GLUTAMIC ACID:
ADVANCES IN BIOCHEMISTRY AND PHYSIOLOGY

MONOGRAPHS OF THE MARIO NEGRI INSTITUTE FOR PHARMACOLOGICAL RESEARCH, MILAN

SERIES EDITOR: SILVIO GARATTINI

Amphetamines and Related Compounds
Edited by E. Costa and S. Garattini, 1970, 976 pp.

Basic and Therapeutic Aspects of Perinatal Pharmacology
Edited by P. L. Morselli, S. Garattini, and F. Sereni, 1975, 456 pp.

The Benzodiazepines
Edited by S. Garattini, E. Mussini, and L. O. Randall, 1973, 707 pp.

Central Mechanisms of Anorectic Drugs
Edited by S. Garattini and R. Samanin, 1978, 502 pp.

Chemotherapy of Cancer Dissemination and Metastasis
Edited by S. Garattini and G. Franchi, 1973, 400 pp.

Drug Interactions
Edited by P. L. Morselli, S. Garattini, and S. N. Cohen, 1974, 416 pp.

Factors Affecting the Action of Narcotics
Edited by M. W. Adler, L. Manara, and R. Samanin, 1978, 797 pp.

Glutamic Acid: Advances in Biochemistry and Physiology
Edited by L. J. Filer, Jr., S. Garattini, M. R. Kare, W. A. Reynolds, and R. J. Wurtman, 1979, 416 pp.

Insolubilized Enzymes
Edited by M. Salmona, C. Saronio, and S. Garattini, 1974, 236 pp.

Interactions Between Putative Neurotransmitters in the Brain
Edited by S. Garattini, J. F. Pujol, and R. Samanin, 1978, 380 pp.

Isolated Liver Perfusion and Its Applications
Edited by I. Bartošek, A. Guaitani, and L. L. Miller, 1973, 303 pp.

Mass Spectrometry in Biochemistry and Medicine
Edited by A. Frigerio and N. Castagnoli, Jr., 1974, 379 pp.

Pharmacology of Steroid Contraceptive Drugs
Edited by S. Garattini and H. W. Berendes, 1977, 391 pp.

Platelets: A Multidisciplinary Approach
Edited by G. de Gaetano and S. Garattini, 1978, 500 pp.

MONOGRAPHS OF
THE MARIO NEGRI INSTITUTE FOR
PHARMACOLOGICAL RESEARCH, MILAN

Glutamic Acid:
Advances in Biochemistry and Physiology

Edited by

L. J. Filer, Jr.
Professor of Pediatrics
University of Iowa
College of Medicine
Iowa City, Iowa

Silvio Garattini
Director
Istituto di Ricerche Farmacologiche
  "Mario Negri"
Milan, Italy

Morley R. Kare
Professor of Physiology and
  Director
Monell Chemical Senses Center
University of Pennsylvania
Philadelphia, Pennsylvania

W. Ann Reynolds
Associate Vice Chancellor for Research
Dean, The Graduate College
Professor of Anatomy
University of Illinois at the Medical
  Center
Chicago, Illinois

Richard J. Wurtman
Professor of Endocrinology and Metabolism
Laboratory of Neuroendocrine Regulation
Department of Nutrition and Food Science
Massachusetts Institute of Technology
Cambridge, Massachusetts

Raven Press ■ New York

Raven Press, 1140 Avenue of the Americas, New York,
New York 10036

© 1979 by Raven Press Books, Ltd. All rights reserved. This book is protected by copyright. No part of it may be reproduced, stored in a retrieval system, or transmitted, in any form or by any means, electronic, mechanical, photocopying, recording, or otherwise, without the prior written permission of the publisher.

Made in the United States of America

Library of Congress Cataloging in Publication Data

Main entry under title:

Glutamic acid.

(Monographs of the Mario Negri Institute for Pharmacological Research, Milan)
Includes bibliographies and index.
1. Glutamic acid—Metabolism. 2. Glutamic acid—Physiological effect. 3. Sodium glutamate—Metabolism. 4. Sodium glutamate—Physiological effect. I. Filer, Mario Negri. Monographs. [DNLM: 1. Glutamates. 2. Glutamates—Physiology. QU60 G566]
QP562.G5G56   599'.01'9245   78-56782
ISBN 0-89004-356-6

Preface

Glutamic acid, one of the most common amino acids found in nature, is present in virtually all proteins as well as in its "free form," that is, nonprotein-linked. Although there has been extensive research on the physiology, biochemistry, and toxicology of glutamic acid, it has been more than 30 years since a text devoted primarily to this substance has been published. We therefore felt that the time was right for the publication of a volume dealing with these multiple aspects of research on glutamic acid.

This volume [based on the International Symposium on the Biochemistry and Physiology of Glutamic Acid held in Milan, May 1978] is divided into five sections. The first group of chapters deals with the sensory and dietary sources of glutamate with particular reference to its natural occurrence in food and to its commercial use as a flavor enhancer. The second section includes chapters on the metabolism of glutamate from both endogenous and added sources. The third section, Glutamate in the Central Nervous System, features reports of studies on glutamic acid biosynthesis, uptake, and metabolism in the brain. The fourth section, Safety Evaluation and Experimental Application, discusses glutamate as a tool for toxicological research and a variety of methods useful in its safety evaluation. The final section covers various clinical aspects, ranging from biochemical and metabolic studies to surveys involved with food use.

Our intention in this volume is to bring to the scientific community both a discussion of our present knowledge about the sensory and dietary aspects of glutamate use in man and reports on those more basic studies on its metabolism in mammals. Also highlighted are studies on the mechanism of action of glutamate as a putative neurotransmitter and the inaccessibility of exogenous glutamate to the brain. The book will be useful reading to a wide range of scientists, from basic physiologists and neurochemists to those who are interested in glutamate from the standpoint of food safety and nutrition.

The Editors

Contents

Sensory and Dietary Aspects of Glutamate

1   Biochemical Studies of Glutamate Taste Receptors: The Synergistic Taste Effect of L-Glutamate and 5′-Ribonucleotides
Robert H. Cagan, Kunio Torii, and Morley R. Kare

11   Self-Selection of Food and Water Flavored with Monosodium Glutamate
Michael Naim

25   Free and Bound Glutamate in Natural Products
T. Giacometti

35   Psychometric Studies on the Taste of Monosodium Glutamate
Shizuko Yamaguchi and Akimitsu Kimizuka

Glutamate Metabolism in Mammals

55   Factors in the Regulation of Glutamate Metabolism
Hamish N. Munro

69   Biochemistry of Glutamate: Glutamine and Glutathione
Alton Meister

85   Comparative Metabolism of Glutamate in the Mouse, Monkey, and Man
L. D. Stegink, W. Ann Reynolds, L. J. Filer, Jr., G. L. Baker, T. T. Daabees, and Roy M. Pitkin

103   Glutamate Metabolism and Placental Transfer in Pregnancy
Roy M. Pitkin, W. Ann Reynolds, L. D. Stegink, and L. J. Filer, Jr.

111   Factors Influencing Dicarboxylic Amino Acid Content of Human Milk
G. L. Baker, L. J. Filer, Jr., and L. D. Stegink

Glutamate in the Central Nervous System

125   Regulation of Amino Acid Availability to Brain: Selective Control Mechanisms for Glutamate
William M. Pardridge

139   Biochemical Aspects of the Neurotransmitter Function of Glutamate
R. P. Shank and M. H. Aprison

151   Glutamic Acid as a Transmitter Precursor and as a Transmitter
E. Costa, A. Guidotti, F. Moroni, and E. Peralta

163 Problems in the Evaluation of Glutamate as a Central Nervous System Transmitter
D. R. Curtis

177 Central Nervous System Receptors for Glutamic Acid
Graham A. R. Johnston

187 Glutamate in the Striatum
E. G. McGeer, P. L. McGeer, and T. Hattori

Glutamate: Safety Evaluation and Experimental Application

203 Glutamate Toxicity in Laboratory Animals
R. Heywood and A. N. Worden

217 Morphology of the Fetal Monkey Hypothalamus After *In Utero* Exposure to Monosodium Glutamate
W. Ann Reynolds, Naomi Lemkey-Johnston, and Lewis D. Stegink

231 *In Utero* and Dietary Administration of Monosodium L-Glutamate to Mice: Reproductive Performance and Development in a Multigeneration Study
K. Anantharaman

255 Toxicological Studies of Monosodium L-Glutamate in Rodents: Relationship Between Routes of Administration and Neurotoxicity
Yutaka Takasaki, Yoshimasa Matsuzawa, Seinosuke Iwata, Yuichi O'hara, Shinobu Yonetani, and Masamichi Ichimura

277 Effects of Glutamate Administration on Pituitary Function
Alan F. Sved and John D. Fernstrom

287 Excitotoxic Amino Acids: Research Applications and Safety Implications
John W. Olney

321 Attempts to Establish the Safety Margin for Neurotoxicity of Monosodium Glutamate
L. Airoldi, A. Bizzi, M. Salmona, and S. Garattini

Clinical Aspects of Glutamate Utilization

333 Factors Affecting Plasma Glutamate Levels in Normal Adult Subjects
Lewis D. Stegink, L. J. Filer, Jr., G. L. Baker, S. M. Mueller, and M. Y-C. Wu-Rideout

353 Metabolism of Free Glutamate in Clinical Products Fed Infants
L. J. Filer, Jr., G. L. Baker, and L. D. Stegink

363 Placebo-Controlled Studies of Human Reaction to Oral Monosodium L-Glutamate
Richard A. Kenney

375 Food-Symptomatology Questionnaires: Risks of Demand-Bias Questions and Population-Biased Surveys
George R. Kerr, Marion Wu-Lee, Mohamed El-Lozy, Robert McGandy, and Frederick J. Stare

389 Summary
R. J. Wurtman

395 Subject Index

Contributors

L. Airoldi
Istituto di Ricerche Farmacologiche
  "Mario Negri"
Via Eritrea, 62
20157 Milan, Italy

K. Anantharaman
Experimental Biology Laboratory
Nestlé Products
Technical Assistance Co., Ltd.
CH-1350 Orbe, Switzerland

M. H. Aprison
Institute of Psychiatric Research and
Departments of Psychiatry and
  Biochemistry
Indiana University Medical Center
Indianapolis, Indiana 46101

G. L. Baker
Departments of Pediatrics and Biochemistry
The University of Iowa College of Medicine
Iowa City, Iowa 52242

A. Bizzi
Istituto di Ricerche Farmacologiche
  "Mario Negri"
Via Eritrea, 62
20157 Milan, Italy

Robert H. Cagan
Veterans Administration Hospital and
  Monell Chemical Senses Center
University of Pennsylvania
3500 Market Street
Philadelphia, Pennsylvania 19104

E. Costa
Laboratory of Preclinical Pharmacology
National Institute of Mental Health
Saint Elizabeths Hospital
Washington, D.C. 20032

D. R. Curtis
Department of Pharmacology
John Curtin School of Medical Research
Australian National University
P. O. Box 334
Canberra City, A.C.T., 2601 Australia

T. T. Daabees
Departments of Pediatrics and Biochemistry
The University of Iowa College of Medicine
Iowa City, Iowa 52242

Mohamed El-Lozy
School of Public Health
Harvard University
665 Huntington Avenue
Boston, Massachusetts 02115

John D. Fernstrom
Laboratory of Brain and Metabolism
Program in Neural and Endocrine
  Regulation
Massachusetts Institute of Technology
Cambridge, Massachusetts 02139

L. J. Filer, Jr.
Departments of Pediatrics and Biochemistry
The University of Iowa College of Medicine
Iowa City, Iowa 52242

CONTRIBUTORS

S. Garattini
Istituto di Ricerche Farmacologiche
"Mario Negri"
Via Eritrea, 62
20157 Milan, Italy

T. Giacometti
22, Avenue de Traménaz
1814 La Tour de Peilz, Switzerland

A. Guidotti
Laboratory of Preclinical Pharmacology
National Institute of Mental Health
Saint Elizabeths Hospital
Washington, D.C. 20032

T. Hattori
Kinsmen Laboratory of Neurological
 Research
Department of Psychiatry
University of British Columbia
Vancouver, British Columbia, Canada
 V6T 1W5

R. Heywood
Huntingdon Research Centre
Huntingdon, Cambridgeshire
PE18 6ES, England

Mrasamichi Ichimura
Life Science Laboratory
Central Research Laboratories
Ajinomoto Co., Inc.
214, Maeda-cho, Totsuka-ku
Yokohama-shi, Japan

Seinosuke Iwata
Life Science Laboratory
Central Research Laboratories
Ajinomoto Co., Inc.
214, Maeda-cho, Totsuka-ku
Yokohama-shi, Japan

Graham A. R. Johnston
Department of Pharmacology
John Curtin School of Medical Research
Australian National University
P. O. Box 334
Canberra City, A.C.T., 2601 Australia

Morley R. Kare
Veterans Administration Hospital and
 Monell Chemical Senses Center
University of Pennsylvania
3500 Market Street
Philadelphia, Pennsylvania 19104

Richard A. Kenney
Department of Physiology
George Washington University Medical
 Center
2300 Eye Street, N.W.
Washington, D.C. 20037

George R. Kerr
School of Public Health
The University of Texas Health Science
 Center
P. O. Box 20186
Houston, Texas 77025

Akimitsu Kimizuka
Central Research Laboratories
Ajinomoto Co., Inc.
Suzuki-cho, Kawasaki
210 Japan

Naomi Lemkey-Johnston
Illinois Institute for Developmental
 Disabilities
Chicago, Illinois 60612

Yoshimasa Matsuzawa
Life Science Laboratory
Central Research Laboratories
Ajinomoto Co., Inc.
214, Maeda-cho, Totsuka-ku
Yokohama-shi, Japan

Robert McGandy
School of Public Health
Harvard University
665 Huntington Avenue
Boston, Massachusetts 02115

CONTRIBUTORS

E. G. McGeer
Kinsmen Laboratory of Neurological
  Research
Department of Psychiatry
University of British Columbia
Vancouver, British Columbia, Canada
  V6T 1W5

P. L. McGeer
Kinsmen Laboratory of Neurological
  Research
Department of Psychiatry
University of British Columbia
Vancouver, Canada V6T 1W5

Alton Meister
Department of Biochemistry
Cornell University Medical College
1300 York Avenue
New York, New York 10021

F. Moroni
Laboratory of Preclinical Pharmacology
National Institute of Mental Health
Saint Elizabeths Hospital
Washington, D.C. 20032

S. M. Mueller
Departments of Pediatrics, Biochemistry
  and Neurology
The University of Iowa College of Medicine
Iowa City, Iowa 52242

Hamish N. Munro
Physiological Chemistry Laboratories
Department of Nutrition and Food Science
Massachusetts Institute of Technology
Cambridge, Massachusetts 02139

Michael Naim
Department of Nutrition
Hebrew University-Hadassah Medical
  School
P. O. Box 1172
Jerusalem, Israel

Yuichi O'hara
Life Science Laboratory
Central Research Laboratories
Ajinomoto Co., Inc.
214, Maeda-cho, Totsuka-ku
Yokohama-shi, Japan

John W. Olney
Departments of Psychiatry and
  Neuropathology
Washington University School of
  Medicine
4940 Audubon Avenue
St. Louis, Missouri 63110

William M. Pardridge
Department of Medicine
Division of Endocrinology and Metabolism
UCLA School of Medicine
Los Angeles, California 90024

E. Peralta
Laboratory of Preclinical Pharmacology
National Institute of Mental Health
Saint Elizabeths Hospital
Washington, D.C. 20032

Roy M. Pitkin
Department of Obstetrics and Gynecology
The University of Iowa
Iowa City, Iowa 52242

W. Ann Reynolds
Departments of Anatomy and Obstetrics
  and Gynecology
University of Illinois at the Medical
  Center
Chicago, Illinois 60680

M. Salmona
Istituto di Ricerche Farmacologiche
  "Mario Negri"
Via Eritrea, 62
20157 Milan, Italy

R. P. Shank
Department of Physiology
Temple University School of Medicine
Philadelphia, Pennsylvania 19140

CONTRIBUTORS

Frederick J. Stare
School of Public Health
Harvard University
665 Huntington Avenue
Boston, Massachusetts 02115

Lewis D. Stegink
Departments of Pediatrics and
 Biochemistry
The University of Iowa College of Medicine
Iowa City, Iowa 52242

Alan F. Sved
Laboratory of Brain and Metabolism
Program in Neural and Endocrine
 Regulation
Massachusetts Institute of Technology
Cambridge, Massachusetts 02139

Yutaka Takasaki
Life Science Laboratory
Central Research Laboratories
Ajinomoto Co., Inc.
214, Maeda-cho, Totsuka-ku
Yokohama-shi, Japan

Kunio Torii
Monell Chemical Senses Center
University of Pennsylvania
3500 Market Street
Philadelphia, Pennsylvania 19104

A. N. Worden
Huntingdon Research Centre
Huntingdon, Cambridgeshire
PE18 6ES, England

Marion Wu-Lee
School of Public Health
Harvard University
665 Huntington Avenue
Boston, Massachusetts 02115

M. Y-C. Wu-Rideout
Departments of Pediatrics, Biochemistry
 and Neurology
The University of Iowa College of Medicine
Iowa City, Iowa 52242

R. J. Wurtman
Laboratory of Neuroendocrine
 Regulation
Department of Nutrition and Food Science
Massachusetts Institute of Technology
Cambridge, Massachusetts 02139

Shizuko Yamaguchi
Central Research Laboratories
Ajinomoto Co., Inc.
Suzuki-cho, Kawasaki
210 Japan

Shinobu Yonetani
Life Science Laboratory
Central Research Laboratories
Ajinomoto Co., Inc.
214, Maeda-cho, Totsuka-ku
Yokohama-shi, Japan

Glutamic Acid: Advances in Biochemistry and Physiology, edited by L. J. Filer, Jr., et al.
Raven Press, New York © 1979.

Biochemical Studies of Glutamate Taste Receptors: The Synergistic Taste Effect of L-Glutamate and 5'-Ribonucleotides

Robert H. Cagan, Kunio Torii, and Morley R. Kare

Veterans Administration Hospital and Monell Chemical Senses Center, University of Pennsylvania, Philadelphia, Pennsylvania 19104

Unique features of the taste effects of monosodium glutamate (MSG) could make MSG taste an important system for a better understanding of the biochemical basis of taste sensation. Our approach to the glutamate taste at the biochemical level derives, on the one hand, from observations of its taste effects in food systems and, on the other, from previous studies in our laboratory on the biochemical basis of taste. The approach we are employing is essentially biochemical, utilizing measures of the initial interaction of the stimulus with its receptor sites.

Dried bonito (*katsuo-bushi*), black mushroom (*shiitake*), and the seaweed sea tangle (*Laminaria sp.*) have been used extensively as condiments in Japanese cuisine. These materials impart a characteristic taste that is called *umami* in Japanese (49,52), meaning "delicious" or "savory" taste (27). Chemical studies have shown that the taste-active substances from these foods belong to two separate classes of chemical compounds. In the early 1900s, the taste-active substance was isolated from sea tangle and identified by Ikeda (20) as a salt of L-glutamic acid. In contrast, the taste-active ingredient from dried bonito was found by Kodama (26) to be the histidine salt of IMP (29,30), and that from black mushroom was more recently shown to be GMP (37,45). The taste properties of MSG and ribonucleotides have been investigated by psychologists and food scientists (3,25,27,29–31,36,45,49,50,52), but the mechanism of action of these compounds in exciting the taste receptors has not been studied. Investigators consistently describe the "distinctive" or "unique" taste of MSG, and certain other amino acids have also been noted (23) to possess some of this characteristic taste.

The fact that these condiments are typically used in combination in Japanese cookery has a certain theoretical interest. Psychophysical taste evaluations have conclusively shown that a powerful synergism exists in mixtures of certain 5'-ribonucleotides and MSG; the taste intensity of such a mixture is greater than the sum of the tastes of the two components. Yamaguchi (49) initially used mixtures of IMP and MSG to study this synergism, and subsequently (50) extended the work to include several additional purines and their derivatives. Because of the potent

1

synergistic effect, Cagan (11) recently noted that: "to understand the mechanism of action of MSG we may actually need to understand the synergistic action of MSG and ribonucleotides rather than the effect of MSG alone."

The use of new approaches and methodologies has led to a number of advances in our basic understanding of the initial events underlying taste sensation. The stimulus-receptor interaction in taste has been investigated in our laboratory (7–10,12–15,28) and elsewhere (32,40). Our recent studies have included the development of improved methods enabling the direct measurement of the initial interaction, which presumably reflects the complex formed (15,28,46). In addition, several testable hypotheses were recently proposed (11) that focused attention on the possible sites of action and the underlying mechanisms responsible for the taste effects of MSG and of MSG-ribonucleotide mixtures. The loci noted were (a) receptor site interactions, (b) transduction processes within receptor cells, (c) peripheral synaptic transmission, and (d) neural transmission and central processing. When considering item (a), receptor site interactions, two fundamentally different types of effects at this level were described: first, effects on binding affinities and, second, effects on the accessibility of receptor sites.

The results of the studies noted above (7–10,12–15,28) have made it feasible to undertake a biochemical investigation of the glutamate taste. We have focused our attention on the binding interaction of L-glutamate to receptor sites in taste tissue and on the effects of GMP and other ribonucleotides on this interaction. Specifically, we wished to establish if the synergism could be accounted for by a peripheral mechanism, and if this could be measured biochemically using the binding of a radioactively labeled ligand. Our evidence demonstrates that L-glutamate binds preferentially to taste receptor tissue, and that GMP and certain other 5'-ribonucleotides cause a marked enhancement of L-glutamate binding (46). We propose that the basis of the synergistic effect is due to a change in the receptor properties, caused by the 5'-ribonucleotide, that affects the binding interaction of L-glutamate with the taste receptors. A corollary of this hypothesis is that the flavor enhancement of foods by MSG is due to a combination of its unique taste character coupled with the enhancement of its taste effect by low levels of 5'-ribonucleotides, such as those that may be present endogenously in the food.

AMINO ACID BINDING TO TASTE RECEPTORS

In order to provide a background for our approach, previous studies from our laboratory on the taste system of the catfish are reviewed. The catfish *Ictalurus* has taste buds distributed over its body surface. The barbels ("whiskers") have the highest density of buds, and the rostral, dorsal, and dorsolateral surfaces also contain appreciable numbers (4,18). Behavioral studies (6,18) showed that the catfish uses its sense of taste to locate food, and electrophysiological measurements (5,16) from the barbel nerve established that the taste system is sensitive to a number of amino acids. In our laboratory (28) we measured the binding of amino acid taste stimuli to the catfish taste receptors *in vitro* by utilizing radioactively labeled amino acids as ligands. In order to carry out these experiments, the taste epithelium is homogenized and fractionated by differential centrifugation to yield a sedimentable fraction (Fraction P2). This fraction, which contains plasma membranes (although not pure), is enriched in binding activity for taste stimulus amino acids (12,13,28). In recent studies (14) we further purified Fraction P2 to isolate the plasma membranes and demonstrated that the binding activity is in fact associated with this plasma membrane fraction. These results show that the initial discrimination of taste stimuli occurs at the outer surface of the receptor cells.

L-[^{3}H]Alanine has been used extensively as a ligand in our studies because the catfish is highly sensitive to this amino acid as a taste stimulus (16). This has allowed us to develop a convenient, reliable assay system by which to measure binding. We thereby established that binding is a measure of an early event in taste sensation, an event which we propose to be the initial discrimination step in the recognition of taste stimulus compounds. Several lines of evidence support the hypothesis: (a) binding is saturable and reversible, (b) the distribution of binding activity corresponds with the known distribution of taste buds, (c) amino acids that are electrophysiological taste stimuli also show binding activity (albeit in different orders of relative effectiveness), and (d) denervation, which is known to cause the degeneration of the taste buds (33,39,47), results in a decrease in binding activity of the denervated preparation.

Of the ligands we employ, the binding properties of L-alanine have been studied in greatest detail (28). The value of the K_D (dissociation constant) for L-alanine is 4.8×10^{-6} M, and the optimal pH for binding is 7.8. The reversibility of the binding was demonstrated both by measuring the dissociation rate of the ligand into ligand-free medium, and by displacement of the bound ^{3}H-ligand with a large excess of unlabeled ligand. The preparation from catfish taste tissue therefore contains receptor sites with which the taste stimulus L-alanine interacts in a reversible binding step.

ENHANCEMENT EFFECT AND "HIDDEN" RECEPTORS

During the course of studies to attempt to stabilize the binding activity of Fraction P2, a striking effect was observed (12,13) following the frozen storage of Fraction P2 in L-alanine. When Fraction P2 was maintained in a relatively high concentration (10 mM) of L-alanine (unlabeled) and then washed to remove the ligand prior to the assay for binding activity, the binding activity actually increased several fold. This enhancement effect shows an interesting degree of stereospecificity with respect to L- and D-alanine; it is similar to the stereospecificity of these isomers as taste stimuli measured electrophysiologically. D-Alanine averages 57% as effective a taste stimulus as L-alanine (16); D-alanine treatment of Fraction P2 enhanced the binding of L-[^{3}H]alanine to 50 to 60% of the level caused by L-alanine treatment (12,13). By analyzing the enhancement phenomenon using Scatchard plots, it was established (13) that the K_D for L-alanine is unchanged while the maximal binding of L-alanine increases by several fold. The effect, therefore, is not on the affinity of the taste receptors for L-alanine, but rather on the number of sites available for binding this amino acid.

These results led to the hypothesis (12,13) that the enhancement due to L-alanine treatment is a result of exposing alanine receptor sites that were previously "hidden" or "buried," such as within the cell membrane. It was further postulated that the high ligand concentration causes a perturbation, presumably through a cooperative effect, of the receptor membrane complex such that additional binding sites become accessible to the ligand. Whether this involves a conformational change of the receptor molecules only, or a larger conformational change involving the membrane structure, will require further study.

MSG AND RIBONUCLEOTIDES

MSG is a taste stimulus in humans, being effective in the millimolar concentration range (25,31,36). MSG also shows the remarkable property of a synergistic taste effect in mixtures with certain 5'-ribonucleotides. Of the naturally occurring nucleotides studied, GMP is the most potent in evoking this effect (35,50).

Relatively few behavioral animal studies with MSG have been carried out, but those reported do agree in certain respects with human psychophysical studies. Although in an early study the food preference of rats was not affected by 1% MSG in the diet (44), rats do prefer solutions of MSG to water (19,38). Weanling calves ate more of a diet when it contained 0.2% MSG (48), and preliminary reports suggest that MSG affects the dietary intake of pigs (17,24). Pygmy goats appeared to have altered taste acuity to certain stimuli in the presence of MSG (34).

Electrophysiological experiments with rats (2,21,41,43,51) and with cats (1,22) showed that MSG is a taste stimulus and, in addition, that mixtures of MSG and ribonucleotides are synergistic when recording from the chorda tympani nerve. As in humans, electrophysiological studies in rats (41) showed GMP to be the most effective of the naturally occurring nucleotides; IMP, UMP, and CMP were also effective.

Structure-activity relationships with nucleotides in human studies (35,50) have revealed several structural features that are necessary for their synergistic effect. For example, the importance of the oxygen function at position 6 of the purine ring has been established, as well as the importance of the phosphate ester being located at the 5'-position of the ribose ring. In addition, certain 2-substituted purine derivatives were shown to be even more potent than GMP. Electrophysiological studies in the rat employed chorda tympani recordings to establish that GMP and IMP were effective potentiators of the MSG response (2,21,41,43,51). Single-fiber recordings (43) showed potentiation by GMP, IMP, and, surprisingly, by AMP; UMP and CMP were relatively ineffective. One of the derivatives of IMP substituted at the 2-position (50) was also tested with rats and found to be highly effective (42).

The results from the animal studies, although not extensive, show that some species studied respond to MSG as a taste stimulus and to mixtures of MSG and ribonucleotides. Little is known, however, of such responses in the species of fish,

Ictalurus punctatus, which we have used extensively in our biochemical research on taste receptors. In addition, the catfish shows relatively small electrophysiological responses to L-glutamate (16).

Although the catfish has provided an experimental methodology and has yielded results leading to the hypothesis of "hidden" or "buried" receptors, it did not appear to be the most suitable experimental model for initiating direct studies of the binding of L-glutamate and of the synergistic effect with ribonucleotides. An additional animal model is available in our laboratory, dating from our earlier research on taste-binding interactions of sugars (9). The binding of ^{14}C-labeled sugars had been measured in bovine taste papillae preparations and in control tissue devoid of taste buds. More binding of the sugars occurred in the taste tissue than in the controls. The findings were recently confirmed and extended to show the involvement of the plasma membrane in the binding interactions with sugars (32,40).

In the intervening years, we have refined the approach, including methods of obtaining preparations with a higher enrichment of taste receptors (7,8), as well as adaptation of the binding methodology that we used for amino acid binding (28) to enable us to measure the binding of ^3H-labeled monellin to taste tissue preparations (15). Using the improved preparative procedures (7,8,46), coupled with the filtration binding assay, we have directly measured the initial interaction of L-[^3H]glutamate to bovine taste tissue preparations. We believe that our results may offer an explanation both of the site of action of MSG and of the mechanistic basis of the synergistic effect of MSG and ribonucleotides. The following is a preliminary report of the major findings, which will be published in detail elsewhere (46).

(a) The binding of L-[^3H]glutamate to the circumvallate (CV) preparation is several fold higher than to the control epithelium (EP) preparation devoid of taste receptors. This observation is consistent in the many experiments we have carried out, and representative data are shown in Table 1. Further, several features indicate

TABLE 1. *Binding of L-[^3H]glutamate to bovine circumvallate (taste) and epithelial (nontaste) tongue tissues*[a]

Preparation	L-Glutamate concentration	L-[^3H]Glutamate bound (nmoles/mg protein)		
		1.4 mM	6.9 mM	14 mM
Circumvallate	—	1.14	4.99	6.49
Epithelium	—	0.20	0.60	1.78

[a] The sidewall epithelium is peeled away from bovine circumvallate papillae and a homogenate is prepared (Circumvallate). The control preparation (Epithelium) consists of pieces of tongue epithelium taken from tongue regions devoid of taste buds and prepared in the same fashion. Differential centrifugation removes the low-speed pellet, and then the pellet that sediments at 7,000 × g (30 min) is used in each case. Binding is measured using L-[^3H]glutamate with a rapid filtration method. All samples are run in duplicate. The data for circumvallate are mean values from three experiments; those for the epithelium are from a single experiment.

that the binding to EP is qualitatively different than that to CV, and we suggest that the lower level of binding to EP is nonspecific. We conclude that the bovine circumvallate papilla has receptor sites for L-glutamate. Quantitatively, however, the amounts of ligand bound suggest that they reflect entrapment of fluid (containing radioactive ligand) in addition to direct complex formation with receptor sites.

(b) The binding of L-[^{3}H]glutamate to CV appears to be saturable. Under our conditions of measurement, binding begins to show saturation, but does not fully reach a plateau as the concentration of L-[^{3}H]glutamate increases. This is undoubtedly due to the interaction being weak; therefore, the concentrations of L-[^{3}H]glutamate required to saturate the binding sites are considerably higher than we have employed experimentally (up to 14 mM). The value of the K_D was estimated using both a double-reciprocal plot (Lineweaver-Burk) and a Scatchard plot, yielding a K_D for L-glutamate in the range of 17 to 20 mM.

(c) The addition of 5'-GMP dramatically increases the amount of L-[^{3}H] glutamate bound to CV and has no effect whatever on the already low level of L-[^{3}H]glutamate bound to EP. An example of the effect is illustrated in Table 2,

TABLE 2. Enhancement by 5'-GMP of L-[^{3}H]glutamate binding to bovine circumvallate (taste) tissue[a]

Preparation	L-[^{3}H]Glutamate bound (nmoles/mg protein)	
	No GMP	+ GMP
Circumvallate	3.44	18.2
Epithelium	0.60	0.80

[a] The tissues were prepared and assays carried out as described in Table 1. L-[^{3}H]Glutamate was present at 6.9 mM and 5'-GMP at 1.4 mM. The data shown are taken from a single experiment, in which samples were run in duplicate.

where a several-fold enhancement of the binding of L-[^{3}H]glutamate to CV is demonstrated. In further experiments, we have used Scatchard analyses to estimate the K_D for L-glutamate in the absence and presence of GMP. The K_D was unchanged by adding GMP, whereas the maximal amount of L-glutamate bound increased by sixfold. The latter observation can be explained by a hypothesis similar to that advanced to explain the ligand enhancement of binding with the L-alanine taste receptors from catfish (12,13). In the present case, GMP is postulated to cause the exposure of "hidden" or "buried" receptor sites for L-glutamate.

(d) The response to ribonucleotides shows a high degree of specificity. The 5'-ribonucleotides GMP, IMP, and UMP are each effective in enhancing the binding of L-[^{3}H]glutamate to the CV. AMP and CMP are ineffective. Also ineffective are guanine, GDP, GTP, adenine, ADP, and ATP. The specificity with respect to the nucleotide therefore shows a high degree of similarity, although not absolute, with human psychophysical responses.

SUMMARY

The ability of MSG to evoke a "distinctive" or "unique" taste sensation is well known. The taste is called *umami* in Japanese, which is translated as "delicious" or "savory." Furthermore, the remarkable synergistic effect of certain 5'-ribonucleotides in enhancing this taste is also clearly documented. Neither the site of action nor the biochemical mechanism is known. We have therefore investigated this question as an extension of our research into the initial steps in taste stimulus recognition; the binding interaction of stimulus molecules with sites in the receptor membranes of the taste cells appears to be critically involved in the initial discrimination of taste stimuli.

We have measured the binding of MSG by means of ^{3}H-labeled L-glutamate. Substantially greater binding of L-[^{3}H]glutamate occurs to the bovine circumvallate papillae preparations, which contain taste receptors, than to the epithelial preparations devoid of taste receptors. The low level of binding to the epithelium is qualitatively as well as quantitatively different than that to the circumvallate, and we conclude that the binding to the epithelial preparation is nonspecific. Our data demonstrate directly that 5'-GMP causes a marked increase in the binding of L-[^{3}H]glutamate to the circumvallate preparation. Furthermore, this effect occurs only with the taste tissue and not with the control, nontaste tongue epithelium. This response shows a high degree of specificity with respect to the nucleotide; GMP, IMP, and UMP are each effective, but AMP and CMP, as well as guanine, GDP, GTP, adenine, ADP, and ATP are ineffective. This system appears to be a useful experimental model for MSG taste and for the MSG-ribonucleotide synergistic taste phenomenon observed in humans and other species.

We propose that the site of action of the synergistic effect of ribonucleotides and MSG is at the peripheral level, acting on the taste receptor cell membrane. We further propose that the mechanism of action of the ribonucleotide is to expose additional receptor sites for L-glutamate. A major corollary of the hypothesis, therefore, is that MSG evokes its characteristic taste by binding to specific taste receptor sites, thereby itself acting as a taste stimulus. The effect of MSG becomes markedly enchanced by low levels of certain 5'-ribonucleotides, such as certain of those that occur endogenously in various foods.

ACKNOWLEDGMENT

This research was supported in part by NIH Research Grant No. NS-08775 (to R.H.C.) from the National Institute of Neurological and Communicative Disorders and Stroke.

REFERENCES

1. Adachi, A. (1964): Neurophysiological study of taste effectiveness of seasoning. *J. Physiol. Soc. Jpn.*, 26:347–355 (English summary).
2. Adachi, A., Okamoto, J., Hamada, T., and Kawamura, Y. (1967): Taste effectiveness of mixtures

of sodium 5'-inosinate and various amino acids. *J. Physiol. Soc. Jpn.*, 29:65–71 (English summary).
3. Amerine, M. A., Pangborn, R. M., and Roessler, E. B. (1965): *Principles of Sensory Evaluation of Food*, pp. 115–120. Academic Press, New York.
4. Bardach, J. E., and Atema, J. (1971): The sense of taste in fishes. In: *Handbook of Sensory Physiology*, Vol. 4, Part 2, edited by L. M. Beidler, pp. 293–336. Springer-Verlag, New York.
5. Bardach, J., Fujiya, M., and Holl, A. (1967): Investigations of external chemoreceptors of fishes. In: *Olfaction and Taste*, Vol. 2, edited by T. Hayashi, pp. 647–665. Pergamon Press, New York.
6. Bardach, J. E., Todd, J. H., and Crickmer, R. (1967): Orientation by taste in fish of the genus *Ictalurus*. *Science*, 155:1276–1278.
7. Brand, J. G., and Cagan, R. H. (1976): Biochemical studies of taste sensation III. Preparation of a suspension of bovine taste bud cells and their labeling with a fluorescent probe. *J. Neurobiol.*, 7:205–220.
8. Brand, J. G., Zeeberg, B. R., and Cagan, R. H. (1976): Biochemical studies of taste sensation V. Binding of quinine to bovine taste papillae and taste bud cells. *Int. J. Neurosci.*, 7:37–43.
9. Cagan, R. H. (1971): Biochemical studies of taste sensation I. Binding of ^{14}C-labeled sugars to bovine taste papillae. *Biochim. Biophys. Acta*, 252:199–206.
10. Cagan, R. H. (1974): Biochemistry of sweet sensation. In: *Sugars in Nutrition*, edited by H. L. Sipple and K. W. McNutt, pp. 19–36. Academic Press, New York.
11. Cagan, R. H. (1977): A framework for the mechanisms of action of special taste substances: The example of monosodium glutamate. In: *The Chemical Senses and Nutrition*, edited by M. R. Kare and O. Maller, pp. 343–359. Academic Press, New York.
12. Cagan, R. H. (1977): Enhancement of taste receptor binding by exposure to high stimulus concentration. *Soc. Neurosci. Abstr.*, 3:77 (Abstr. 222).
13. Cagan, R. H. (1979): Biochemical studies of taste sensation. Enhancement of taste stimulus binding to a catfish taste receptor preparation by prior exposure to the stimulus. *J. Neurobiol. (in press)*.
14. Cagan, R. H., and Boyle, A. G. (1978): Plasma membranes from taste receptor tissue: Isolation and binding activity. *Fed. Proc.*, 37:1818 (Abstr. 3006).
15. Cagan, R. H., and Morris, R. W. (1979): to be published.
16. Caprio, J. (1975): High sensitivity of catfish taste receptors to amino acids. *Comp. Biochem. Physiol.* [A], 52:247–251.
17. Henson, J. N., Bogdonoff, P. D., and Thrasher, G. W. (1962): Levels of monosodium glutamate in pig starter preference. *J. Anim. Sci.*, 21:999–1000 (Abstract).
18. Herrick, C. J. (1904): The organ and sense of taste in fishes. *Bull. U.S. Fish. Comm.*, 22:237–272.
19. Hiji, Y., and Sato, M. (1967): Preference-aversion function for sodium monoaminodicarboxylates in rats. *J. Physiol. Soc. Jpn.*, 29:168–169.
20. Ikeda, K. (1912): On the taste of the salt of glutamic acid. *Orig. Commun. 8th Int. Congr. Appl. Chem.* 18:147 (Abstract).
21. Kasahara, Y., Kawamura, Y., and Ikeda, S. (1970): Neurophysiological analysis of relations between taste effectiveness and chemical structures of 5'-ribonucleotides. *J. Physiol. Soc. Jpn.*, 32:748–755 (English summary).
22. Kawamura, Y., and Adachi, A. (1965): Single taste nerve responses to the chemical taste enhancers. *J. Physiol. Soc. Jpn.*, 27:279–284. (English summary).
23. Kirimura, J., Shimizu, A., Kimizuka, A., Ninomiya, J., and Katsuya, N. (1969): The contribution of peptides and amino acids to the taste of foodstuffs. *J. Agric. Food Chem.*, 17:689–695.
24. Klay, R. F. (1964): Monosodium glutamate in pig creep rations. *J. Anim. Sci.*, 23:598 (Abstract).
25. Knowles, D., and Johnson, P. E. (1941): A study of the sensitiveness of prospective food judges to the primary tastes. *Food Res.*, 6:207–216.
26. Kodama, S. (1913): On a procedure for separating inosinic acid. *J. Tokyo Chem. Soc.*, 34:751–757.
27. Komata, Y. (1976): Utilization of nucleic acid-related substances. Utilization in foods. In: *Microbial Production of Nucleic Acid-Related Substances*, edited by K. Ogata, S. Kinoshita, T. Tsunoda, and K. Aida, pp. 299–319. Wiley, New York.
28. Krueger, J. M., and Cagan, R. H. (1976): Biochemical studies of taste sensation IV. Binding of L-[^{3}H]alanine to a sedimentable fraction from catfish barbel epithelium. *J. Biol. Chem.*, 251:88–97.
29. Kuninaka, A. (1967): Flavor potentiators. In: *Symposium on Foods: The Chemistry and Physiology of Flavors*, edited by H. W. Schultz, E. A. Day, and L. M. Libbey, pp. 515–535. Avi Publishing Co., Westport., Conn.
30. Kuninaka, A., Kibi, M., and Sakaguchi, K. (1964): History and development of flavor nucleotides. *Food Technol.*, 18:287–293.

31. Lockhart, E. E., and Gainer, J. M. (1950): Effect of monosodium glutamate on taste of pure sucrose and sodium chloride. *Food Res.*, 15:459–464.
32. Lum, C. K. L., and Henkin, R. I. (1976): Sugar binding to purified fractions from bovine taste buds and epithelial tissue. Relationships to bioactivity. *Biochim. Biophys. Acta*, 421:380–394.
33. May, R. M. (1925): The relation of nerves to degenerating and regenerating taste buds. *J. Exp. Zool.*, 42:371–410.
34. Mehren, M. J., and Church, D. C. (1976): Influence of taste-modifiers on taste responses of pygmy goats. *Anim. Prod.*, 22:255–260.
35. Mizuta, E., Toda, J., Suzuki, N., Sugibayashi, H., Imai, K.-I., and Nishikawa, M. (1972): Structure-activity relationship in the taste effect of ribonucleotide derivatives. *Chem. Pharm. Bull.*, 20:1114–1124.
36. Mosel, J. N., and Kantrowitz, G. (1954): Absolute sensitivity to the glutamic taste. *J. Gen. Psychol.*, 51:11–18.
37. Nakajima, N., Ichikawa, K., Kamada, M., and Fujita, E. (1961): Food chemical studies on 5'-ribonucleotides I. On the 5'-ribonucleotides in foods (1) Determination of the 5'-ribonucleotides in various stocks by ion exchange chromatography. *J. Agric. Chem. Soc. Jpn.*, 35:797–803.
38. Ohara, I., and Naim, M. (1977): Effects of monosodium glutamate on eating and drinking behavior in rats. *Physiol. Behav.*, 19:627–634.
39. Olmsted, J. M. D. (1920): The results of cutting the seventh cranial nerve in *Amiurus nebulosus* (Lesueur). *J. Exp. Zool.*, 31:369–401.
40. Ostretsova, I. B., Safarian, E. K., and Etingof, R. N. (1975): On the presence and localization of glucose-binding proteins in the tongue. *Proc. Acad. Sci. U.S.S.R.*, 223:1484–1487.
41. Sato, M., and Akaike, N. (1965): 5'-Ribonucleotides as gustatory stimuli in rats. Electrophysiological studies. *Jpn. J. Physiol.*, 15:53–70.
42. Sato, M., Ogawa, H., and Yamashita, S. (1971): Comparison of potentiating effect on gustatory response by disodium 2-methyl mercapto-5'-inosinate with that by 5'-IMP. *Jpn. J. Physiol.*, 21:669–679.
43. Sato, M., Yamashita, S., and Ogawa, H. (1970): Potentiation of gustatory response to monosodium glutamate in rat chorda tympani fibers by addition of 5'-ribonucleotides. *Jpn. J. Physiol.*, 20:444–464.
44. Scott, E. M., and Quint, E. (1946): Self selection of diet II. The effect of flavor. *J. Nutr.*, 32:113–119.
45. Shimazono, H. (1964): Distribution of 5'-ribonucleotides in foods and their application to foods. *Food Technol.*, 18:294–303.
46. Torii, K., and Cagan, R. H. (1979): to be published.
47. Torrey, T. W. (1934): The relation of taste buds to their nerve fibers. *J. Comp. Neurol.*, 59:203–220.
48. Waldern, D. E., and Van Dyk, R. D. (1971): Effect of monosodium glutamate in starter rations on feed consumption and performance of early weaned calves. *J. Dairy Sci.*, 54:262–265.
49. Yamaguchi, S. (1967): The synergistic taste effect of monosodium glutamate and disodium 5'-inosinate. *J. Food Sci.*, 32:473–478.
50. Yamaguchi, S., Yoshikawa, T., Ikeda, S., and Ninomiya, T. (1971): Measurement of the relative taste intensity of some L-α-amino acids and 5'-nucleotides. *J. Food Sci.*, 36:846–849.
51. Yamashita, S., Ogawa, H. and Sato, M. (1973): The enhancing action of 5'-ribonucleotide on rat gustatory nerve fiber response to monosodium glutamate. *Jpn. J. Physiol.*, 23:59–68.
52. Yoshida, M., and Saito, S. (1969): Multidimensional scaling of the taste of amino acids. *Jpn. Psychol. Res.*, 11:149–166.

Glutamic Acid: Advances in Biochemistry and Physiology, edited by L. J. Filer, Jr., et al.
Raven Press, New York © 1979.

Self-Selection of Food and Water Flavored with Monosodium Glutamate

Michael Naim

Monell Chemical Senses Center, University of Pennsylvania, Philadelphia, Pennsylvania 19104; and *Department of Nutrition, Hebrew University-Hadassah Medical School, Jerusalem, Israel

The recognition and acceptance of food mainly depends on its chemostimulatory characteristics. Therefore, taste and flavor perception are important in food discrimination (2,5,24). The flavor can be said to motivate ingestion and may serve as a protector against the consumption of foods of questionable quality. Experimental evidence suggests that the flavor characteristic of a nutrient may be paired with its nutritive value (13,36,49), helping the animal to regulate his diet selection according to physiological needs (18,35,48,56).

Taste and oral stimulation can also reflexively initiate certain digestive and metabolic processes even before food reaches the stomach (39,42,44). This includes the stimulation of gastric acid secretion (16,47), release of gastrin (43,47), increase in pancreatic exocrine output (3,38,40,44,53), and mobilization of insulin from the endocrine pancreas (20,60).

Monosodium glutamate (MSG) is widely used as a flavor in cooking, presumably as a taste stimulant and a flavor enhancer. Possible mechanisms for the ability of MSG to stimulate taste receptors have recently been discussed by Cagan (6). L-Glutamic acid, the free acid of MSG, is a constituent of protein, and also occurs as a free amino acid (0.01 to 0.2%) in a variety of vegetables, meat, and seafood (17,31). Milk from lactating human females contains 0.007 to 0.02% free L-glutamate (59). It has been suggested by Fagerson (10) that only the form of glutamate with both carboxyl groups ionized is active as a taste stimulant in humans. Taste thresholds of MSG for humans are in millimolar concentrations (30). Electrophysiological measurements in rats showed that MSG is effective in stimulating the chorda tympani nerve (54). The taste of some other amino acids has been referred to as an MSG-like taste (25). A favorable effect of MSG as a flavor enhancer for meat, poultry, seafood, and vegetables has been reported (15,58). A synergistic effect between MSG and 5'-ribonucleotides in enhancing food flavor has been demonstrated (27).

Relatively little consideration has been given to MSG preference in animals,

---

*Present address.

although animal models are commonly employed in the analysis of food intake mechanisms. Scott and Quint (55) showed that 1% MSG did not influence the diet preference of rats. Hiji and Sato (19) demonstrated that female rats select 5×10^{-2} to 1.5×10^{-1} M MSG-flavored water over water in two-choice, long-term preference tests. This preference for MSG solutions was observed only in a situation where solutions containing MSG were offered in an ascending concentration sequence. No such preference was shown in the descending concentration sequence. Weanling calves, beginning at 3 weeks of age, selected more of a 0.2% MSG-flavored diet than a nonflavored diet (61). The taste responses of pygmy goats to sucrose and quinine hydrochloride solutions were moderated by adding 5, 50, 500 ppm MSG (32).

In this chapter we will consider both the sensory and possible postingestional properties of MSG in an attempt to understand the consummatory responses to MSG by rats. Experimental data of a recent study by Ohara and Naim (41) will be discussed in some detail.

FACTORS AFFECTING THE SELECTION OF NUTRIENTS

The control of the ingestion of nutrients is known to be subject to the influences of taste, smell, and texture, as well as postingestional factors (18,23,33,48,56). Postingestional factors refer to those effects occurring after materials are ingested, including gastrointestinal tract processes or postabsorptive metabolic effects (9,22,33).

To ensure that any ingested food will not be poisonous, the animal uses at least two mechanisms. First, there are learning processes that make use of flavor as a cue, such that any postingestional consequences can be associated with food previously eaten. The rat accepts novel food slowly. This usually assures him of enough time to make a quality assessment. This type of learning has been termed "conditioned taste aversion" (13,14). Second, as a result of innate response, when an animal is given a choice of two foods, one of which is adulterated with an aversive taste stimulus, the animal will prefer the unadulterated (more "palatable") food. Many aversive taste stimuli are also poisons. This type of protective mechanism (i.e., simple rejection of adulterated food without apparent learning) might thus have had survival value in the phylogenetic history of vertebrates and, therefore, was genetically retained. Innate, learned, and postingestional responses can therefore influence the preference-aversion curves derived from preference tests.

It appears that, in a choice situation, animals are often able to select nutrients according to their physiological needs (18,48,56). This is specially true for rats. The immediate recognition of sodium by sodium-deficient rats has been argued to be an innate response via taste mechanisms (for review, see ref. 36). More common are the mechanisms wherein the animal associates the sensory quality of the food with postingestional consequences. For example, a 24-hr preference test indicated that glucose is preferred over sucrose (50), whereas during a brief exposure (i.e., 1 hr), sucrose is preferred over glucose (57). Harris et al. (18), and later Scott and Verney (56), reported that, when given several choices, rats deficient in the B vitamins were able to select the food containing these vitamins. Figure 1 illustrates a recent study (37) in which rats were given a choice between two diets differing not only in flavor, but also in nutritional potential (46). One diet contained defatted raw soybeans as a protein source with the addition of 0.35% (w/w) sodium saccharin (appealing taste). The other diet contained defatted heated soybeans with the addition of aversive stimulus of either 2.0% (w/w) sucrose octaacetate or 0.02% quinine sulfate. Raw soybeans have a lower nutritive value than the cooked soybeans because the raw contain digestive enzyme inhibitors and other undesirable physiological factors (4,28,29,46). Given a choice between these two diets, the rats initially preferred the diet with the better taste (bad nutrition). However, after 6 to 7 days, the rats changed their preference to the diet with the "better nutrition," even though it contained sucrose octaacetate (Fig. 1A). This may be explained by assuming that postingestional factors eventually influenced the animals to choose the diet offering better nutrition. If so, the rats associated the appealing taste with "bad nutrition" and the aversive taste with "good nutrition." When quinine sulfate was substituted for sucrose octaacetate in the above regimen (Fig. 1B), this change in preference was not seen. This might be explained by suggesting that no sensory habituation

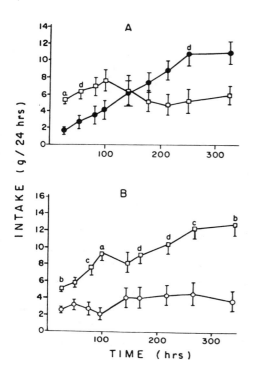

FIG. 1. Preference tests for diets containing unground raw soybean flakes (RS) mixed with an appealing taste stimulus vs. diets containing heated soybean meal (HS) mixed with an aversive taste stimulus. **A**: HS + 2.0% (w/w) sucrose octaacetate (●) vs. RS + 0.35% sodium saccharin (□). **B**: HS + 0.02% quinine sulfate (○) vs. RS + 0.35% sodium saccharin (□). Values are average food intake in grams ± SEM of 12 to 14 rats [$a = p < 0.001$, $b = p < 0.01$, $c = p < 0.02$, $d = p < 0.05$]. Diets were present at all times in the cages. Position of the diet cups was alternated daily. Water was available at all times. (From Naim, et al., ref. 37.)

occurred to the aversive taste of quinine. Alternatively, quinine might have produced postingestional information, e.g., pharmacological effects (52), alerting the rats that the combination of heated soybeans with quinine is less nutritious or, perhaps, even harmful.

Preference tests are often used for making qualitative sensory judgements on the properties of a specific flavor. The brief-exposure experiments, where the taste stimuli are available for only a few minutes, are considered to be more valid measurements of taste, with little confounding by postingestional factors (45).

SELF-SELECTION OF MSG-FLAVORED WATER

As mentioned above, comparisons between the brief-exposure preference tests and long-term exposure tests can often separate sensory effects from postingestional ones. In our study (41) the brief-exposure, two-choice preference test was carried out according to the method described by Cagan and Maller (7). Forty-two rats (200 to 250 g) were individually housed. Animals were divided into three equal groups. After 8 hr of water deprivation, rats were subjected to a preference test between a single concentration of MSG in water versus water for a period of 10 min. Rats were tested three times during the week on alternate days. Each group of rats was subjected sequentially from lower to higher concentrations of MSG in the following percentages:

Group 1: 0.005, 0.05, 0.5, 5.0, 50.0 (2.7×10^{-5} to 2.7 M).
Group 2: 0.001, 0.02, 0.1, 1.0, 8.0, 16.0 (5.3×10^{-5} to 8.6×10^{-1} M).
Group 3: 0.005, 0.25, 3.0, 12.0 (2.7×10^{-4} to 6.4×10^{-1} M).

In long-term preference tests, 126 weanling rats (40 to 50 g) were used. Rats were divided into nine groups. Each group was subjected to a two-choice preference test between a single concentration of MSG in water versus water for a period of 14 days. During this period, the rats were maintained *ad lib.* on rat chow. The following percentage concentrations of MSG in water were used: 0.005, 0.02, 0.05, 0.1, 0.25, 0.5, 1.0, 3.0, and 5.0 (2.7×10^{-4} to 2.7×10^{-1} M).

Figures 2 and 3 indicate that rats show a strong preference-aversion response to the taste of aqueous solutions containing MSG, in both short- and long-term testing. These results are qualitatively compatible with electrophysiological measurements of the chorda tympani nerve in rats (54). The results of long-term tests using male rats are similar quantitatively to those obtained by Hiji and Sato (19) with female rats.

Positive responses to MSG solutions were observed when MSG served as a single taste stimulus in deionized water, suggesting that under these circumstances MSG was a taste stimulant rather than a taste enhancer (15). The Na^+ alone cannot simply account for the sensory response for MSG (1). In a different experiment (41), rats

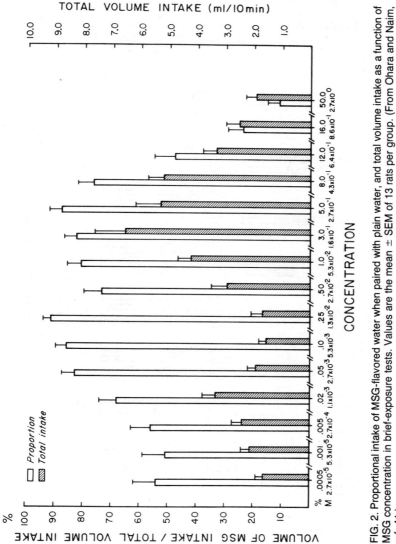

FIG. 2. Proportional intake of MSG-flavored water when paired with plain water, and total volume intake as a function of MSG concentration in brief-exposure tests. Values are the mean ± SEM of 13 rats per group. (From Ohara and Naim, ref. 41.)

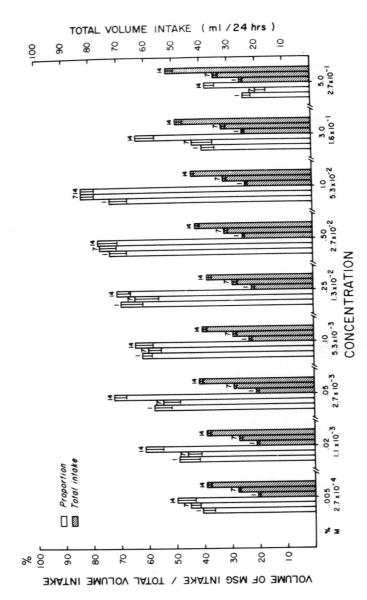

FIG. 3. Proportional intake of MSG-flavored water when paired with plain water, and total volume intake as a function of MSG concentration on days 1, 7, and 14 of long-term test. (From Ohara and Naim, ref. 41.)

could detect the difference between MSG, monosodium aspertate, sodium acetate, and sodium glutamate in situations where the Na^+ content and pH were kept equal.

As shown by the short-term tests (Fig. 2), the preference for solutions containing MSG occurred for a wide range of the offered concentrations. For brief-exposure experiments, there was a significant stimulation of the total liquid intake (MSG solution and water) over the range between 1 and 8% MSG as compared to lower concentration (0.05 to 1%). This suggests that solutions containing 1 to 8% MSG were most preferred in brief-exposure experiments. A solution containing 16% MSG was the minimum concentration required to cause a significant reduction of intake compared to water.

In long-term tests (Fig. 3), the total intake of liquid from both choices was slightly increased at high concentrations of MSG. This might be due to the necessity of diluting the concentrated solutions, since the proportional intake (intake of MSG solution as percentage of total intake) of solutions containing MSG was significantly reduced in that range. The proportion data suggested that in long-term tests solutions of 0.1 to 1% MSG are consistently preferred over water. A solution that contained 5% MSG was consistently less preferred than water.

Comparisons of the proportion data between brief-exposure and long-term testing suggest that solutions of 3 to 5% MSG might have produced, in long-term tests, postingestional feedback that reduced the amount of MSG intake. In short-term testing, these concentrations were still highly preferred. However, any analysis of the postingestional effects that might be caused by the ingestion of MSG requires a determination of the absolute amounts consumed. A relationship that expresses the ingestion of milliliters of taste stimulus solution versus milliliters of deionized water does not always give meaningful information. Therefore, the number of moles of MSG ingested as a function of the molar concentration of MSG in the solutions offered was calculated. Figure 4 shows that for brief exposure experiments the total number of moles of MSG consumed increased as the concentration of MSG solution rose, peaking at a concentration of about 3 to 5×10^{-1} M MSG. The total number of moles consumed was reduced when higher concentrations of MSG were presented. In the long-term experiments, the total number of moles of MSG consumed (Fig. 5) increased as a function of increasing MSG concentration. This includes the solution of 5% (2.7×10^{-1} M) MSG, which was less preferred than water in terms of milliliters ingested (Fig. 3).

Measurements of MSG intake were taken at concentrations that ranged 10^5-fold for brief-exposure tests and 10^3-fold for long-term tests. The data indicate that the total moles of MSG ingested peaked for both brief and long exposure at a similar concentration of MSG (Figs. 4 and 5). This suggests that the maximum intake of moles of MSG occurs at similar concentrations of MSG, whether or not the experiments were conducted over a 10-min period or over a 2-week period. It is therefore concluded that rats did not change their preference for ingestion of moles of MSG when exposed to long-term preference tests as compared to the short-term.

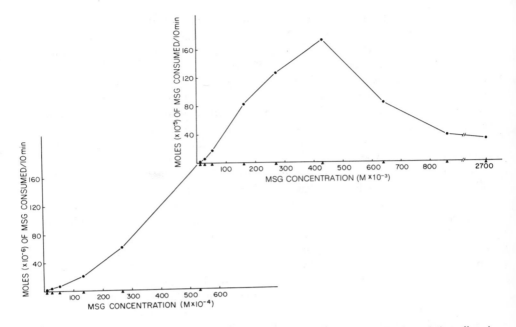

FIG. 4. The intake of moles of MSG as a function of MSG concentration in the solution offered during brief-exposure tests. Values are the mean of 13 rats per group. (From Ohara and Naim, ref. 41.)

This result thus rules out postingestional effects in long-term tests with solutions containing up to 5% MSG. The curve shape suggests sensory response (26).

At low concentrations of offered MSG, the curve is ascending linearly, whereas higher concentrations bring about the saturation of taste receptors (Figs. 4 and 5). In the very high MSG concentrations (tested only for brief exposure), the total moles of MSG consumed is significantly reduced. This cannot be explained by viscosity (texture) changes when the solutions became very concentrated, since the measured viscosity value of solutions containing 50% MSG (2.7 M) was 5.5 centistokes lower than those values necessary for preference by rats (C. M. Christensen, unpublished observation).

SELF-SELECTION OF MSG-FLAVORED DIET

Weanling rats were divided into two groups of 70 each. Each group was subdivided into five subgroups. Rats were given a two-choice preference test for a period of 7 days between a casein-based diet and a single concentration of MSG in the same diet. For one group the casein diet contained 9% protein, while for the second group it contained 18%. Each subgroup received a single concentration of MSG in their casein diet as a choice along with the diet without MSG. The following diet concentrations of MSG were used: 0.1, 0.5, 1.0, 3.0, and 7.0%.

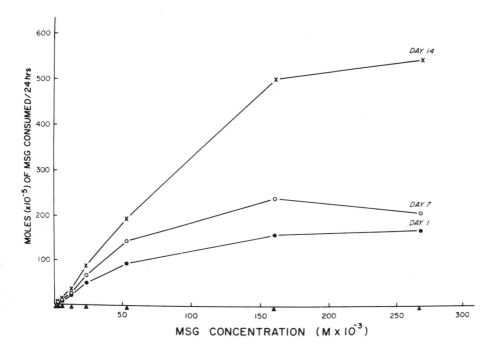

FIG. 5. The intake of moles of MSG as a function of MSG concentration in the solution offered during long-term tests. Values are the mean of 14 rats per group. (From Ohara and Naim, ref. 41.)

Table 1 indicates that rats do not show a preference for MSG-flavored diets over unadulterated diets. This was true for both low- or high-protein diets. This part of the study confirms the data of Scott and Quint (55). Since no preference for MSG-flavored diet was observed, the taste-enhancing phenomenon cannot be concluded from this study. Diets containing a high (7%) level of MSG were, in general, less preferred than unadulterated diets.

The lack of preference for MSG in solid food as compared to a strong preference in solution may be explained in two ways. Perhaps there is a competition between MSG and dietary components for the gustatory receptors that masks the sensory properties of MSG. Another possible reason for the lack of preference for an MSG-flavored diet might be an incomplete ionization of the MSG carboxyl groups in the diet compared to its aqueous state. This ionization has been reported by Fagerson (10) as necessary for taste sensation. A competition hypothesis is supported by the data showing a lower proportional intake (intake of MSG-flavored diet as a percentage of the total intake) of 9% protein—7% MSG diets than 18% protein—7% MSG. It is possible that, in the competition between protein components and MSG for gustatory receptors, MSG molecules were favored when the diet containing a low protein level. As a result, the rats could more easily detect the

TABLE 1. *Proportional intake of MSG-flavored diet when paired with plain diet, both containing either 9 or 18% protein**

		MSG concentrations in diet (%)				
	%	0.1	0.5	1.0	3.0	7.0
Day 1	9	48 ± 4[a,†]	54 ± 4[a]	58 ± 6[a]	60 ± 6[a]	46 ± 5[a]
	18	54 ± 5[a]	53 ± 5[a]	51 ± 4[a]	51 ± 4[a]	41 ± 5[a]
Days 2–3	9	43 ± 4[a]	43 ± 5[a]	50 ± 5[a]	39 ± 6[a]	22 ± 3[b]
	18	47 ± 3[a]	51 ± 3[a]	53 ± 4[a]	51 ± 3[a]	38 ± 3[a]
Days 4–7	9	56 ± 6[a]	55 ± 7[a]	60 ± 5[a]	40 ± 8[a,b]	12 ± 2[c]
	18	47 ± 3[a]	51 ± 3[a]	57 ± 4[a]	46 ± 3[a,b]	30 ± 3[b]

*Intake of MSG-flavored diet expressed as a percentage of total intake. Values are mean ± SEM of 13 to 14 rats per group.
† Comparisons following analysis of variance were made for each block of days across concentrations. All values not designated by the same superscript letter are different at least at the 5% level.
From Ohara and Naim, ref. 41.

aversive taste of 7% MSG. No differences were found for the total intake of diets, nor for body weight gains, as a function of MSG concentration in the diet. This might indicate that no amino acid imbalances occurred. Depression in food intake is known to occur within 3 hr in response to feeding a diet with amino acid imbalances (51). Thus, the thesis that the selection of MSG was by sensory means rather than postingestional feedback regulation is further supported in the diet experiments.

CONCLUSIONS

Different analyses resulted in two different expressions of MSG preference in rats. The expression of MSG acceptance by volume intake of MSG solutions, compared to deionized water, was different from that which expressed the intake of moles of MSG as a function of its concentration in solution offered. To separate the sensory properties of MSG from possible postingestional effects, it was suggested that acceptance is best expressed as the number of moles of MSG ingested for each solution.

It can be concluded from the data that in a two-choice situation, rats, by using sensory input, will not select high levels of glutamate. This may be an example of the sensory quality limiting consumption without the necessity of invoking postingestional feedback mechanisms. The suggestions that sensory, rather than postingestional factors, regulate the intake of MSG is compatible with other studies showing that an increase of glutamate in the diet does not usually lead to an increase of glutamate in the circulatory system (8,12,34). Since the rate of free amino acids crossing the intestinal wall diminishes with increasing electrical charge (11), glutamic and aspartic acids are slowly absorbed. These acids were also found to be the most rapidly excreted (21). Further confirmation is afforded by Windmueller and Spaeth (62), who demonstrated that the intestine metabolizes nearly all absorbed dietary glutamate.

ACKNOWLEDGMENT

I thank Dr. Judith R. Ganchrow for her valuable help in the preparation of this manuscript.

REFERENCES

1. Bartoshuk, L. M., Cain, W. S., Cleveland, C. T., Grossman, L. S., Marks, L. E., Stevens, J. C., and Stolwijk, J. A. J. (1974): Saltiness of monosodium glutamate and sodium intake. *JAMA*, 230:670.
2. Beauchamp, G. K., and Maller, O. (1977): The development of flavor preferences in humans: A review. In: *The Chemical Senses and Nutrition,* edited by M. R. Kare and O. Maller, pp. 291–310, Academic Press, New York.
3. Behrman, H. R., and Kare, M. R. (1968): Canine pancreatic secretion in response to acceptable and aversive taste stimuli. *Proc. Soc. Exp. Biol. Med.,* 129:343–346.
4. Birk, Y. (1961): Purification and some properties of a highly active inhibitor of trypsin and α-chymotrypsin from soybeans. *Biochim. Biophys. Acta,* 54:378–381.
5. Brand, J. G., Naim, M., and Kare, M. R. (1978): Taste and nutrition. In: *Handbook of Nutrition and Food,* edited by M. Recheigl, CRC Press, Cleveland, Ohio (*in press*).
6. Cagan, R. H. (1977): A framework for the mechanisms of action of special taste substances: The example of monosodium glutamate. In: *The Chemical Senses and Nutrition,* edited by M. R. Kare and O. Maller, pp. 343–359. Academic Press, New York.
7. Cagan, R. H., and Maller, O. (1974): Taste of sugars: Brief exposure single-stimulus behavioral method. *J. Comp. Physiol. Psychol.,* 87:47–55.
8. Delhumeau, G., Pratt, G. V., and Gitler, C. (1962): The absorption of amino acid mixtures from the small intestine of the rat. I. Equimolar mixtures and those simulating egg albumin, casein and zein. *J. Nutr.,* 77:52–60.
9. Epstein, A., and Teitelbaum, P. (1962): Regulation of food intake in the absence of taste, smell, and other oropharyngeal sensation. *J. Comp. Physiol. Psychol.,* 55:753–759.
10. Fagerson, I. S. (1954): Possible relationship between the ionic species of glutamate and flavor. *J. Agric. Food Chem.,* 2:474–476.
11. Fauconneau, G., and Michael, M. C. (1970): The role of the gastrointestinal tract in the regulation of protein metabolism. In: *Mammalian Protein Metabolism,* Vol. 4, edited by H. N. Munro, pp. 481–522. Academic Press, New York.
12. Finch, L. R., and Hird, F. J. R. (1960): The uptake of amino acids by isolated segments of rat intestine. I. A survey of factors affecting the measurement of uptake. *Biochim. Biophys. Acta,* 43:268–277.
13. Garcia, J., Hankins, W. G., and Rusinak, K. W. (1974): Behavioral regulation of the milieu interne in man and rat. *Science,* 185:824–831.
14. Garcia, J., and Brett, L. P. (1977): Condition responses to food odor and taste in rats and wild predators. In: *The Chemical Senses and Nutrition,* edited by M. R. Kare and O. Maller, pp. 277–289. Academic Press, New York.
15. Girardot, N. F., and Peryam, D. R. (1954): MSG's power to perk up foods. *Food Eng.,* 182:71–72, 182–183.
16. Grossman, M. I. (1967): Neural and hormonal stimulation of gastric secretion. In: *Handbook of Physiology,* Vol. 2, edited by C. F. Code, Sec. 6, pp. 835–863. American Physiological Society, Washington, D.C.
17. Hac, L. R., Long, M. L., and Blish, M. J. (1949): The occurrence of free L-glutamic acid in various foods. *Food Technol.,* 3:351–354.
18. Harris, L. J., Clay J., Hargreaves, F., and Ward, A. (1933): Appetite and choice of diet. The ability of the vitamin B deficient rat to discriminate between diets containing and lacking the vitamin. *Proc. Roy. Soc. Lond.* [*Biol.*], 113:161–190.

19. Hiji, Y., and Sato, M. (1967): Preference-aversion function for sodium monoamino-dicarboxylates in rats. *J. Physiol. Soc. Jpn.*, 29:168–169.
20. Hommel, H., Fischer, U., Retzlaff, K., and Knöfler, H. (1972): The mechanism of insulin secretion after oral glucose administration. II. Reflex insulin secretion in conscious dogs bearing fistulas of the digestive tract by sham-feeding of glucose or tap water. *Diabetologia*, 8:111–116.
21. Jacobs, F. A., and Lang, A. H. (1965): Dynamics of amino acid transport in the intact intestine. *Proc. Soc. Exp. Biol. Med.*, 118:772–776.
22. Jacobs, H. L. (1962): Some physical, metabolic and sensory components in the appetite for glucose. *Am. J. Physiol.*, 203:1043–1054.
23. Janowitz, H. D., and Grossman, M. (1949): Some factors affecting food intake of normal dogs and dogs with esophagostomy and gastric fistulae. *Am. J. Physiol.*, 159:143–148.
24. Kare, M. R. (1970): Taste, smell and hearing. In: *Dukes' Physiology of Domestic Animals*, edited by M. J. Swenson, pp. 1160–1185. Cornell University Press, Ithaca, New York.
25. Kirimura, J., Shimizu, A., Kimizuka, A., Ninomiya, T., and Katsuya, N. (1969): The contribution of peptides and amino acids to the taste of foodstuffs. *J. Agric. Food Chem.*, 17:689–695.
26. Krueger, J. M., and Cagan, R. H. (1976): Biochemical studies of taste sensation. Binding of L-[^{3}H]alanine to a sedimentable fraction from catfish barbel epithelium. *J. Biol. Chem.*, 251:88–97.
27. Kuninaka, A., Kibi, M., and Sakaguchi, K. (1964): History and development of flavor nucleotides. *Food Technol.*, 18:287–293.
28. Kunitz, M. (1947): Crystalline soybean trypsin inhibitor. II. General properties. *J. Gen. Physiol.*, 30:291–310.
29. Liener, I. E., and Kakade, M. L. (1969): Protease inhibitors. In: *Toxic Constituents of Plant Foodstuff*, edited by I. E. Liener, pp. 7–68. Academic Press, New York.
30. Lockhard, E. E., and Gainer, J. M. (1950): Effect of monosodium glutamate on taste of pure sucrose and sodium chloride. *Food Res.*, 15:459–464.
31. Maeda, S., Eguchi, S., and Sasaki, H. (1958): The content of free L-glutamic acid in various foods. *J. Home Econ. Jpn.*, 9:163–167.
32. Mehren, M. J., and Church, D. C. (1976): Influence of taste-modifiers on taste responses of pygmy goats. *Anim. Prod.*, 22:255–260.
33. Mook, D. G. (1963): Oral and postingestional determinants of the intake of various solutions in rats with esophageal fistulas. *J. Comp. Physiol. Psychol.*, 56:645–659.
34. Munro, H. N. (1970): Free amino acid pools and their role in regulation. In: *Mammalian Protein Metabolism*, Vol. 4, edited by H. N. Munro, pp. 299–386. Academic Press, New York.
35. Nachman, M. (1963): Taste preferences for lithium chloride by adrenalectomized rats. *Am. J. Physiol.*, 205:219–221.
36. Nachman, M., and Cole, L. P. (1971): Role of taste in specific hungers. In: *Handbook of Sensory Physiology*, Vol. 4, edited by L. M. Beidler, pp. 337–362. Springer-Verlag, Berlin.
37. Naim, M., Kare, M. R., and Ingle, D. E. (1977): Sensory factors which affect the acceptance of raw and heated defatted soybeans by rats. *J. Nutr.*, 107:1653–1658.
38. Naim, M., Kare, M. R., and Merritt, A. M. (1978): Effects of oral stimulation on the cephalic phase of pancreatic exocrine secretion in dogs. *Physiol. Behav.*, 20:563–570.
39. Nicolaidis, S. (1969): Early systemic responses to orogastric stimulation in the regulation of food and water balances: Functional and electrophysiological data. *Ann. N.Y. Acad. Sci.*, 157:1176–1203.
40. Novis, B. H., Banks, S., and Marks, I. N. (1971): The cephalic phase of pancreatic secretion in man. *Scan. J. Gastroenterol.*, 6:417–421.
41. Ohara, I., and Naim, M. (1977): Effects of monosodium glutamate on eating and drinking behavior in rats. *Physiol. Behav.*, 19:627–634.
42. Pavlov, I. P. (1910): *The Work of the Digestive Glands*, 2nd ed., translated by W. H. Thompson. Griffin, London.
43. PeThein, M., and Schofield, B. (1959): Release of gastrin from the pyloric antrum following vagal stimulation by sham feeding in dogs. *J. Physiol. (Lond.)*, 148:291–305.
44. Preshaw, R. M., Cook, A. R., and Grossman, M. I. (1966): Sham feeding and pancreatic secretion in the dog. *Gastroenterology*, 50:171–178.
45. Puerto, A., Deutsch, J. A., Molina, F., and Roll, P. L. (1976): Rapid discrimination of rewarding nutrient by the upper gastrointestinal tract. *Science*, 192:485–487.
46. Rackis, J. J. (1974): Biological and physiological factors in soybean. *J. Am. Oil Chem. Soc.*, 51:161A–174A.
47. Richardson, C. T., Walsh, J. H., Cooper, K. A., Feldman, M., and Fordtran, J. S. (1977): Studies

on the role of cephalic-vagal stimulation in the acid secretory response to eating in normal human subjects. *J. Clin. Invest.*, 60:435–441.
48. Richter, C. P. (1936): Increased salt appetite in adrenalectomized rats. *Am. J. Physiol.*, 115:155–161.
49. Richter, C. P. (1953): Experimentally produced behavior reactions to food poisoning in wild and domesticated rats. *Ann. N.Y. Acad. Sci.*, 56:225–239.
50. Richter, C. P., and Campbell, K. H. (1940): Taste threshold and taste preferences of rats for five common sugars. *J. Nutr.*, 20:31–46.
51. Rogers, Q. R., and Leung, P. M. B. (1977): The control of food intake: When and how are amino acids involved? In: *The Chemical Senses and Nutrition*, edited by M. R. Kare and O. Maller, pp. 213–248. Academic Press, New York.
52. Rollo, I. M. (1975): Drugs used in the chemotherapy of malaria. In: *The Pharmacological Basis of Therapeutics*, 5th ed., edited by L. S. Goodman and A. Gilman, pp. 1045–1068, Macmillan Publishing Co., New York.
53. Sarles, H., Dani, R., Prezelin, G., Souville, C., and Figarella, C. (1968): Cephalic phase of pancreatic secretion in man. *Gut*, 9:214–221.
54. Sato, M., and Akaike, N. (1965): 5'-Ribonucleotides as gustatory stimuli in rats. Electrophysiological studies. *Jpn. J. Physiol.*, 15:53–70.
55. Scott, E. M., and Quint, E. (1946): Self selection of diet. II. The effect of flavor. *J. Nutr.*, 32:113–119.
56. Scott, E. M., and Verney, E. L. (1947): Self selection of diet. VI. The nature of appetites for B vitamins. *J. Nutr.*, 34:471–480.
57. Shuford, E. H., Jr. (1959): Palatibility and osmotic pressure of glucose and sucrose solutions as determinants of intake. *J. Comp. Physiol. Psychol.*, 52:150–153.
58. Sjöstörm, L. B., and Crocker, E. C. (1948): The role of monosodium glutamate in the seasoning of certain vegetables. *Food Technol.*, 2:317–321.
59. Stegink, L. D., Filer, L. J., Jr., and Baker, G. L. (1972): Monosodium glutamate: Effect on plasma and breast milk amino acid levels in lactating women. *Proc. Soc. Exp. Biol. Med.*, 140:836–841.
60. Steffens, A. B. (1976): Influence of the oral cavity on insulin release in the rat. *Am. J. Physiol.*, 230:1411–1415.
61. Waldem, D. E., and Van Dyk, R. D. (1971): Effect of monosodium glutamate in starter rations on feed consumption and performance of early weaned calves. *J. Dairy Sci.*, 54:262–265.
62. Windmueller, H. G., and Spaeth, A. E. (1975): Intestinal metabolism of glutamine and glutamate from the lumen as compared to glutamine from blood. *Arch. Biochem. Biophys.*, 171:662–672.

Glutamic Acid: Advances in Biochemistry and Physiology, edited by L. J. Filer, Jr., et al.
Raven Press, New York © 1979.

Free and Bound Glutamate in Natural Products

T. Giacometti

22, Avenue de Tramenaz, 1814 La Tour de Peilz, Switzerland

> For almost 90 years, glutamic acid, a ubiquitous component of protein of well defined chemical composition and a nonessential amino acid which the body can synthesize, has attracted relatively little attention. Suddenly, through the remarkable coincidence of many unexpected findings it has been brought into the bright light of an interest which goes well beyond the limits of science. It has now become the object of most vivacious discussions in front of public opinion. (7)

The state of affairs alluded to in the above statement from 1955 by Kuhnau in Klingmüller's *Biochemie, Physiologie und Klinik der Glutaminsaure* (7), a no longer readily available compendium of the scientific knowledge of this important amino acid, should certainly be applicable to our present situation; however, it is not. The opinions in 1955 concerning the dispute on whether glutamate (Glu) could enhance intellectual level could very well apply to the situation in 1978, but Kuhnau would be very surprised to learn the present meaning of public opinion: a heterogeneous mass of listeners and readers who can be easily reached and impressed.

For the vast majority of all known proteins, Glu is the major amino acid. As an essential link in intermediary metabolism, it is also largely present in its free form in animal and plant tissues. Table 1 shows the levels of Glu that have been reported in various human tissues. The small amount in plasma is remarkable, and it

TABLE 1. *Free Glu in the organs of a normal adult*

Tissue	Free Glu (mg)
Muscles	6,000
Brain	2,250
Kidneys	680
Liver	670
Blood plasma[a]	40
Total	9,640

[a] Glu in plasma—total free Glu: 0.41%.
From K. Lang, *unpublished*.

TABLE 2. *Free amino acids in the muscles of a man of 70 kg*

Muscle mass	28 kg
Intracellular muscular water	18.2 liters
Total amino acid intracellular pool consisting of	86.5 g
Glutamine	61%
Glu	13.5%
Alanine	4.4%
8 Essential amino acids	8.4%

does not increase to more than 120 mg even after ingestion in excess of current average dietary amounts of the free form. This underscores the very high metabolic rate (5 to 10 g/hr). According to Bergstrom (1), the concentration in human muscle could be greater than the 6,000 mg shown in Table 1 (Table 2). It is remarkable that 79% of the free amino acids in the muscle consists of three closely related compounds: glutamine, Glu, and alanine.

As the sodium salt, Glu has become a widely used food ingredient. There is hardly a product that could more justifiably be called a food ingredient. Unfortunately, monosodium glutamate (MSG) is considered to be a "synthetic food additive" or a "chemical seasoning." Brillat-Savarin, who 150 years ago, claimed that the future of gastronomy belonged to chemistry, would have liked the definition. Today, "chemical" seems to imply danger and may elicit an emotional response to a given issue.

It is useful to remember the formula of MSG because the water of crystallization tends to be ignored in toxicological evaluations, and this makes a 10% difference in weight (Table 3). For current use, the sodium is insignificant (not more than in a glass of milk), unless total Na restriction is recommended. However, in the case of high-dose experiments, the risk of a concomitant hypernatremia should not be overlooked. Klingmüller (7) felt that it might be misleading to equate Na in MSG to the equimolecular amount in NaCl because Na and Cl follow a very close metabolic path, whereas Na and Glu widely diverge.

The Glu properties of improving the palatability of processed and preserved foods

TABLE 3. *Characterization of MSG:*

$$COOH - CH_2 - CH_2 - CH_2 - COONa \cdot H_2O$$
$$|$$
$$NH_2$$

Constituent	Percent
Glu	78.2
Na	12.2
H_2O	9.6
Molecular weight	187.3
1 g Glu	1.27 g MSG
1 g MSG	0.122 g Na

TABLE 4. *Annual industrial production of Glu*

Country	Production (tons)
Japan	65,000
Europe	40,000
Korea	40,000
Taiwan	25,000
U.S.	20,000
Others	10,000
World production	200,000

and of giving the last touch to high gastronomy were known before its discovery. The origin of soy sauce (containing about 1% Glu) is lost to us. However, the actual beginning of industrial production of acidic protein hydrolysates could have been inspired by Justus von Liebig (1803 to 1873), who remarked that they tasted pleasant and meatlike. An enterprising young Swiss of Italian extraction, Julius Maggi, picked up the idea and developed the industrial production of protein hydrolysate, which became a striking commercial success. He thought, at first, that he had hit on an authentic reproduction of meat extract. He soon found that the composition was very different and that the quality of his product depended on the choice of protein, the best being wheat gluten and casein. He had actually discovered the two essential elements of what we today call a hydrolyzed vegetable protein (HVP): the reacted flavors and Glu. Maggi learned this before his death in 1912. In fact, Ritthausen had isolated Glu in 1866 without reporting its organoleptic properties. The properties of Glu were detected in a natural product where it is present in its free form: in 1908, Ikeda extracted Glu from a seaweed widely used in ancient Japanese cooking (*Dashikombu* or *Laminaria japonica*), which is one of the natural substances with the highest content of free Glu (0.42%). This discovery, coupled with the finding by Ritthausen that wheat gluten contained 25 to 30% of Glu, was the starting point for the industrial production of MSG.

Today, MSG is produced primarily by a fermentation process. Some figures about Glu production are shown in Tables 4 and 5, the daily world metabolic turnover being 450,000 to 900,000 tons per day. Without Ikeda or Admiral Perry, a Frenchman pondering the ideal association of wine and cheese could have found

TABLE 5. *Some natural sources of Glu (as MSG)*

Food source (tons)	MSG contents (tons)
Tomatoes	
U.S. (6,500,000)	11,600
Western Europe (8,000,000)	14,300
Parmesan-type cheese	
Italy (150,000)	2,286

TABLE 6. *MSG consumption in selected countries*

Country	Total tons/yr	g/day
Taiwan	18,000	3
Korea	30,000	2.3
Japan	65,000	1.6
Italy	8,000	0.4
U.S.	28,000	0.35

TABLE 7. *Natural free Glu (as MSG) from other sources in Italy*

Food source (tons)	MSG contents (tons)	g/day
Tomatoes (3,000,000)	5,350	0.26
Parmesan cheese (150,000)	2,286	0.11

that the velvety smoothness was conferred to wine by the high content of free Glu in the cheese. Furthermore, an Italian could have asked why Parmesan cheese was such a popular all-purpose seasoning, why tomatoes are so essential in a sauce, and why they blend so well with Parmesan cheese. In fact, the consistency of many traditional foods seems to be related to the presence of free Glu.

Taiwan has the highest per capita consumption of MSG (3 g/day) (Tables 6 and 7). The consumption of Glu in the United States is marginal.

When speaking of the percentage of bound Glu in the food protein, it is most

TABLE 8. *Current food items for which Glu is not the leading amino acid on total nitrogen*

Food item	Glu (mg/g N)	Asp (mg/g N)
Potato	639	775
Sweet potato	541	825
Beet	946	1,131
Apple	700	1,300
Apricot	372	1,300
Avocado	769	1,413
Banana	575	656
Fig	600	1,500
Orange	760	880
Pear, Japanese	540	2,800
Strawberry	920	1,400
Brewer's yeast	669	678
Candida krusei	675	800
Ansenula anomala	775	779

TABLE 9. Glu in animal proteins

Protein	Glu (g/100 g)
Albumin (human serum)	17.0
Fibrinogen (human serum)	14.5
γ-Globulin (human serum)	11.8
Albumin (egg white)	16.5
α-Casein (milk)	22.5
β-Lactoglobulin (milk)	20.0
Actin (muscle)	14.8
Myosin (muscle)	21.0
Insulin	18.6
Pepsin	11.9
Keratin (human hair)	14.4
Keratin (wool)	11.9
Collagen (tendon)	11.3

practical to start by mentioning the few proteins in which Glu is not the major amino acid. The relevant FAO data do not concern unique proteins. In fruits and leaves, the nitrogen compounds include free amino acids and amides. If we disregard exotic roots and beggarstick and scratchbush leaves, we have the list of the more common items shown in Table 8. Glu and the leading amino acid, which is invariably aspartic acid (Asp), are given as milligrams per gram of nitrogen (4). Glu is also the most abundant amino acid in the proteins that have been isolated by scientific research (Table 9) (2). Various techniques have been used for the separation of proteins, as well as for their classification (Table 10) (2). Finally, for most food products it is possible to indicate their amino acid composition as a percentage of total nitrogen, since this has been done in the FAO tables, with the reservation, already mentioned, that some of the Glu and Asp are actually present in their free form. This figure is again a confirmation of the importance of Glu as a source of nonessential nitrogen (Table 11) (4).

Since the earliest times, cereals have been submitted to fermentation processes

TABLE 10. Glu in isolated plant proteins

Protein	Glu (g/100 g)
Gliadin (wheat)	45.7
Zein (maize)	26.9
Edestin (flax)	20.7
Hordenin (barley)	38.4
Globulin (coconut)	21.8
Arachin (peanut)	20.8
Globulin (cotton seed)	23.6
Glycinin (soybean)	20.5
Glutenin (wheat)	24.7
Lupin (lupine bean)	27.2

TABLE 11. *Glu as percentage of 16 g of total nitrogen in various foods*

Food item	Glu (%)	Food item	Glu (%)
Wheat gluten	37.4	Pea	14.6
Wheat flour (70–80%)	34.2	Lentil	16.7
Barley	23.6	Beet	14.9
Rye, whole meal	24.2	Asparagus	20.8
Rice, milled, polished	19.2	Carrot	19.4
Oats, meal	20.9	Potato	10.2
Maize, whole meal	19.0	Tomato	40.6
		Apple	11.4
		Apricot	6.1
		Grape	20.8
Hazelnut	20.6		
Brazil nut	18.6	Anchovy	15.1
Sunflower seeds	21.8	Haddock	15.8
Soya milk	17.4	Salmon	13.2

involving enzymatic hydrolysis of their proteins. Thus, free Glu became a taste component in human food. This is also evident for milk protein, and free Glu is an important flavor component of cheese (Table 12).

The levels of natural or added free Glu in the diet must be assessed within the framework of metabolic capacity by the normal activation of the digestive system. The latter is best observed in the course of feeding of a 4- to 6-month-old baby. There are basically two possibilities to consider when feeding, for instance, a jar of baby food to an infant. The infant either may accept the food or may react with rebellious sputtering until something is worked up. The digestive secretion is actually being worked up in the process. Figure 1 gives Nasset's sketch (9) of the mechanism of concurring actions that are set off by digestion. The gastrointestinal tract can deliver into its lumen a substantial amount of endogenous nitrogen (including amino acid), diluting the exogenous nitrogen in the small intestine by sevenfold in response to the ingestion of any type of meal. If the ingested protein has an unusual amino acid composition, as in zein, it is effectively obscured after

TABLE 12. *Free Glu in cheese*

Cheese	Glu (g/100 g)	MSG equiv.[a] (g/100 g)
Parmesan	1.2	1.52
Stilton	0.82	0.94
Roquefort	1.28	1.62
Gruyere de Comte	1.05	1.33
Saint Paulin	0.21	0.27
Camembert	0.39	0.49
Danish Blue	0.67	0.85
Gouda	0.46	0.58

[a] Glu × 1.27.

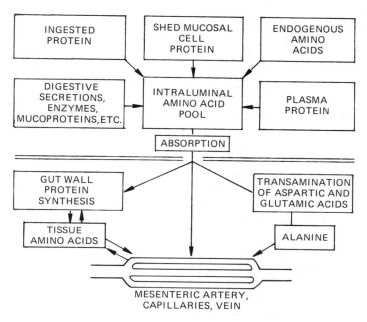

FIG. 1. Amino acid pool in small gut.

being exposed to the normal process of digestion. The amounts of amino acid delivered into the lumen depend on the chemical composition of the lumen. If several amino acids are present together, there is competition among them for sites in the transport system. In other words, there seems to be a mechanism that prevents acute, large fluctuations in the amino acid mixture that is available for absorption (a mechanism that is bypassed by intragastric alimentation). This may also be true for the ingestion of concentrated water solutions of Glu by human volunteers.

The concentration of MSG in baby food, before its voluntary removal by the manufacturers, was reported to be as high as 0.6% by the National Academy of Sciences in July 1970; (but no source of this information was given) (11). In my experience in this field, the addition was never more than 0.3%. Both 0.3 and 0.6% are generally within individual metabolic capacity; however, more than 0.3% in baby food is technologically excessive. The question, then, is whether MSG was a desirable ingredient. It has been said that its addition should have pleased only mothers, since infants are not taste conscious. This opinion is based on the assumption of the so-called cosmetic action of MSG, which should improve the quality of any food—the more the better, thus allowing for a reduction in the use of expensive ingredients. This is totally misleading. MSG cannot make a delicacy of meat and vegetable baby food. It can only make it more acceptable by reducing the occasional coarseness or astringency caused by processing. The favorable influence of Glu is primarily related to its taste-lingering effect, which, along with its stimulation of the salivary glands, is its most easily measurable action. MSG is not a flavor enhancer in the sense that has been purported. For instance, MSG makes saltiness milder, but does not increase the demand for salt. On the contrary, it tends to reduce the consumption of salt because it creates a new, more analytical perception of taste, as observations of patients deprived of salt have shown. This finding can also be applicable to infants. Supplemental baby foods have the advantage of assisting the cultivation of a taste for a variety of foods that, later on, are even more important in the diet. The parents' attitude toward food is unquestionably recognized by infants at an early age.

Protein-rich processed food may contain 0.3% MSG or slightly more. The clear broths (beef or chicken flavored)—which, combined with noodles, rice, or vegetables, constitute a wholesome and inexpensive staple food, not only in the Far East, but in all the Mediterranean countries—may contain 0.5 to 0.8%. It has been suggested that the occasional jar of baby food as well as the current processed food may expose the growing child to an excess of free Glu. From the still incomplete data, the ample supply of free Glu and Asp appears to come from fresh unprocessed foods and traditional foods (3,5,6,11). In Table 13 the contents of free Glu and Asp have been added together and transformed into their MSG equivalent.

Finally, it is noteworthy that a nutriment that nature has conceived for the requirements of the newborn—mother's milk—contains small but significant amounts of free Glu. The highest concentration is measured in the very first days after birth when the infant is supposed to be more sensitive. This may be nutritionally meaningless, although this was not the opinion given in 1971 by Montreuil, a pediatrician who felt that the existing knowledge justified the systematic and

TABLE 13. Free Glu and Asp in natural foods

Food item	Glu (mg/100 g)	Asp (mg/100 g)	MSG equiv.[a] (mg/100 g)
Tomato	140	35	221
Fresh tomato juice	260	60	406
Processed tomato juice	230	60	370
Grapefruit, white meat	11.5	87.1	125
Grapefruit juice	18.6	130	190
Orange juice	21	89	140
Nectarine, fruit	9.6	200	269
Peach, juice	32	212	274
Plum, yellow fruit	7.9	185	243
Prunes (California)	14.4	185.5	254
Prune dry	18.6	518.4	684
Grape, red Malaga	184	12	250
Grape juice	258	16.8	350
Strawberry	44.4	60.1	128
Potato	102	—	129
Broccoli	176	40	274
Parmesan	1,200	—	1,524
Gruyere de Comte	1,050	60	1,460
Mushroom (Psalliota campestris)	180	30	267

[a] Glu + Asp.

TABLE 14. *Free amino acid concentration in human and cow's milk*

Amino acid	Human milk (mg/100 ml)			Cow's milk (mg/100 ml)[a]
	2nd day	3 weeks	2 months	2 months
Glu	12.88	7.66	4.20	0.64
Glutamine	9.48	1.79	1.75	0.41
Asp	2.92	1.40	0.53	0.08

[a] At 2 months.

immediate supplementation of cow's milk with the free amino acid contained in human milk (8). At any rate, we can be sure that the wisdom of maternal physiology is not taking an unnecessary risk when the newborn has a plasma level above average, whereas the mother has only 50% of her basal level. According to the data in Table 14, the 3-kg newborn with a daily intake of 480 g of breast milk is exposed to 20.6 mg/kg body weight Glu and 4.7 mg/kg Asp. The total of the two dicarboxylic amino acids expressed as MSG is 32 mg/kg body weight.

CONCLUSIONS

A vast and long-standing literature indicates the overwhelming presence of bound and free Glu in the food supply. Some of the older figures are probably 25 to 50% too low. Similarly for Asp, some of the values may be only 50% of what would be obtained by the best modern methods.

The free dicarboxylic amino acids are quantitatively important constituents of tomatoes, fruit, cheese, and mushrooms. The relevant data are still incomplete, fragmentary and diverge widely according to variety, ripeness, fertilizing with respect to fruit and tomatoes, etc.

Glu is an important element in natural and traditional ripening processes that achieve fullness of taste. Its industrial use is a logical consequence.

Food preservation and processing tend to decrease free and bound Glu. The industrial use of MSG tends to compensate for this loss.

REFERENCES

1. Bergstrom, J., Fürst, P., Norée, L.-O., and Vinnars, E. (1974): Intracellular free amino acid concentration in human muscle tissue. *J. Appl. Physiol.*, 36:693–697.
2. Block, J. R., and Bolling, D. (1951): *The Amino Acid Composition of Proteins and Food.* Charles C Thomas, Springfield, Ill.
3. Do Ngoc, M., Do Ngoc, M., and Lenoir, J., and Choisy, C. (1971): Les acides libres des frommages affine de Camembert, Saint Paulin et Gruyere de Comte. *Rev. Lait. Fr.*, 288:447–454.
4. FAO/OMS (1970): *Amino Acid Content of Food and Biological Data on Proteins.* FAO/OMS, Rome.
5. Fernandez-Flores, E., et al. (1970): Qualitative GLC analysis of free amino acids in fruits and juices. *J. AOAC*, 536:1203–1208.
6. Kliewer, W. M. (1969): Free amino acids and other nitrogenous substances of table grape varieties. *J. Food Sci.*, 34:274–278.

7. Klingmüller, V. (1955): *Biochemie, Physiologie und Klinik der Glutaminsaure.* Editio Cantor, Aulendorf, Germany.
8. Montreuil, J. (1971): Humanisation du lait de vache. *Nutr. Alim.,* 25:A1–A37.
9. Nasset, E. S. (1965): Role of digestive system in protein metabolism. *Fed. Proc.,* 24:953–958.
10. National Research Council (1970): *Safety and Suitability of Monosodium Glutamate for Use in Baby Foods.* National Academy of Sciences, Washington, D.C.
11. Stadtmen, F. H. (1972): Free amino acids in raw and processed tomato juices. *J. Food Sci.,* 37:944–951.

Psychometric Studies on the Taste of Monosodium Glutamate

Shizuko Yamaguchi and Akimitsu Kimizuka

Central Research Laboratories, Ajinomoto Co., Inc., Suzuki-cho, Kawasaki 210, Japan

Monosodium glutamate (MSG) is generally used as a flavor enhancer. There are several scientific papers on the flavor effect of MSG added to food, but they seem to focus on its practical use (3a,4,5,6,17a,19,25). We conducted a two-part study to define the flavor effect of MSG. In the first part, the aspect of the flavor profile change of food when MSG was added to food was investigated psychometrically. In the second part, the fundamental flavor properties of MSG and other flavor substances were examined.

STUDY ON THE FLAVOR EFFECT OF MSG ON FOODS

Psychometric Approach

The addition of MSG, broth, salt, or sugar to a variety of different foods was investigated to learn how a general population, not specialists in food science, describe the flavor change of food. The flavor profile evaluation for a mass panel used the Semantic Differential Method (20) and the results were analyzed statistically.

Collection of Evaluation Terms

One hundred and fifty adults in our research laboratories and 30 female students studying food science were presented with eight different foods, with or without added MSG (0.1 to 2%) or containing different concentrations of broth, salt, or sugar. Comparing the samples, the subjects expressed freely their impression of the flavor profile of the food using their own expressions. They were not informed of the sample ingredients nor the purpose of the test.

Out of approximately 500 expressions obtained, 32 pairs of the expressions, which appeared most frequently and expressed the differentiated characteristics concretely, were selected. The expression of MSG-like taste or MSG taste appeared frequently. But these terms were consciously eliminated because the purpose of the study was to make a clear flavor profile of MSG itself. The 32 paired terms were listed on the evaluation sheet (Fig. 1) for the following set of experiments described below.

Methods in the Flavor Evaluation Test

The panel comprised 300 people, and the panel size of each experiment was 25 to 50 persons. Panel members received oral instructions from the experimenter. They

FLAVOR PROFILE OF [FOOD NAME]

Date ———————
Name ———————

DIRECTIONS: mark each line in the place that best expresses your feelings of SAMPLE B compared with SAMPLE A.

			certainly	slightly	almost same	slightly	certainly	
			−2	−1	0	1	2	
1	Whole aroma	/ weak	├──┼──┼──┼──┤					strong
2	Meaty aroma	/ weak						strong
3	Aroma derived from (······)	/ weak						strong
4	Whole aroma	/ bad						good
5	Meaty flavor	/ weak						strong
6	Flavor derived from (······)	/ weak						strong
7	Flavor of spice	/ weak						strong
8	Whole taste	/ weak						strong
9	Salty taste	/ weak						strong
10	Salty taste	/ rough						smooth
11	Sweet taste	/ weak						strong
12	Sour taste	/ weak						strong
13	Bitter taste	/ weak						strong
14	Meaty taste	/ weak						strong
15	Taste derived from (······)	/ weak						strong
16	Oily or Fatty	/ weak						strong
17	Foreign flavor	/ weak						strong
18	Continuity	/ short						long
19	Simple							Complex
20	Watery							Concentrated
21	Mouthfulness	/ weak						strong
22	Development	/ narrow						broad
23	Flat							Body
24	Light							Heavy
25	Poor							Rich
26	Thin							Thick
27	Harsh							Mild
28	Crude							Aged
29	Balance	/ bad						good
30	Punch	/ weak						strong
31	Unfavorable							Tasty
32	Palatability	/ bad						good

FIG 1. Evaluation sheet for the food flavor profile.

were to use the five-point rating scale for evaluating the 32 paired terms. The meanings of the terms were not defined for the panel members.

The foods were prepared (2,13,26,27,35) just before the test, and their appearance and temperature were controlled to preserve their best condition at serving (Table 1). Beef or chicken broth was prepared according to Berolzheimer's *Encyclopedic Cook Book* (2). Test samples were standardized at 100 g.

TABLE 1. *Foods presented for flavor profile test*

Item	Main raw material	Test additive
Soup		
beef consommé	Lean beef, vegetables	MSG, beef broth
chicken consommé	Chicken, vegetables	MSG
cream of chicken soup	Chicken, milk	MSG
chicken noodle soup	Chicken, noodle	MSG
cream of vegetable soup	Potato, onion, milk	MSG, chicken broth
vichyssoise	Potato, milk, chicken broth	MSG
onion soup	Onion, butter	MSG
cream of tomato soup	Tomato, milk	MSG, chicken broth
Japanese miso soup	Soybean paste	MSG
Meat		
hamburger	Ground beef, onion	MSG, salt
Poultry and eggs		
seasoned egg custard	Eggs, chicken, mushroom	MSG, bonito broth
Sea food		
coquilles of scallops	Scallops	MSG
Vegetable		
cooked vegetables	Carrot, corn, peas	MSG, chicken broth, beef broth
cooked broccoli	Broccoli	MSG
Dessert		
Bavarian cream	Gelatine, egg, cream	Sugar
caramel custard	Eggs, milk	Sugar

Results of the Flavor Evaluation Test

Classification of Evaluation Terms

The average scores obtained by the test on the evaluation scales were analyzed by principal-component analysis and by cluster analysis. The paired terms were classified into five major groups according to flavor functions, and some paired terms were united by their high coefficient of correlation (Table 2).

This classification was used to arrange the results of the evaluation sheets. The average score given by the subjects on each scale was drawn in a bar diagram to show the flavor profile change induced by the addition of MSG or other flavor

TABLE 2. *Classification of evaluation terms*

Aroma
  Whole aroma
  Aroma derived from material (meaty, vegetable-like, etc.)

Basic taste
  Whole taste
  Salty
  Sweet
  Sour
  Bitter

Flavor character
  Continuity
  Mouthfulness (complex, development, body, rich)
  Impact (concentrated, heavy, punch)
  Mild (aged)
  Thick

Other flavor
  Spicy
  Oily
  Flavor derived from material (meaty, vegetablelike, etc.)

Whole preference
  Preference (palatability, tasty, balance)

substance to the food (Figs. 2–7). Each bar diagram includes a line indicating the 95% confidence level.

Flavor Profile

The addition of MSG to beef consommé had no effect on aroma, but increased the overall taste intensity. Saltiness, sweetness, sourness, and bitterness were not significantly increased (Fig. 2). The addition of MSG increased the characteristics of the flavor (Fig. 2), i.e., continuity, mouthfulness, impact, mildness, and thickness of beef consommé. It also increased the meaty flavor and the overall preference for the beef consommé.

The same pattern was observed with hamburger, chicken consommé, chicken noodle soup, and cream of chicken soup (Figs. 3 and 4), as well as with scallop coquilles, cooked vegetables, and seasoned egg custard.

Doubling the concentration of beef consommé gave the same pattern of change in the flavor profile of beef consommé as did the addition of MSG, but additionally increased the intensity of aroma and the four basic tastes (Fig. 2).

Cream of tomato soup was found to be an exception. The addition of MSG did not change the flavor profile of this food very much. Adding chicken broth to the tomato soup changed only the sourness and continuity of the flavor (Fig. 5).

The beef consommé mentioned above contained 0.8% NaCl. Increasing NaCl levels to 1.2% changed only the saltiness ratings of the food and decreased its palatability (Fig. 6). But an increase in NaCl from 0.2% to 0.8% enhanced

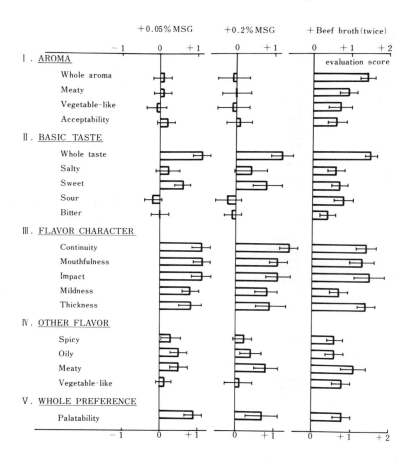

FIG. 2. Effect of MSG and beef broth on the flavor profile of beef consommé.

palatability and increased the flavor characteristics of continuity, mouthfulness, impact, mildness, and thickness (Fig. 6).

When comparing the different sugar contents of Bavarian cream between 10 and 20%, the latter was given larger evaluation scores of continuity, mouthfulness, impact, mildness, and thickness, as well as increased sweetness (Fig. 7).

Thus, in the case of some foods, both salt and sugar not only increased their intrinsic tastes, but also enhanced the flavor characteristics of the flavor character.

Flavor Effects of MSG

In summary, our experiments demonstrated the following pattern of effects of MSG on foods:

1. MSG has no effect on aroma of food.

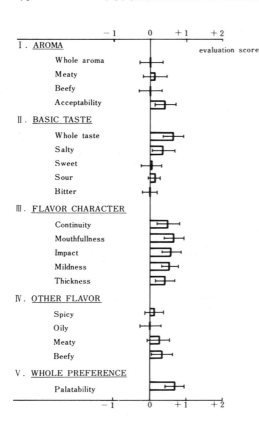

FIG. 3. Effect of MSG (+1%) on the flavor profile of hamburger.

2. MSG increases the total taste intensity of food. The quality of the taste brought about by MSG is different from the four basic tastes.
3. MSG enhances certain flavor characteristics of food: continuity, mouthfulness, impact, mildness, and thickness.
4. MSG enhances the specific flavor of meat and poultry foods.
5. MSG has a flavor effect similar to broth (beef stock), although MSG has no effect on aroma.
6. MSG increases the whole preference or palatability of food.

Discussion

The pattern of flavor effects of MSG on food was obtained by the flavor profile test. The role of MSG in food can be summarized by saying that MSG increases the taste other than the four basic tastes and improves certain flavor characteristics of food.

MSG has two flavor functions. One is that MSG imparts an intrinsic taste different than the four basic tastes. This intrinsic taste corresponds to the Japanese expression *umami,* meaning tastiness.

Another function of MSG is that it intensifies the flavor characteristics of food: continuity, mouthfulness, impact, mildness, and thickness; MSG also increases the whole preference of food. The same effect is brought about by the increase of broth concentration. Both salt and sugar, in some cases, not only increase the intrinsic tastes of foods, but also enhance the flavor characteristics mentioned above. These facts suggest that MSG, as well as salt and sugar, may generally be called a flavor enhancer.

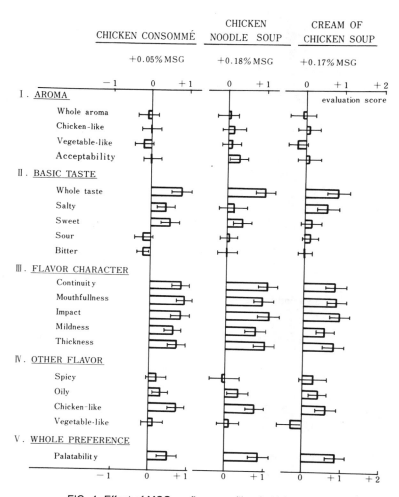

FIG. 4. Effect of MSG on flavor profile of chicken soups.

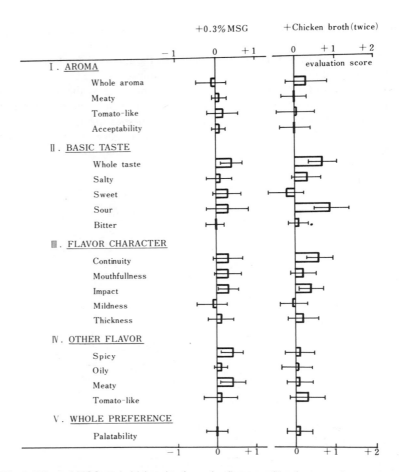

FIG. 5. Effect of MSG and chicken broth on the flavor profile of cream of tomato soup.

STUDY ON THE FUNDAMENTAL FLAVOR PROPERTIES OF MSG

Taste Threshold for MSG

The taste threshold for MSG is reported in a wide range of values from 4×10^{-5} to 4×10^{-3} M (7,12,14,17,18,23,24,36). The value varies depending on the methods of measurement or on the composition of the panel.

The absolute threshold for MSG was measured as follows: The triangle test was administered to a panel composed of 30 persons from our laboratories. The three samples, two pure water and one MSG solution, were presented to the panelist in each trial. For the series of trials, the samples of MSG solution were presented in order of decreasing concentration until subjects reported no difference between the MSG solution and pure water.

The absolute threshold for MSG was found to be 6.25×10^{-4} M. The thresholds for the four basic taste substances were determined in the same way by the same panel (Table 3). The threshold for MSG was higher than for quinine sulfate or tartaric acid, lower than for sucrose, and about the same as for sodium chloride.

Taste Intensity of MSG

The relationship between concentration and the perceived taste intensity of MSG and the four basic tastes were studied. The panel was the same group used for establishing the taste thresholds. The panelists were first trained to rate taste intensity by using a 100-point scale: zero was pure water and 100 was a 3.2×10^{-4} M quinine sulfate. The panelists evaluated 30 samples (5 taste com-

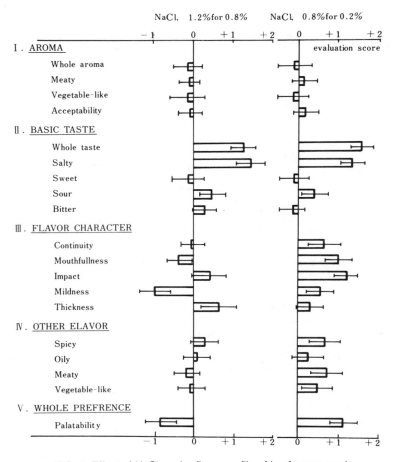

FIG. 6. Effect of NaCl on the flavor profile of beef consommé.

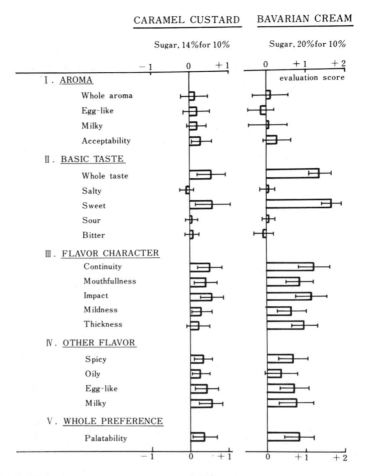

FIG. 7. Effect of sugar on the flavor profile of caramel custard and Bavarian cream.

TABLE 3. *Taste thresholds for selected compounds*

Solvent	Absolute threshold[a]				
	Sucrose	Sodium chloride	Tartaric acid	Quinine sulfate	MSG
Pure water	1.25×10^{-3} M $(4.3 \times 10^{-2}\%)$	6.25×10^{-4} M $(3.7 \times 10^{-3}\%)$	6.25×10^{-5} M $(9.4 \times 10^{-4}\%)$	6.25×10^{-7} M $(4.9 \times 10^{-5}\%)$	6.25×10^{-4} M $(1.2 \times 10^{-2}\%)$
5×10^{-3} M MSG	1.25×10^{-3} M	6.25×10^{-4} M	1.25×10^{-4} M $(1.9 \times 10^{-3}\%)$	6.25×10^{-7} M	—
5×10^{-3} M IMP	1.25×10^{-3} M	6.25×10^{-4} M	2.0×10^{-3} M $(3.0 \times 10^{-2}\%)$	2.5×10^{-6} M $(2.0 \times 10^{-4}\%)$	—

[a] Significant at 5% level.

pounds times 6 concentrations); the order of sample presentation was randomized. All 30 samples were evaluated twice by the same panelist. The panelist was instructed to do the following: (a) hold 10 ml of the sample in the mouth for 10 sec, (b) evaluate and score the taste strength, (c) rinse with tap water, and (d) take 1-min interval before the next tasting. The panelist recorded the subjective rating score of the taste intensity of each sample in the range between 0 and 100 points.

The results showed that Weber-Fechner's law was applicable to the relationship between concentration and the taste intensity of MSG, as well as to the relationship of the other four basic tastes (Fig. 8) (1,8).

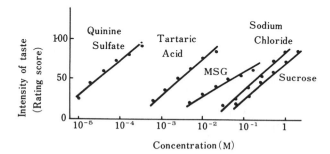

FIG. 8. Relationship between concentration and taste intensity.

Interaction Between MSG and the Four Basic Tastes

The influences of both MSG and inosinate (IMP) on the absolute thresholds of four basic tastes were tested with the same panel (Table 3). The results showed that the threshold for the four basic tastes in 5×10^{-3} M MSG or IMP solution was the same as in pure water. A slightly higher threshold for sourness in the MSG or IMP solution may have been caused by the change of the pH value.

The same panel was used to determine the interactions between MSG and the four basic tastes at suprathreshold levels. Magnitude estimation was used to measure the influence of MSG on the four other basic tastes. For example, the strength of sweetness was reported after the addition of MSG (the sweetness was set as 100 with no addition). The same method was used to measure the influence of the four basic tastes on MSG.

MSG did not increase the intensity of the four basic tastes. Conversely, the four basic taste substances did not increase the intensity of the taste of MSG (Fig. 9). A masking effect was observed more or less between MSG and the four basic tastes; this effect has been reported among the four basic tastes (9,21,22). The taste of MSG seems not to be one of the traditional four basic tastes nor to be composed of them. Multidimensional scaling has shown that the taste of MSG comprises a dimension independent from those of the four basic tastes (37). A number of psychologists who have studied MSG taste call it "distinctive" or "unique" (3).

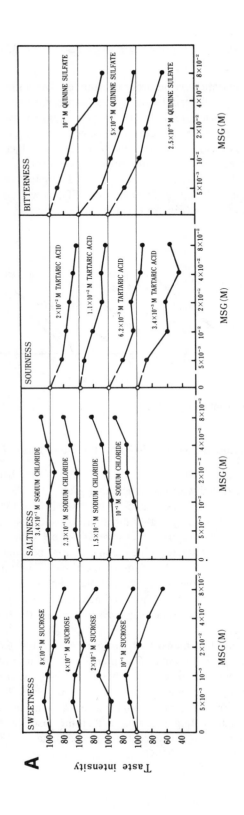

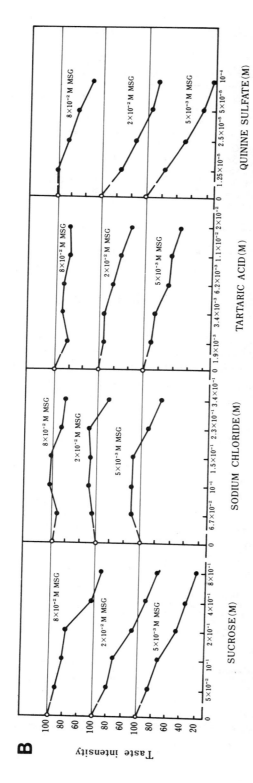

FIG. 9. Interactions between MSG and the four basic tastes by the Magnitude Estimation Method. **A:** Effect of MSG on the four basic tastes. **B:** Effect of the four basic tastes on MSG.

TABLE 4. *Taste substances similar to MSG*

Substance	Relative taste intensity	
	(g/g)	(mole/mole)
Monosodium L-glutamate·H_2O	1	1
Monosodium DL-threo-β-hydroxy glutamate·H_2O	0.86 ± 0.06	0.92
Monosodium DL-homocystate·H_2O	0.77 ± 0.04	0.92
Monosodium L-aspartate·H_2O	0.077 ± 0.007	0.071
Monosodium L-α-amino adipate·H_2O	0.098 ± 0.008	0.10
L-Tricholomic acid (erythro form)[a]	5–30	4.3–26
L-Ibotenic acid[a]	5–30	4.2–25

Table from Yamaguchi et al., ref. 34.
[a] From Terasaki et al., refs. 28 and 29.

Taste Compounds Similar to MSG

Several compounds have a similar taste to MSG: monosodium DL-threo-β-hydroxy glutamate, monosodium DL-homocystate, monosodium L-aspartate, monosodium L-α-amino adipate (10,11) L-tricholomic acid, and L-ibotenic acid (28,29). The taste intensity of these compounds was measured (29,34), and the results are shown in Table 4.

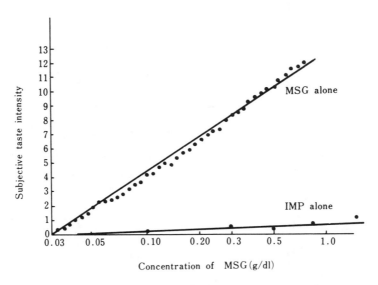

FIG. 10. Relationship between the concentration of MSG or IMP alone and taste intensity. (From Yamaguchi, ref. 31.)

Synergistic Effect of Nucleotides on the Taste of MSG

Inosinate (IMP) (15) alone has only a very weak taste (34) (Fig. 10). It is, however, known that IMP increases the intensity of the taste of MSG synergistically (16,30). The quantitative relationship of IMP to the taste intensity of MSG has been reported in detail (31,32,33).

The main points of those investigations are itemized as follows:

1. When the total concentration of MSG and IMP is constant, the taste intensity of the mixture increases remarkably following an increase of IMP. When the ratio of IMP reaches 50%, the intensity starts decreasing. The relationship between the portion of IMP and the taste intensity of the mixture is shown as a bisymmetric curve in Fig. 11.

2. According to the concentration of the mixture of both MSG and IMP, the taste intensity of the mixture increases acceleratingly, compared with MSG alone (Fig. 12).

3. The relationship between the taste intensity of the mixture and the concentration of the components is expressed as follows:

$$y = u + 1,200\ uv$$

where y is the concentration (% or g/100 ml) of MSG alone, giving the equivalent taste intensity to the mixture;
  u is the concentration (% or g/100 ml) of MSG in the mixture; and
  v is the concentration (% or g/100 ml) of IMP in the mixture.

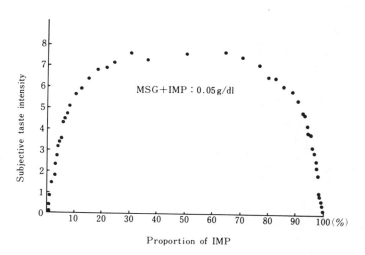

FIG. 11. Relationship between the mixing ratio of MSG and IMP and taste intensity. (From Yamaguchi, ref. 31.)

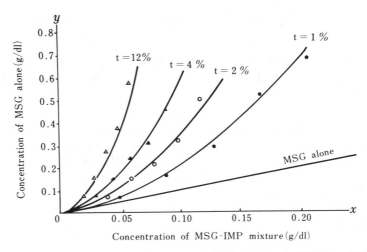

FIG. 12. Relationship of the taste intensity between MSG and mixtures of MSG and IMP (*t* being the proportion of IMP in the mixture). (From Yamaguchi, ref. 31.)

TABLE 5. *Nucleotides having synergistic effect on MSG*

Substance (disodium salt)	Relative intensity of taste activity	
	(g/g)	(mole/mole)
5'-Inosinate·7.5 H$_2$O	1	1
5'-Guanylate·7 H$_2$O	2.3 ± 0.07	2.3
5'-Xanthylate·3 H$_2$O	0.61 ± 0.04	0.53
5'-Adenylate	0.18 ± 0.03	0.13
Deoxy 5'-guanylate·3 H$_2$O	0.62 ± 0.07	0.52
2-Methyl-5'-inosinate·6 H$_2$O	2.3 ± 0.16	2.2
2-Ethyl-5'-inosinate·1.5 H$_2$O	2.3 ± 0.14	2.0
2-Methylthio-5'-inosinate·6 H$_2$O	8.0 ± 0.97	8.2
2-Ethylthio-5'-inosinate·2 H$_2$O	7.5 ± 0.75	6.9
2-Methoxy-5'-inosinate·H$_2$O	4.2 ± 0.33	3.5
2-Chloro-5'-inosinate·1.5 H$_2$O	3.1 ± 0.25	2.7
2-N-Methyl-5'-guanylate·5.5 H$_2$O	2.3 ± 0.15	2.3
2-N-Dimethyl-5'-guanylate·2.5 H$_2$O	2.4 ± 0.13	2.2
N^1-Methyl-5'-inosinate·H$_2$O	0.74 ± 0.09	0.59
N^1-Methyl-5'-guanylate·H$_2$O	1.3 ± 0.13	1.1
N^1-Methyl-2-methylthio-5'-inosinate	8.4 ± 0.75	7.4
6-Chloropurine riboside 5'-phosphate·H$_2$O	2.0 ± 0.20	1.6
6-Mercaptopurine riboside 5'-phosphate·6 H$_2$O	3.4 ± 0.35	3.3
2-Methyl-6-mercaptopurine riboside 5'-phosphate·H$_2$O	8.0 ± 0.83	6.7
2-Methylthio-6-mercaptopurine riboside 5'-phosphate·2.5 H$_2$O	7.9 ± 0.69	7.5
2',3'-O-Isopropylidene 5'-inosinate	0.21 ± 0.06	0.16
2',3'-O-Isopropylidene 5'-guanylate	0.35 ± 0.06	0.28

From Yamaguchi et al., ref. 34.

In case of guanylate (GMP), instead of IMP, the constant number is 2,800, which is 2.3 times larger than the 1,200 of IMP.

4. MSG has this synergistic effect with many other kinds of 5'-ribonucleotides (Table 5) (32).

Relationship Between Flavor Preference of Food and MSG

In order to show the relationship between the flavor preference of food and its MSG content, the y value, mentioned above, was calculated by substituting both u and v with the analytical values obtained for glutamate, inosinate, and guanylate in each food presented in the flavor evaluation test of Table 6.

The y value of beef consommé with no added MSG was 0.15; however, the value became 0.59 by increasing beef broth concentration, and the whole preference score was 0.80. The addition of MSG (0.05%) to beef consommé gave 0.91 of the y value by the effect of both IMP and GMP, which were contained naturally in the food itself, and gave 0.85 of the whole preference score. The addition of very little MSG gave a larger y value to this food, as did the increase of beef broth concentration, and increased the whole preference score. One observes the close relationship between the whole palatability of food and the total concentration (the y value) of MSG and nucleotides, whether they are added intentionally to food or contained naturally in food.

Discussion

The above-mentioned experiments suggest that MSG has an intrinsic taste independent of the four basic tastes: saltiness, sweetness, sourness, and bitterness. The taste intensity of MSG is shown to be increased synergistically by 5'-ribonucleotides, which alone have a very weak taste intensity. The effect of 5'-ribonucleotides on MSG is confirmed by the phenomena that the addition of a small amount of MSG (0.05%) greatly increases the palatability of food like beef consommé, which contains large amounts of 5'-ribonucleotide.

CONCLUSION

Psychometric methods were used to clarify the flavor function of MSG added to food. The flavor profile changes of foods produced by the addition of MSG and the taste properties of MSG itself were evaluated by panels of subjects.

MSG not only imparts an intrinsic taste of its own, but also enhances several specific flavor characteristics such as continuity, mouthfulness, impact, mildness, and thickness of the food. Furthermore, it improves the overall preference for a food. A similar change of the flavor characteristics is observed with beef consommé by increasing the concentration of the consommé stock.

These studies have provided the experimental basis for understanding the use of MSG as a flavor enhancer and for the reported improvement MSG adds to the flavor

TABLE 6. Relationship between y value and palatability of food

Item	Sample A Analysis (%)			y	Sample B Additive		y	Difference of B to A Evaluation score of the whole preference
	MSG	IMP	GMP		Name	Quantity (%)		
Beef consommé	0.010	0.0113	0.0002	0.15	Beef broth	—	0.59	0.80
Beef consommé				0.15	MSG	0.05	0.91	0.85
Beef consommé				0.15	MSG	0.1	1.67	0.62
Beef consommé				0.15	MSG	0.2	3.19	0.67
Beef consommé				0.15	MSG	0.4	6.08	0.03
Chicken consommé	0.023	0.0097	0.0005	0.31	MSG	0.05	1.01	0.54
Cream of chicken soup	0.010	0.0023	0.0006	0.05	MSG	0.17	0.83	0.85
Chicken noodle soup	0.008	0.0014	0.0001	0.02	MSG	0.18	0.63	0.87
Cream of vegetable soup	0.026	0.0005	0.0002	0.06	MSG	0.05	0.16	0.49
Cream of vegetable soup				0.06	Chicken broth	—	0.21	0.71
Vichyssoise	0.011	n.d.	0.0003	0.02	MSG	0.18	0.30	0.58
Onion soup	0.012	n.d.	n.d.	0.01	MSG	0.50	0.51	0.85
Cream of tomato soup	0.122	n.d.	0.0006	0.32	MSG	0.30	1.11	0.17
Cream of tomato soup				0.32	Chicken broth	—	0.92	0.08
Japanese miso soup	0.074	n.d.	n.d.	0.07	MSG	0.3	0.37	0.56
Hamburger	0.009	0.0579	0.0011	0.68	MSG	1.0	74.2	0.68
Seasoned egg custard	0.029	0.0005	0.0002	0.06	MSG	0.3	0.68	0.75
Seasoned egg custard				0.06	Bonito broth	—	0.37	0.42
Cooked mixed vegetables	0.069	n.d.	0.0005	0.17	MSG	0.5	1.39	0.40
Cooked broccoli	0.061	n.d.	0.0004	0.13	MSG	0.25	0.65	−0.20

of food. The addition of salt or sugar to certain foods increases the flavor characteristics mentioned above. Therefore, in this context, salt and sugar may also be called flavor enhancers.

The results of the studies lend support to the notion that MSG has an intrinsic taste—*umami* or tastiness—independent of the other four basic tastes, and these studies demonstrate that MSG improves the flavor of food.

ACKNOWLEDGMENT

The authors gratefully acknowledge useful discussions with Drs. J. Kirimura, Y. Sugita, and M. Aoki. Thankfulness is also extended to Drs. M. R. Kare, R. H. Cagan, and C. M. Christensen of Monell Chemical Senses Center for their kind reading of our manuscript. We are indebted to Ms. M. Oh-ishi, Ms. M. Itoh, and Ms. S. Kanno for their assistance in carrying out much of the experimental work. Our appreciation is also extended to Dr. Y. Komata for his valuable advice.

REFERENCES

1. Beebe-Center, J. G. (1949): Standards for use of the gust scale. *J. Psychol.*, 28:411–419.
2. Berolzheimer, R., ed. (1950): *Encyclopedic Cook Book*. Culinary Arts Institute, Chicago.
3. Cagan, R. H. (1977): A framework for the mechanisms of action of special taste substances: The example of monosodium glutamate. In: *Chemical Senses and Nutrition*, edited by M. R. Kare and Q. Maller, pp. 343–360. Academic Press, New York.
3a. *Flavor and Acceptability of Monosodium Glutamate*, (Proceedings of the First Symposium on Monosodium Glutamate, Chicago (1948): The Quartermaster Food and Container Institute for the Armed Forces and Associates, Food and Container Institute, Inc., Chicago.
4. Girardot, N. F., and Perryan, D. R. (1954): MSG's power to perk up foods. *Food Eng.*, 26:71–72,182,185.
5. Glutamate Manufacturers' Technical Committee (1954): Monosodium glutamate, unique flavor enhancer. *Food Eng.*, 26:76–77.
6. Hanson, H. L., Brushway, M. J., and Lineweaner, H. (1960): Monosodium glutamate studies. 1. Factors affecting detection of and for added glutamate in foods. *Food Technol.*, 14:320–327.
7. Ikeda, K. (1909): On a new seasoning. *J. Tokyo Chem. Soc.*, 30:820–836.
8. Indow, T. (1966): A general equi-distance scale of the four qualities of taste. *Jpn. Psychol. Res.*, 8:136–150.
9. Indow, T. (1969): An application of the scale of taste: Interaction among the four qualities of taste. *Percept. psychophys.*, 5:347–351.
10. Kaneko, T., Yoshida, R., and Katsura, H. (1959): The configuration of -hydroxy glutamic acid. *J. Chem. soc. Jpn.*, 80:316–321.
11. Kaneko, T., Yoshida, R., and Takano, I. (1961): Synthesis of cysteic acid homologs and their testing property. *14th Annual Meeting of the Chemical Society of Japan, Tokyo*, April 1, 1961.
12. Kirimura, J., Shimizu, A., Kimizuka, A., Ninomiya, T., and Katsuya, N. (1969): The contribution of peptides and amino acids to the taste of food stuffs. *J. Agric. Food Chem.*, 17:689–695.
13. Kirk, D., ed. (1955): *Woman's Home Companion Cook Book*. P. F. Collier and Son, New York.
14. Knowles, D., and Johnson, P. E. (1941): A study of the sensitiveness of prospective food judges to the primary tests. *Food Res.*, 6:207–216.
15. Kodama, S. (1912): On the isolation of inosinic acids. *J. Tokyo Chem. Soc.*, 34:751–757.
16. Kuninaka, A., Kibi, M., and Sakaguchi, K. (1964): History and development of flavor nucleotides. *Food Technol.*, 18:287–293.
17. Lockhart, E. E., and Gainer, J. M. (1950): Effect of monosodium glutamate on taste of pure sucrose and sodium chloride. *Food Res.*, 15:459–464.
17a. *Monosodium Glutamate: A Second Symposium*, (Proceedings of the Second Symposium on

Monosodium Glutamate, Chicago (1955): The Research and Development Associates, Food and Container Institute, Inc., Chicago.
18. Mosel, J. N., and Kantrowitz, G. (1952): The effect of monosodium glutamate on activity of the primary taste. *Am. J. Psychol.*, 65:573–579.
19. Norton, K. B., Tressler, D. K., and Farkas, L. P. (1952): The use of monosodium glutamate in frozen foods. *Food Technol.*, 6:405–411.
20. Osgood, C. E., Suci, G. J., and Tannenbaum, P. E. (1957): *The Measurement of Meaning.* University of Illinois Press, Chicago, Ill.
21. Pangborn, R. M. (1961): Taste interrelationships. II. Suprathreshold solutions of sucrose and citric acid. *J. Food Sci.*, 26:648–655.
22. Pangborn, R. M. (1962): Taste interrelationships. III. Suprathreshold solutions of sucrose and sodium chloride. *J. Food Sci.*, 27:495–500.
23. Sakaguchi, T., and Fukumizu, H. (1957): Threshold test. V. "umamai." *J. Brewing Soc. Jpn.*, 52:108–109.
24. Sanders, R. (1948): The significance of taste activity in seasoning with glutamate. In: *Flavor and Acceptability of Monosodium Glutamate*, pp. 70–72. Quartermaster Food and Container Institute for the Armed Forces and Associates, Food and Container Institute, Chicago, Ill.
25. Sjostrom, L. B., and Crocker, E. C. (1948): The role of monosodium glutamate in the seasoning of certain vegetables. *Food Technol.*, 2:317–321.
26. Steinbert, R. (1969): *Foods of the World—The Cooking of Japan.* Time-Life Books, New York.
27. Sunset Books (1967): *Favorite Recipes for Soups and Stews.* Lane Publishing Co., California.
28. Terasaki, M., Fujita, E., Wada, S., Takemoto, T., Nakajima, T., and Yokobe, T. (1965): Studies on taste of tricholomic acid and ibotenic acid. 1. *J. Jpn. Soc. Food Nutr.*, 18:172–175.
29. Terasaki, M., Wada, S. Takemoto, T. Nakajima, T., Fujita, E., and Yokobe, T. (1965): Studies on taste of tricholomic acid and ibotenic acid. 2. Taste identifications with adenine nucleotides, uridylic acid and cytidylic acid. *J. Jpn. Soc. Food Nutr.*, 18:222–225.
30. Toi, B., Maeda, S., Ikeda, S., and Furukawa, H. (1960): On the taste of disodium inosinate. *General Meeting of Kanto Branch, the Agricultural Chemical Society of Japan, Tokyo,* November 12, 1960.
31. Yamaguchi, S. (1967): The synergistic taste effect of monosodium glutamate and disodium 5'-inosinate. *J. Food Sci.*, 32:473–478.
32. Yamaguchi, S., Yoshikawa, T., Ikeda, S., and Ninomiya, T. (1968): Synergistic taste effect of some new ribonucleotide derivatives. *Agric. Biol. Chem.*, 32:797–802.
33. Yamaguchi, S., Yoshikawa, T., Ikeda, S. and Ninomiya, T. (1968): The synergistic taste effect of monosodium glutamate and disodium 5' guanylate. *J. Agric. Chem. Soc. Jpn.*, 42:378–381.
34. Yamaguchi, S., Yoshikawa, T., Ikeda, S., and Ninomiya, T. (1971): Measurement of the relative taste intensity of some L-amino acids and 5'-nucleotides. *J. Food Sci.*, 26:846–849.
35. Yamazaki, K., and Shimada, K. (1967): *Cookery and Its Theory.* Dobun Shoin Co., Tokyo.
36. Yoshida, M., Ninomiya, T., Ikeda, S., Yamaguchi, S., Yoshikawa, T., and O'hara, M. (1966): Studies on the taste of amino acids. 1. Determination of threshold values of various amino acids. *J. Agric. Chem. Soc. Jpn.*, 40:295–299.
37. Yoshida, M., and Saito, S. (1969): Multidimensional scaling of the taste of amino acids. *Jpn. Psychol. Res.*, 11:149–166.

Glutamic Acid: Advances in Biochemistry and Physiology, edited by L. J. Filer, Jr., et al. Raven Press, New York © 1979.

Factors in the Regulation of Glutamate Metabolism

Hamish N. Munro

Physiological Chemistry Laboratories, Department of Nutrition and Food Science, Massachusetts Institute of Technology, Cambridge, Massachusetts 02139

The amino acids of the body are in a dynamic state in which input comes from the food in our dietary protein and output takes the form of excreted nitrogenous end-products. Between these two, intermediary metabolism of amino acids consists of reactions, many of which irreversibly remove the amino acids from the body pool. However, protein turnover is a major metabolic pathway in which the majority of amino acids used for incorporation into proteins are eventually returned to the free amino acid pool, except for a small proportion resulting from posttranslational modification, such as 3-methylhistidine and hydroxyproline, which are liberated by protein breakdown as derived amino acids and do not become available for reincorporation.

The purpose of this review is to analyze the metabolism of glutamic acid in order to identify some of the mechanisms regulating its abundance in the free amino acid pools of the body. Accordingly, this article begins by considering the amount of glutamic acid consumed daily, its absorption, and its fate as it passes through the intestinal wall. Next, the roles of glutamic acid and glutamine in the exchange of nitrogenous compounds between organs and tissues will be explored. Then, a picture of the magnitude of the pools of free glutamate in different body compartments, and the fluxes through them, will be described. Finally, the effects of certain hormones and various dietary factors on glutamate metabolism will be summarized.

Two factors impede the assembly of a detailed quantitative picture of the metabolism of glutamic acid. First, glutamine is converted to glutamic acid during the acid hydrolysis of food and tissue proteins prior to analysis, and, in consequence, the reported glutamic acid content of dietary and tissue proteins is therefore inflated through glutamine breakdown (7). Second, the central position of glutamic acid in nitrogen metabolism makes it virtually impossible on the basis of current data to compute the contributions of the various intracellular metabolic pathways to additions to or removal of glutamic acid. We shall therefore have to content ourselves with quantitation of glutamic acid and glutamine exchange between tissues and from dietary sources.

INTAKE AND ABSORPTION OF GLUTAMIC ACID

Glutamic and aspartic acid and their amines are major constituents of most proteins. As mentioned above, the amides are converted to the corresponding dicarboxylic acids when the protein is hydrolyzed in strong acid preparatory to amino acid analysis (7), so that most tabulations for food protein composition provide *total* glutamic and aspartic acids, including the amides. From the amount of ammonia liberated by acid hydrolysis, an approximate estimate of the proportion as amides can be obtained. In the case of proteins of animal origin, this turns out to be about 50%, whereas for plant proteins this can be as high as 80% of the total dicarboxylic acid content (7). On this basis, liver has about 12% total glutamic acid and 9% total aspartic acid, whereas muscle contains about 16 and 10%, respectively, and wheat, 33 and 4% (7,27). These figures can be used to estimate the total dicarboxylic amino acid intake from the diet.

During this century, the average American daily protein intake of 100 g has changed little, despite that the sources of dietary protein have increasingly emphasized animal foods over cereal sources in this period (25). The analysis of the intake of individual amino acids shows that the eight essential amino acids account for 38 g of this daily protein consumption (25). Of the remaining 63 g of nonessential amino acids, total intake of glutamic acid is about 20 g and total intake of aspartic acid is some 8 g on the basis of the food protein composition given in the preceding paragraph. About 50% of these intakes are likely to be in the form of the corresponding amide.

Protein metabolism commences with digestion and absorption, areas in which interest has been renewed in recent years. As is well known, digestion of dietary protein is dependent on hydrolysis by gastric pepsin followed by the proteolytic enzymes secreted by the pancreas and by the intestinal mucosa. The secretion of proteolytic enzymes by the pancreas is known to be regulated by the presence of dietary protein in the gut contents (Fig. 1) through a system of feedback regulation of pancreatic enzyme secretion (16). The enzymes of digestion resolve the dietary proteins into small molecules for absorption through the mucosal cells of the small intestine. Although some dietary protein is hydrolyzed to free amino acids prior to absorption, small peptides have recently been shown to play a significant role in the assimilation of dietary protein. Nevertheless, because of the presence of peptide hydrolases in the brush border and cytosol of the mucosal cells (20), these peptides undergo hydrolysis to free amino acids as they enter the mucosal cells, and, in consequence, only free amino acids are transferred to the portal vein (Fig. 1).

Another significant area is the secretion of protein into the gut. This includes digestive enzymes added to the gut contents, and epithelial cells shed from the gut mucosa in the process of being replaced by cell division in the crypts at the base of the villi. The magnitude of this endogenous protein output into the gut lumen is controversial, perhaps amounting to some 70 g of protein (23). Together with the average of 100 g protein consumed by an adult on a Western-type diet, some 170 g protein would be provided for absorption. Since fecal nitrogen output is commonly equivalent to 10 g protein daily, the efficiency of digestion and absorption of both dietary and endogenous protein must normally be high (Fig. 1).

Extensive work has been done on the absorption of free amino acids by the

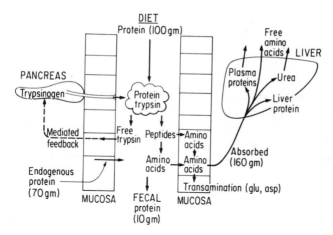

FIG. 1. The digestion of dietary protein, the shedding of protein into the gut, and the transport of amino acids and peptides across the mucosa. (From Crim and Munro, ref. 9.)

mucosa of the small intestine, where transport across the brush border is facilitated by three major carrier-mediated processes for neutral, dibasic, and diacidic amino acids (14). Studies made on rats by injecting mixtures of amino acids into loops of intestine show that the absorption of free aspartic acid and glutamic acid are much slower than that of other amino acids (14). This has been confirmed in humans (1) by using an intubation technique to perfuse the jejunum with equimolar amounts of free amino acids. Once more, the diacidic amino acids aspartic and glutamic acids were the slowest to be taken up. Silk et al. (36) extended this work to the absorption of peptides containing aspartic and glutamic acids. They confirmed that free glutamic acid and especially free aspartic acid are only slowly absorbed, whereas the same amino acids from peptides present in a tryptic digest of casein appeared to be better absorbed. However, as discussed above, much of the glutamic and aspartic acids estimated in acid hydrolysates are present in the original protein as amides. When Silk et al. (36) examined absorption of mixtures of equal proportions of the diacidic acids and their amides, the rate of uptake was equal to that of peptide glutamic and aspartic acids. The findings are thus equivocal regarding the more rapid uptake of aspartic and glutamic acids in peptide form, but do show that both dicarboxylic amino acids are at least as readily available from small peptides as in the form of free amino acids. Since the amino acids in peptide form do not generally appear in the portal blood, it can be assumed that absorbed peptides are efficiently hydrolyzed to free amino acids within the mucosal cells, where they join amino acids absorbed in the free form (Fig. 1).

The mucosal cells of the small intestine are active in the transamination of the dicarboxylic amino acids. Neame and Wiseman (30) demonstrated that absorbed glutamate is extensively transaminated with pyruvate to form alanine, which then appears in increasing amounts in the portal blood. This has been confirmed by a variety of other studies (10,31,34,37,40). The concentrations of two transaminases, glutamate-oxalacetate and glutamate-pyruvate transaminase, have been examined in rat intestinal mucosal cell at various ages by Wen and Gershoff (38). In suckling rats, activity was low but rose steeply after weaning, provided that an adequate amount of the cofactor vitamin B_6 was present in the diet (Fig. 2).

The synthesis of alanine from glutamine and glutamate by mucosal cells and its release into the portal circulation has been extensively studied by Windmueller and Spaeth (39–41) and by Hanson and Parsons (17). Using an isolated preparation of rat small intestine perfused with blood, Windmueller and Spaeth (39) first showed

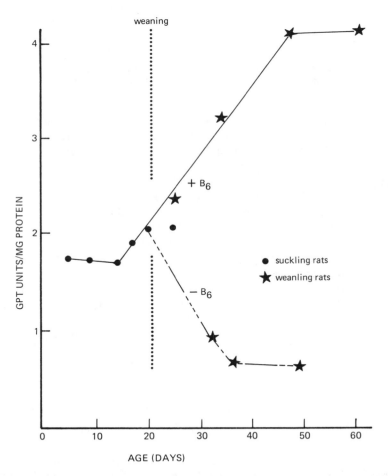

FIG. 2. Effect of age and of dietary vitamin B_6 on the glutamate-pyruvate transaminase content of the intestinal mucosa in suckling (●) and weanling (★) rats. (From Wen and Gershoff, ref. 38.)

that large amounts of glutamine were taken up from the incoming blood, and some one-third of its nitrogen could be accounted for as released alanine, whereas more than 50% of the carbon of the glutamine appeared as CO_2. In subsequent studies (40), they found that glutamate and glutamine administered by way of the lumen of the intestine underwent a similar fate. Thus, there is a single metabolic pool for glutamine entering the mucosal cell from the blood or from the lumen of the intestine. The first step in its utilization is hydrolysis to glutamate, through the action of glutaminase present in abundance in the mucosal cells, followed by transamination to yield alanine and other products of glutamate metabolism, such as proline, ornithine, and citrulline.

A significant question is the extent to which transamination in the mucosa reduces transfer of various loads of glutamate and glutamine to the portal and systemic bloodstreams. The use of perfused intestinal segments by Windmueller and Spaeth (40) showed that small concentrations (6 mM) of glutamine are not only more rapidly absorbed, but a larger proportion (34%) can be transferred to the portal blood than for glutamate (2%). When the intraluminal concentration was raised to 45 mM, the output of glutamine rose to 70%, but glutamate transfer remained negligible. Thus, deamidation is rate limiting for glutamine metabolism by the gut wall. However, large doses of glutamic acid force-fed to rats do, in fact, raise both portal and systemic plasma levels, as well as elevating alanine and glutamine blood levels (34).

EXCHANGE OF GLUTAMATE AND GLUTAMINE BETWEEN ORGANS

The liver is the main or exclusive site of oxidation of seven of the essential amino acids, the branched-chain amino acids being oxidized mostly in muscle and kidney (10). On the other hand, the metabolism of the nonessential amino acids, including glutamate, is widespread in the tissues of the body. The liver is subjected to an extensive increase in amino acid supply through the portal vein, often leading to a 10-fold increment in the levels of some amino acids in the portal blood (24). The liver is nevertheless able to monitor these large loads. Thus, Elwyn (10) reports a study on dogs in which cannulas were implanted in the portal vein and hepatic artery, both providing blood for the liver, and in the hepatic vein, removing blood from the liver. In this way amino acid exchange and urea output by the liver could be monitored over a 12-hr period after feeding a large meal of meat (Fig. 1). It was found that 57% of the absorbed amino acid load was converted to urea as it passed through the liver and 6% to plasma proteins, whereas only 23% of the absorbed amino acids entered the general circulation as free amino acids; the remaining 14% not accounted for was presumed to be temporarily retained in the liver as hepatic protein (enzymes). These findings indicate that the systemic circulation is protected against excessive changes in free amino acid concentrations by temporary adaptive responses within the liver. To be physiologically useful, the hepatic response would need to be sensitive to the needs of the body, and, since the liver is the major site of degradation of many essential amino acids, it has to discriminate between suboptimal and superoptimal amounts; and this has indeed been shown to be the case. For example, Harper (18) fed young rats different levels of dietary casein from insufficient up to quantities exceeding their needs for growth. Threonine-serine dehydratase activity in the liver remained low until the casein content of the diet reached 20%, which is optimal for growth of the rat; at intakes above 20%, the activity of this enzyme rose sharply. In contrast, the activity of two transaminases handling glutamic acid rose progressively with dietary protein intake.

Elwyn's study (10) provides us with a balance sheet of the exchange of glutamic acid, glutamine, and alanine across the gut and liver, both in the fed and fasting dog. Table 1 shows that the fed dog transferred more of the intake of total glutamic acid (glutamate plus glutamine in meat protein) into the portal vein as glutamine than as glutamic acid, in agreement with the studies of Windmueller and Spaeth (40) cited earlier; however, a large output of glutathione into the red cells could account for much of the missing glutamic acid. There was also a considerable output of alanine from the gut. In contrast, the liver removed glutamic acid, glutamine, and alanine from the portal blood, so that the net result of passage of blood through the splanchnic area was a small reduction in the levels of all three amino acids. During the postabsorptive period (not shown in Table 1), more glutamine and alanine were extracted from the blood as it passed across the splanchnic area.

TABLE 1. *Exchanges of glutamic acid, glutamine, and alanine across the viscera and limb muscles*

Species	Total exchange during period		
	Glutamic acid	Glutamine	Alanine
Fed dog (mmoles/12 hr)[a]			
Absorbed from gut lumen	———— 160 ————		100
Gut output	+10	+50	+230
Liver output	−15	−70	−260
Total splanchnic output	−5	−20	−30
Fed sheep (mmoles/hr)[b]			
Portal viscera output	−0.2	−1.5	+2.3
Liver output	+1.1	−2.1	−3.2
Total splanchnic output	+0.9	+3.6	−0.9
Hind-quarters output	−11	+13	+14
Fasted human (μmoles/min)[c]			
Total splanchnic output	+48	−59	−60
Leg output	−24	+50	+26

[a] Computed from Elwyn (10). "Glutamic acid" intake from meat protein (160 mmoles) is probably 50% glutamine. Note that glutathione output was 80 mmoles from the gut and 35 mmoles from the liver.

[b] From Bergman and Heitmann (4). Note that the hind-quarters output is given as μmoles/liter, estimated blood flow being unavailable.

[c] From Felig et al. (13), with the figure for alanine output from muscle taken from Felig et al. (12).

These findings are amplified by studies on sheep (4) and man (13), in which plasma was measured instead of whole blood. This is unfortunate, since Elwyn et al. (11) reported that glutamate is present in higher concentration in the red cells and that glutamate in red cells and plasma behave differently during transit across the splanchnic area. In the case of fed sheep (Table 1), Bergman and Heitmann (4) have demonstrated by cannulation of the appropriate vessels that glutamine and a small amount of glutamate are removed from the plasma as it perfuses the gut and other portal viscera; the liver removes further glutamine, but adds a significant amount of glutamate to the plasma. On the other hand, the portal viscera put out alanine, whereas the liver takes it up. This general picture of the exchange across the splanchnic area, including the liver (uptake of glutamine and alanine, with a smaller output of glutamic acid), is confirmed by studies on the plasma of human subjects (13) (Table 1).

These patterns of amino acid metabolism in the viscera are dependent on exchanges of the same amino acids in the peripheral tissues, notably muscle. Muscle represents the major depot within the body of free amino acids (24) and is also the largest single component of body protein. Changes in muscle amino acid flux can thus have a considerable effect on their concentrations in the blood and their availability to other organs. In the fasting human, it has been shown that muscle releases large amounts of alanine (35) and glutamine (13), and, as shown above, these two amino acids are removed by the viscera. This is most clearly seen in the fasting subject (Table 1), who demonstrates a net uptake of glutamic acid and a release of glutamine and alanine as blood passes through the leg muscles. A similar picture is seen for sheep (Table 1). In the case of both humans (13) and sheep (4), the output of glutamine and alanine is greater in the fasting than in the fed state, thus providing a source of carbon for gluconeogenesis during fasting. Details of the reactions involved in the formation of glutamine and alanine are shown in Fig. 3. It will be noted that the state of protein synthesis or breakdown can affect the availability of the free amino acid pool within muscle, and that excretion of the nonreutilizable amino acid 3-methylhistidine allows the investigator to have an independent measure of the contribution of the breakdown of muscle protein to this pool (8).

In summary, Fig. 4 shows the overall picture of glutamic acid, glutamine, and alanine in relation to their uptake and release by the visceral and peripheral organs. The net output of amino acids by the arm or leg muscles of fasting human subjects has been used to compute total muscle protein breakdown, assessed at 75 g per day in the fasting subject (35). We (8) have used the urinary output of the nonreutilized amino acid 3-methylhistidine as a specific index of myofibrillar muscle protein breakdown and obtain a comparable figure (50 g daily).

MAGNITUDE OF GLUTAMATE POOLS AND FLUXES

Beginning with the preceding information, we can assemble a composite picture of the flux of glutamic acid and glutamine in the body. Figure 5 reconstructs amino

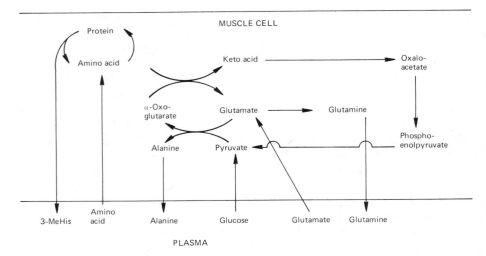

FIG. 3. Interactions of certain amino acids between plasma and muscle. Note that the identification (15) of the enzyme phosphoenolpyruvate carboxykinase in muscle allows the utilization of keto-acids for synthesis of pyruvate as an acceptor of amino groups.

acid turnover in a 70-kg man. The customary protein intake in Western countries is 100 g daily, and some 70 g protein is secreted or shed into the lumen of the gut. In consequence, about 160 g protein are absorbed. The daily turnover of body protein is computed to be about 300 g. The difference between intake of 100 g and turnover of 300 g thus represents the recycling of amino acids and implies a dynamic state for the free amino acid pool. The free amino pools in the tissues constitute at least 70 g, of which the greater part consists of four nonessential amino acids, namely, alanine, glutamic acid, glutamine, and glycine (24).

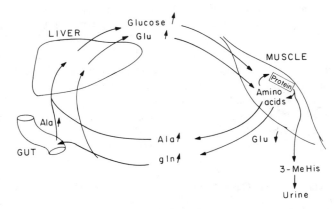

FIG. 4. Interchange of glutamic acid, glutamine, alanine, and glucose between muscle, intestinal mucosa, and liver.

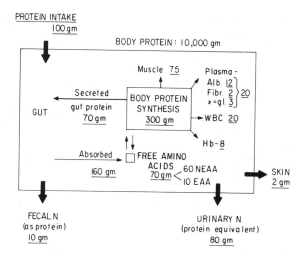

FIG. 5. Diagram illustrating the daily flux of amino acids in the body of a 70-kg man. (From Munro, ref. 26.)

This general picture can be made specific for glutamic and aspartic acids (Table 2). Using data for *total* dicarboxylic acids (glutamic acid and glutamine and aspartic acid and asparagine), estimates of some aspects of glutamic acid and aspartic acid flux can be computed. The most striking features are the large amounts of glutamic and aspartic acids released from body protein in the course of turnover and the very small amounts of these dicarboxylic amino acids in the plasma. If the plasma is to act as an effective channel for transport, the turnover of these plasma amino acids must be very rapid, and, indeed, Elwyn et al. (11) have estimated their plasma half-life to be less than 1 min.

TABLE 2. *Daily flux of glutamic and aspartic acids in a 70-kg man*[a]

Source of amino acids	Total amino acids (g)	Glutamic acid (g)	Aspartic acid (g)
Dietary protein	101	20	8
Endogenous gut protein	70	8	6
Body protein			
Total	10,000	1,600	900
Daily turnover	300	48	27
Free amino acids			
Plasma	—	0.02	0.004
Muscle	86	20	4

[a] Flux of total amino acids taken from Munro (26). Flux of glutamic and aspartic acids (including their amides) based on their abundance in the proteins of the average diet (7) and in tissue proteins (27). Pools of plasma and muscle free glutamic and aspartic acid are computed from data of Bergstrom et al. (5).

These data for man can be viewed in relation to other mammalian species. Elsewhere, an account of the influence of the size of species on intensity of metabolism has been given (22). The analysis of various parameters shows that the intensity of metabolism is about five times greater in the rat than in man. Table 3 illustrates this by showing the turnover of plasma albumin in a number of mammals. Whereas the renewal rate for plasma albumin is 59% in the mouse, it is only 4% in man, species of intervening size having intermediate rates of turnover. The amount of RNA per cell in the liver (the site of albumin synthesis) can be correlated with this decline in turnover rate. However, Table 3 shows that the levels of essential and nonessential amino acids in liver and muscle, and the levels of glutamic and aspartic acids and glutamine in plasma, show no consistent relationship to the size of a species. Thus, when comparisons are made between glutamate metabolism in man and small species such as the rat, these differences in metabolic intensity should be recognized.

HORMONES AND GLUTAMATE METABOLISM

Hormones affect the levels of free glutamic acid and glutamine through changes in the entry and exit of these amino acids between body compartments. For example, the administration of corticosteroids to rats (6) has been shown to increase the levels of glutamic and aspartic acids and of alanine in muscle, plasma, and liver. This is probably mediated by an increase in muscle protein breakdown resulting from corticosteroid administration, as evidenced by an increased output of 3-methylhistidine from the breakdown of muscle protein (28). Plasma glutamate is also elevated by hyperthyroidism in man, but administration of an oral load of glutamate is tolerated normally (3). Urinary excretion of glutamate is also increased in the hyperthyroid state.

RESPONSES TO CHANGES IN DIET

Enzymes of glutamic acid metabolism respond to changes in protein intake, due to the involvement of glutamate as a nitrogen donor in many reactions, including urea synthesis. In a recent study, McGivan et al. (21) have shown that transport of glutamate across the mitochondrial membrane can be rate limiting for urea synthesis in the liver and that this intracellular transport mechanism is enhanced in carrier capacity by raising the protein intake. In addition, liver transaminases involving glutamate undergo increased activity as the intake of dietary protein is increased (18).

Studies of the toxicity levels of dietary free amino acids have been reviewed by Harper (19). Based on a variety of factors, but mainly on rate of growth of young rats, methionine was most toxic (above 1.5% of the diet) and glutamic acid least toxic (7% of diet). In a study on young rats force-fed a meal containing 5% glutamic acid, Peng et al. (33) observed a doubling of plasma glutamate and a 60% increase in plasma alanine, but no change in the levels of either amino acid in brain. A few

TABLE 3. Effect of size of species of mammal on body protein turnover, liver RNA content, and free amino acid pools[a]

Species	Body weight (kg)	Albumin synthesis (fractional rate/day (%))	Liver RNA/DNA	Liver amino acids[b] (μmoles/g)		Muscle amino acids[b] (μmoles/g)		Plasma amino acids[b] (μmoles/ml)				
				Essential	Nonessential	Essential	Nonessential	Essential	Nonessential	Aspartic acid	Glutamic acid	Glutamine
Mouse	0.03	59	4.5	1.2	6.6	1.1	5.5	0.5	0.8	0.01	0.04	0.49
Rat	0.2	28	3.1	0.8	4.2	0.8	4.0	0.7	1.1	0.02	0.07	0.62
Rabbit	2	12	2.4	1.4	3.9	0.6	4.0	0.5	1.6	0.02	0.10	0.59
Dog	30	9	1.7	0.8	2.6	0.7	2.9	0.6	0.8	0.01	0.04	—
Man	70	4	—	—	—	1.4	8.0	0.7	0.8	0.01	0.05	0.51

[a] Data from Munro (22), with data for man from Bergstrom et al. (5).
[b] Essential amino acids = sum of isoleucine, leucine, phenylalanine, and valine; nonessential amino acids = sum of alanine, aspartic acid, glutamic acid, glycine, and serine.

human studies also suggest a high tolerance for glutamic acid. In human studies of the amino acid needs of boys, Nakagawa et al. (29) observed no toxic effects when 12.75 g free glutamic acid were fed daily. This picture of the efficient metabolic removal of orally administered excess glutamate in humans is reflected in the finding that free glutamic acid levels in the plasma are the least affected by meal-related diurnal rhythms (42) and by infusion of glutamate into the intestine (2). A study (32) in which a large meal of protein was given to human subjects also failed to raise plasma glutamate, but did cause an increase in glutamine concentration. This may represent the more extensive absorption of glutamine than of glutamic acid across the gut mucosa noted earlier.

CONCLUSION

The metabolism of glutamic acid and glutamine involves cooperation between tissues. The daily intakes of these amino acids from the diet are large; the body pools are also extensive; nonetheless, the total amounts present in the plasma are small. Thus, the regulation of the plasma concentrations of glutamic acid and glutamine is an important aspect of their metabolism. This is achieved by restricting the passage of these amino acids across the intestinal mucosa, and by efficient removal by organs such as the intestine. It is, however, still too early to assemble a quantitative picture of the overall metabolism of these amino acids in the body because of the need for more quantitative information on their transport and metabolism at the subcellular level.

REFERENCES

1. Adibi, S. A., Gray, S. J., and Menden, E. (1967): The kinetics of amino acid absorption and alteration of plasma composition of free amino acids after intestinal perfusion of amino acid mixtures. *Am. J. Clin. Nutr.*, 20:24–33.
2. Adibi, S. A., Modesto, T. A., Morse, E. L., and Amin, P. M. (1973): Amino acid levels in plasma, liver and skeletal muscle during protein deprivation. *Am. J. Physiol.*, 225:408–415.
3. Bélanger, R., Chandramohan, N., Misbin, R., and Rivlin, R. S. (1972): Tyrosine and glutamic acid in plasma and urine of patients with altered thyroid function. *Metabolism,* 21:855–864.
4. Bergman, E. N., and Heitmann, R. N. (1978): Metabolism of amino acids by the gut, liver, kidneys and peripheral tissues. *Fed. Proc.*, 37:1228–1232.
5. Bergström, J., Fürst, P., Noree, L. O., and Vinnars, E. (1974): The intracellular free amino acid concentration in human muscle tissue. *J. Appl. Physiol.*, 36:693–697.
6. Betheil, J. J., Feigelson, M., and Feigelson, P. (1965): The differential effects of glucocorticoid on tissue and plasma amino acid levels. *Biochim. Biophys. Acta,* 104:92–97.
7. Bigwood, E. J. (1972): Amino acid patterns of animal and vegetable proteins. In: *International Encyclopaedia of Food and Nutrition, Vol. 11; Protein and Amino Acid Functions,* edited by E. J. Bigwood, pp. 215–257. Pergamon Press, Oxford.
8. Bilmazes, C., Uauy, R., Haverberg, L. N., Munro, H. N., and Young, V. R. (1978): Muscle protein breakdown rates in humans based on N^r-methylhistidine (3-methylhistidine) content of mixed proteins in skeletal muscle and urinary output of N^r-methylhistidine. *Metabolism,* 25:525–530.
9. Crim, M. C., and Munro, H. N. (1977): Protein and amino acid requirements and metabolism in relation to defined formula diets. In: *Defined Formula Diets for Medical Purposes,* edited by M. E. Shils, pp. 5–15. American Medical Association, Chicago, Ill.
10. Elwyn, D. H. (1970): The role of the liver in regulation of amino acid and protein metabolism. In:

Mammalian Protein Metabolism, Vol. 4, edited by H. N. Munro, pp. 523–557, Academic Press, New York.
11. Elwyn, D. H., Launder, W. J., Parikh, H. C., and Wise, E. M., Jr. (1972): Roles of plasma and erythrocytes in interorgan transport of amino acids in dogs. *Am. J. Physiol.,* 222:1333–1341.
12. Felig, P., and Wahren, J. (1971): Amino acid metabolism in exercising man. *J. Clin. Invest.,* 50:2703–2714.
13. Felig, P., Wahren, J., Karl, I., Cerasi, E., Luft, R., and Kipnis, D. M. (1973): Glutamine and glutamate metabolism in normal and diabetic subjects. *Diabetes,* 22:573–576.
14. Gitler, C. (1964): Protein digestion and absorption in nonruminants. In: *Mammalian Protein Metabolism,* edited by H. N. Munro and J. B. Allison, pp. 35–69. Academic Press, New York.
15. Goldstein, L., and Newsholme, E. A. (1976): The formation of alanine from amino acids in diaphragm muscle of the rat. *Biochem. J.,* 154:555–558.
16. Green, G. M., Olds, B. A., Matthews, G., and Lyman, R. L. (1973): Protein as a regulator of pancreatic enzyme secretion in the rat. *Proc. Soc. Exp. Biol. Med.,* 142:1162–1167.
17. Hanson, P. J., and Parsons, D. S. (1977): Metabolism and transport of glutamine and glucose in vascularly perfused small intestine rat. *Biochem. J.,* 166:509–519.
18. Harper, A. E. (1968): Diet and plasma amino acids. *Am. J. Clin. Nutr.,* 21:358–366.
19. Harper, A. E. (1973): Amino acids of nutritional importance. In: *Toxicants Occurring Naturally in Foods,* edited by Committee on Food Protection, Food and Nutrition Board, National Research Council, pp. 130–152. National Academy of Sciences, Washington, D.C.
20. Kim, Y. S., and Freeman, H. J. (1977): The digestion and absorption of protein. In: *Clinical Nutrition Update: Amino Acids,* edited by H. L. Greene, M. A. Holliday, and H. N. Munro, pp. 135–141. American Medical Association, Chicago, Ill.
21. McGivan, J. D., Bradford, N. M., and Chappell, J. B. (1974): Adaptive changes in the capacity of systems used for the synthesis of citrulline in rat liver mitochondria in response to high- and low-protein diets. *Biochem. J.,* 142:359–364.
22. Munro, H. N. *(1969): Evolution of protein metabolism in mammals. In: *Mammalian Protein Metabolism,* Vol. 3, edited by H. N. Munro, pp. 133–182. Academic Press, New York.
23. Munro, H. N. (1969): A general survey of techniques used in studying protein metabolism in whole animals and intact cells. In: *Mammalian Protein Metabolism,* Vol. 3, edited by H. N. Munro, pp. 237–262. Academic Press, New York.
24. Munro, H. N. (1970): Free amino acid pools and their role in regulation. In: *Mammalian Protein Metabolism,* Vol. 4, edited by H. N. Munro, pp. 299–386. Academic Press, New York.
25. Munro, H. N. (1976): Health-related aspects of animal products for human consumption. In: *Fat Content and Composition of Animal Products,* edited by National Research Council, pp. 24–44. National Academy of Sciences, Washington, D.C.
26. Munro, H. N. (1977): Parenteral nutrition: metabolic consequences of bypassing the gut and liver. In: *Clinical Nutrition Update: Amino Acids,* edited by H. L. Greene, M. A. Holliday, and H. N. Munro, pp. 141–146. American Medical Association, Chicago, Ill.
27. Munro, H. N., and Fleck, A. (1969): Analysis of tissues and body fluids for nitrogenous constituents. In: *Mammalian Protein Metabolism,* Vol. 3, edited by H. N. Munro, pp. 423–525. Academic Press, New York.
28. Munro, H. N., Tomas, F. M., Randall, R., Bilmazes, C., and Young, V. R. (1978): Effect of glucocorticoids on myofibrillar protein breakdown measured by N^r-methylhistidine output. *Fed. Proc.,* 37:751.
29. Nakagawa, I., Takahashi, T., and Suzuki, T. (1960): Amino acid requirements of children. *J. Nutr.,* 71:176–181.
30. Neame, K. D., and Wiseman, G. (1957): The transamination of glutamic and aspartic acids during absorption by the small intestine of the dog *in vivo. J. Physiol.,* 135:442–450.
31. Neame, K. D., and Wiseman, G. (1958): The alanine and oxo acid concentrations in mesenteric blood during the absorption of L-glutamic acid by the small intestine of dog, cat and rabbit *in vivo. J. Physiol.,* 140:148–155.
32. Palmer, T., Rossiter, M. A., Levin, B., and Oberholzer, V. G. (1973): The effect of protein loads on plasma amino acid levels. *Clin. Sci. Mol. Med.,* 45:827–830.
33. Peng, Y., Gubin, J., Harper, A. E., Vavich, M. G., and Kemmerer, A. R. (1973): Food intake regulation: Amino acid toxicity and changes in rat brain and plasma amino acids. *J. Nutr.,* 103:608–617.
34. Peraino, C., and Harper, A. E. (1962): Concentrations of free amino acids in blood plasma of rats force-fed L-glutamic acid, L-glutamine or L-alanine. *Arch. Biochem. Biophys.,* 97:442–448.

35. Pozefsky, T., Felig, P., Tobin, J., Soeldner, J. S., and Cahill, G. F. (1969): Amino acid balance across tissues of the forearm in post-absorptive man. Effects of insulin at two dose levels. *J. Clin. Invest.*, 48:2273–2282.
36. Silk, D. B. A., Marrs, T. C., Addison, J. M., Burston, D., Clark, M. L., and Matthews, D. M. (1973): Absorption of amino acids from an amino acid mixture simulating casein and a tryptic hydrolysate of casein in man. *Clin. Sci. Mol. Med.*, 45:715–719.
37. Stegink, L. D., Filer, L. J., Jr., and G. L. Baker (1973): Monosodium glutamate metabolism in the neonatal pig: Effect of load on plasma, brain, muscle and spinal fluid free amino acid levels. *J. Nutr.*, 103:1138–1145.
38. Wen, C.-P., and Gershoff, S. N. (1972): Effects of dietary vitamin B_6 on the utilization of monosodium glutamate by rats. *J. Nutr.*, 102:835–840.
39. Windmueller, H. G., and Spaeth, A. E. (1974): Uptake and metabolism of plasma glutamine by the small intestine. *J. Biol. Chem.*, 249:5070–5079.
40. Windmueller, H. G., and Spaeth, A. E. (1975): Intestinal metabolism of glutamine and glutamate from the lumen as compared to glutamine from blood. *Arch. Biochem. Biophys.*, 171:662–672.
41. Windmueller, H. G., and Spaeth, A. E. (1978): Identification of ketone bodies and glutamine as the major respiratory fuels *in vivo* for postabsorptive rat small intestine. *J. Biol. Chem.*, 253:69–76.
42. Wurtman, R. J. (1970): Diurnal rhythms in mammalian protein metabolism. In: *Mammalian Protein Metabolism*, Vol. 4, edited by H. N. Munro, pp. 445–479. Academic Press, New York.

Glutamic Acid: Advances in Biochemistry and Physiology, edited by L. J. Filer, Jr., et al. Raven Press, New York © 1979.

Biochemistry of Glutamate: Glutamine and Glutathione

Alton Meister

Department of Biochemistry, Cornell University Medical College, New York, New York 10021

Glutamate and two of its γ-linked derivatives, glutamine and glutathione, play central roles in the metabolism of amino acids and ammonia. In this chapter, I will present an outline of the metabolism of glutamate in mammalian tissues and review some recent studies on glutamate metabolism and on the function of the γ-glutamyl moiety that have been carried out in our laboratory.

OUTLINE OF GLUTAMATE METABOLISM

The scheme given in Fig. 1 summarizes what appear to be the major enzyme-catalyzed reactions involved in the metabolism of glutamate. α-Ketoglutarate may be converted to glutamate by reductive amination catalyzed by glutamate dehydrogenase (reaction 2) and by transamination reactions involving a number of amino acids (reaction 3) (26). The quantitative significance of the glutamate dehydrogenase reaction in the formation of glutamate in mammals requires additional study. If the relatively high K_m value for ammonia of mitochondrial liver glutamate dehydrogenase reflects its *in vivo* affinity for ammonia, one may seriously question a biosynthetic role for this enzyme. On the other hand, glutamate dehydrogenase might be linked *in vivo* with an enzyme such as glutaminase (reaction 5) so as to make the amide nitrogen atom of glutamine directly available for the synthesis of glutamate and thus also for the formation of the α-amino groups of amino acids. [There is at this time no evidence for the presence of glutamate synthase in mammalian tissues, but the presence in mammals of this enzyme or of a complex representing its catalytic equivalent is not excluded (17)]. Liver carbamyl phosphate synthetase I (reaction 7) is a major catalyst for ammonia utilization in this organ (19,42). This enzyme requires N-acetylglutamate for activity and the rate and extent of synthesis of this cofactor may play a role in the regulation of carbamyl phosphate synthesis. Glutamate is also formed in the degradation of arginine, ornithine, proline, and histidine (26). Ammonia is also formed in other reactions, for example, in the deamination of adenosine 5'-monophosphate (24).

Glutamate is used directly for protein synthesis; thus, the α-carboxyl group of this amino acid is activated by a specific aminoacyl tRNA synthetase (8,22). A quantita-

GLUTAMATE, GLUTAMINE, AND GLUTATHIONE

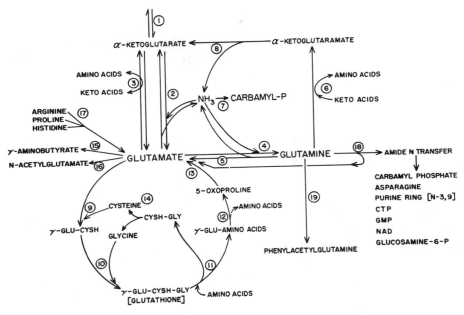

FIG. 1. An outline of glutamate metabolism in mammalian tissues. 1, Reactions of the citric acid cycle; 2, glutamate dehydrogenase; 3, glutamate transaminases; 4, glutamine synthetase; 5, glutaminase; 6, glutamine transaminase; 7, carbamyl phosphate synthetase (liver); 8, α-keto acid ω-amidase; 9, γ-glutamyl cysteine synthetase; 10, glutathione synthetase; 11, γ-glutamyl transpeptidase; 12, γ-glutamyl cyclotransferase; 13, 5-oxoprolinase; 14, cysteinylglycinase; 15, glutamate decarboxylase; 16, glutamate N-acylase; 17, various enzymes involved in the degradation of these amino acids; 18, glutamine amidotransferases known to occur in mammalian tissues; and 19, phenylacetyl glutamine synthetase (Acyl-CoA-L-glutamine N-acyltransferase).

tively minor, but physiologically significant, pathway of glutamate metabolism involves decarboxylation to γ-aminobutyrate, a putative neurotransmitter. There is growing evidence that both γ-aminobutyrate and glutamate are of importance as inhibitory and excitatory neurotransmitters, respectively (12). As discussed below, glutamate is a precursor of two γ-glutamyl compounds of major biochemical importance: glutamine and glutathione. Glutamine occurs both intra- and extracellularly, but glutathione has a predominantly intracellular localization.

METABOLISM OF GLUTAMINE

The conversion of glutamate to glutamine, catalyzed by glutamine synthetase, is of great significance in the utilization of ammonia. This reaction takes place in a number of mammalian tissues, e.g., liver, brain, kidney, muscle, and intestine (28). Glutamine, which is widely distributed in mammalian tissues, is not only an essential building block of proteins, but is a central compound in nitrogen metabolism. Glutamine functions in the uptake, storage, and formation of ammonia; the homeostatic control of amino acid balance; the synthesis of the purine and pyrimidine moieties of nucleic acids, ATP and other nucleotides, and the amide groups of the pyridine nucleotide coenzymes; the formation of amino sugars; and the biosynthesis of a number of amino acids and other compounds of biological importance. Although glutamine is not a dietary essential amino acid, it is nevertheless usually the amino acid present in highest concentration in mammalian blood plasma. Glutamine is also present in high concentrations in mammalian tissues. Glutamine, which crosses cell membranes more readily than glutamate, appears to serve as a transport form of both glutamate and ammonia. The physiological function of glutamine varies depending on the tissue or cell. For example, in the kidney, glutamine is a major source of energy and of urinary ammonia (3). The role of glutamine in the brain seems related to the synaptic functions of glutamate and γ-aminobutyrate. In certain plants, glutamine is a major storage form of nitrogen, and in a number of microorganisms glutamine functions as an essential intermediate in the assimilation of nitrogen. It seems notable that in man (and in some of his close relatives) glutamine is coupled with phenylacetate to form phenylacetylglutamine, a so-called detoxication reaction that does not occur in most mammals (37,56).

The conversion of glutamine to glutamate is effected by several enzymes, including glutamine synthetase, which catalyzes a reversible reaction (23); such reversal is probably not of physiological importance since the equilibrium of this reaction lies distinctly in the direction of synthesis. Glutamine is hydrolyzed to glutamate and ammonia by glutaminase, a reaction of particular importance in the kidney, but one which takes place in other tissues as well. There is also a rather heterogeneous group of enzyme-catalyzed reactions in which the amide nitrogen atom of glutamine is utilized (with concomitant formation of glutamate) in reactions leading to the formation of new compounds. These reactions are catalyzed by the glutamine amidotransferases, and 13 such catalytic activities have thus far been identified (6,31). Glutaminase may be considered as a glutamine amidotransferase in which the amide group is transferred to a hydrogen ion. Glutamine amidotransferases are involved in the biosynthesis of both the purine and pyrimidine rings and in the introduction into these rings of certain amino groups. The amidotransferases also catalyze reactions leading to the synthesis of several amino acids, including asparagine, arginine, and (in bacteria) of glutamate (glutamate synthase), tryptophan, histidine, and p-aminobenzoate. A glutamine amidotransferase is also involved in the conversion of fructose-6-phosphate to glucosamine-6-phosphate and of deamido-NAD to NAD.

Although recent biochemical studies tend to emphasize the reactions catalyzed by the glutamine amidotransferases, reactions involving the α-amino group of glutamine are probably also of considerable metabolic and physiological significance. Reactions of the latter type are catalyzed by the physiologically coupled enzymes glutamine transaminase and α-keto acid ω-amidase (9–11,35). There are several separate glutamine transaminases that exhibit high affinity for glutamine and certain α-keto acids. Glutamate-aspartate and glutamate-alanine transaminases, and the transaminases that catalyze various α-ketoglutarate-amino acid transamination reactions (including those that involve branched-chain and aromatic amino acids) do not act on glutamine at significant rates. The glutamine transaminase reactions, like other transamination reactions, are freely reversible, but in contrast to reactions such as those catalyzed by glutamate-aspartate transaminase, glutamine transaminases catalyze reactions that proceed, under physiological conditions, in the direction of glutamine utilization rather than its synthesis. The steady-state concentrations of α-ketoglutaramate in mammalian tissues are relatively low (13), and it appears that, in the presence of α-keto acid ω-amidase, the open-chain form of α-ketoglutaramate formed in the transamination of glutamine undergoes rapid enzyme-catalyzed deamidation yielding α-ketoglutarate. Therefore, the transamination of glutamine is essentially irreversible *in vivo,* and it follows that its metabolic role must be associated with the utilization of glutamine, formation of ammonia, and the utilization of certain α-keto acids for the synthesis of the corresponding amino acids. Although there is relatively little evidence that the glutamine transaminase-ω-amidase pathway has a major role in ammoniagenesis, it may contribute to ammonia formation to some extent. The most plausible idea concerning the physiological role of glutamine transaminases is that they function in the conversion of α-keto acids to amino acids. The major degradative pathways of many amino acids lead initially to the formation of the corresponding α-keto acids, but there are some notable exceptions, including, for example, phenylalanine and methionine (26). The concentrations of amino acids in mammalian tissues probably fluctuate depending on nutritional and other factors, and it therefore appears likely that certain amino acids are temporarily accumulated in amounts that exceed those necessary for the synthesis of proteins and for the formation of other products. When such amino acids accumulate, they may undergo transamination to the corresponding α-keto acids catalyzed by α-ketoglutarate-amino acid transamination reactions; indeed, it has long been known that such reactions occur in a number of mammalian tissues. Under these circumstances, the carbon chains of amino acids, such as phenylalanine, tyrosine, and methionine, might be lost by excretion or degradation and therefore would become unavailable for protein synthesis. It is of metabolic importance to retain these essential carbon chains, and it appears that the glutamine transaminases function in this salvage process. The transamination reactions between glutamine and α-keto acids are driven by the removal of the α-keto acid product (α-ketoglutaramate), and the metabolic balance tends to be further stabilized by the formation of ammonia, which can be used for glutamine synthesis. Thus, one can picture a mechanism by which the glutamine transaminases serve as part of a homeostatic mechanism for the preservation of amino acid balance. Such a process may be illustrated by the following example:

$$\text{phenylpyruvate} + \text{glutamine} \rightarrow \alpha\text{-ketoglutaramate} + \text{phenylalanine}$$
$$\alpha\text{-ketoglutaramate} \rightarrow \alpha\text{-ketoglutarate} + NH_3$$
$$\text{glutamate} + NH_3 + ATP \rightarrow \text{glutamine} + ADP + P_i$$

Sum: phenylpyruvate + glutamate + ATP → phenylalanine + α-ketoglutarate + ADP + P_i

The overall reaction involves a large free energy change and the formation of phenylalanine. These considerations emphasize the metabolic importance of glutamine transaminase and glutamine synthetase in amino acid formation and in the recovery of the carbon chains of certain amino acids.

It has also been suggested that the glutamine transaminases function in transport phenomena. Possibly, certain amino acids are transported into or out of cells or intracellular organelles as the corresponding α-keto acids; in such a system, the α-keto acid might be formed on one side of the membrane and reaminated on the other. This concept is, of course, consistent with a large body of data showing that there is extensive deamination and reamination of most of the amino acids in mammalian species. The initial studies in this area were performed by Schoenheimer and colleagues, who found that when $^{15}NH_4^+$ or ^{15}N-labeled amino acids were given to rats, the isotope appeared in almost all of the amino acids (52). It is also known that the α-keto analogs of most of the amino acids can replace the corresponding L-amino acids in the diet (26). In studies on patients with phenylketonuria, it was found that the administration of glutamine led to substantially decreased urinary excretion of phenylpyruvate (36). This result can be explained by the occurrence of glutamine-phenylpyruvate transamination and suggests that a metabolic abnormality accompanied by an accumulation of an α-keto acid can be corrected, at least partially, by a mechanism involving transamination. Recent attempts have been made to treat patients with "nitrogen accumulation diseases" by the administration of α-keto acids (4,7,25,46,49–51,65–67). Thus, patients with chronic renal failure, who accumulate urea and other nitrogen-containing compounds, were improved by administration of mixtures containing the α-keto acid analogs of the essential amino acids. Similarly, studies on obese patients undergoing a starvation therapy showed that administration of α-keto acids decreased the loss of nitrogen and apparently improved the efficiency of utilization of amino acids. It seems probable that therapy with α-keto acids decreases the formation of urea by diverting nitrogen from the carbamyl phosphate synthetase pathway into the pathways leading to formation of glutamate and glutamine. The accumulated evidence indicates that mammalian tissues have a homeostatic metabolic mechanism for preserving amino acid balance in which the dietary nonessential amino acids, especially glutamine, function to maintain the tissue levels of the other amino acids by preventing the loss of essential carbon chains. The glutamine transaminase-ω-amidase system seems to be a physiologically significant catalyst in this process.

METABOLISM OF GLUTATHIONE

The findings reviewed above support the view that conversion of glutamate to glutamine is of crucial importance in various biosynthetic processes, transport phenomena, preservation of amino acid balance, and ammonia metabolism. However, glutamate is also converted to another γ-glutamyl compound of major metabolic significance, namely, glutathione (γ-glutamylcysteinylglycine). Glutamate is used, together with cysteine and glycine for the biosynthesis of glutathione. The enzymatic synthesis and degradation of glutathione take place by a cyclic metabolic pathway, the γ-glutamyl cycle (Fig. 1; reactions 9–14). This pathway was elucidated in our laboratory through studies on the two enzymes that catalyze the synthesis of glutathione and by experimental work that demonstrated that 5-oxoproline is a quantitatively significant metabolite of glutathione; 5-oxoproline was also found in mammalian tissues and body fluids (27,33,38,61–64). Although 5-oxoproline was previously known to be formed in the degradation of glutathione by tissue preparations, the significance of this finding was not apparent, and the possibility was seriously considered that the formation of 5-oxoproline under these conditions might be an artifact. Although mammalian tissues and body fluids contain a measurable steady-state concentration of 5-oxoproline, this compound does not normally accumulate to an appreciable extent. The discovery of the enzyme 5-oxoprolinase (Fig. 1, reaction 13) showed that there is a metabolic link between the reactions involved in the degradation of glutathione and those that catalyze its synthesis, making it possible to visualize the γ-glutamyl cycle.

Glutathione is synthesized in many mammalian tissues, including liver, kidney, brain, intestine, lens, muscle, and the erythrocyte (30). Highly purified preparations of γ-glutamylcysteine synthetase and glutathione synthetase have been obtained from several sources, and the mechanisms of actions of these enzymes have been extensively examined (29). The initial step in the breakdown of glutathione is catalyzed by γ-glutamyl transpeptidase, which catalyzes transfer of the γ-glutamyl moiety of glutathione (and other γ-glutamyl compounds) to amino acid and other acceptors. γ-Glutamyl transpeptidase is bound to membranes of various epithelial cells, for example, proximal renal tubules, jejunal villi, choroid plexus, ciliary body, visual receptor cells, retinal epithelium, and cerebral astrocytes and their capillaries (34). Highly purified preparations of γ-glutamyl transpeptidase have been obtained from kidney and other tissues (33). The cysteinylglycine formed in the transpeptidase reaction is cleaved by widely distributed peptidase activity to form cysteine and glycine. The γ-glutamyl amino acids formed by transpeptidation may be substrates for additional transpeptidation reactions and may be converted to 5-oxoproline and the corresponding amino acids by the soluble enzyme, γ-glutamyl cyclotransferase. Highly purified preparations of the cyclotransferase have been obtained from brain (40), liver (1,39), and kidney (57). The 5-oxoproline formed in the reaction catalyzed by the γ-glutamyl cyclotransferase is converted to glutamate by 5-oxoprolinase:

$$\text{5-oxo-L-proline} + \text{ATP} + 2\text{H}_2\text{O} \rightarrow \text{L-glutamate} + \text{ADP} + \text{P}_i$$

5-Oxoprolinase activity is widely distributed, having been found in kidney, liver, brain, and other mammalian tissues, and a highly purified preparation of the enzyme has been obtained from rat kidney (60).

Although the initial formulation of the γ-glutamyl cycle was derived largely from enzyme data, there is now excellent evidence that the reactions of the γ-glutamyl cycle take place *in vivo*. Indeed, each of the reactions has been demonstrated by metabolite-labeling studies or through use of specific enzyme inhibitors. It is notable that the turnover of glutathione, as measured by the incorporation of labeled glutamate or labeled 5-oxoproline into glutathione, is substantially higher in mouse kidney than in liver (54); this result reflects the relatively higher activities of the γ-glutamyl cycle enzymes in kidney than in the liver. The rate of incorporation of 5-oxoproline into glutathione is similar to that of glutamate incorporation. After administration of labeled glutamate, labeled 5-oxoproline is found in kidney and liver (55). Further evidence for the *in vivo* function of the cycle has come from studies in which animals were treated with a competitive inhibitor of 5-oxoprolinase (L-2-imidazolidone-4-carboxylate) (63,64). Treatment of mice with this inhibitor decreases their ability to convert labeled 5-oxoproline to labeled respiratory carbon dioxide. Such mice accumulate 5-oxoproline in several tissues, including liver, kidney, and brain. Mice treated with the inhibitor together with various amino acids accumulate substantial amounts of 5-oxoproline in their tissues. An increase in the accumulation of 5-oxoproline after administration of amino acids can be explained in terms of the γ-glutamyl cycle, since it would be expected that the increase in amino acid concentration produced by administering amino acids would be accompanied by increased utilization of glutathione in transpeptidation and thus by increased formation of 5-oxoproline. Additional evidence that the γ-glutamyl cycle functions *in vivo* has come from studies on patients with the inborn error of glutathione metabolism, 5-oxoprolinuria; in this condition, there is a modified γ-glutamyl cycle associated with a marked deficiency of glutathione synthetase activity (32,68).

In an effort to further elucidate the *in vivo* function of the γ-glutamyl cycle, we have looked for specific inhibitors of the several reactions of the cycle and also for analogs of the substrates that would function in some, but not all, of the reactions of the cycle. Although most of these reactions involve the γ-carboxyl group of glutamate, we reasoned that the several enzymatic reaction pathways must be different and that they are probably therefore associated with differences in the enzyme-bound conformations of the glutamate carbon chain at the active sites of the various enzymes. We attempted to exploit such expected differences among the enzymes by variation of substrate structure. We found that suitable modification of the glutamyl moiety of the substrates can indeed produce the desired results, i.e., effective and specific inhibitors or nonmetabolizable analogs at each of the steps of the γ-glutamyl cycle (15).

The enzymes required for the two-step synthesis of glutathione also catalyze the synthesis of the naturally occurring analogs, ophthalmic and norophthalmic acids, compounds in which the cysteine moiety is replaced by α-aminobutyrate and alanine, respectively. As indicated in Fig. 2, glutathione synthetase exhibits a much broader specificity towards substrates in which the glutamyl group is modified than does γ-glutamylcysteine synthetase. It is notable that glutathione synthetase is active towards D-γ-glutamyl-L-α-aminobutyrate; in contrast, D-glutamate is not an

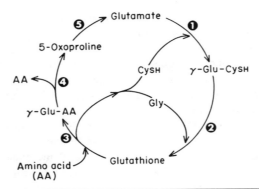

Substrate Analogs	ENZYMES				
	❶ γ-Glu-CySH Synthetase	❷ GSH Synthetase	❸ γ-Glu Trans-peptidase	❹ γ-Glu Cyclo-transferase	❺ 5-Oxoprolinase
D-Glu	0,↓	+	+	0,↓	0
α-Methyl-Glu	(+)	+	+	0	0
N-Methyl-Glu	+	+	0	0	0
β-Glu	+	+	0	0,↓	
β-Methyl-Glu	(+),↓	+,↓	0	+	0,↓
γ-Methyl-Glu	0,↓	+	0	0	(+),↓

FIG. 2. Interaction of glutamate and glutamyl analogs with enzymes of the γ-glutamyl cycle. CySH, cysteine; GSH, glutathione; ↓, > 50% inhibition; +, > 10% as active as L-Glu; (+), 1–10% as active as L-Glu; 0, < 1% as active as L-Glu. (From Griffith and Meister, ref. 15.)

appreciable substrate for peptide synthesis by γ-glutamylcysteine synthetase. γ-Glutamylcysteine synthetase is markedly inhibited by β-methylglutamate and γ-methylglutamate, as well as by D-glutamate. Studies on γ-glutamyl transpeptidase have shown that this enzyme is significantly active with analogs containing D-glutamyl and α-methylglutamyl residues, but is not active with substrates that have N-methylglutamyl, β-glutamyl, β-methylglutamyl, or γ-methylglutamyl groups. We found previously that D-γ-glutamyl-p-nitroanilide is a good substrate of the transpeptidase; because D-amino acids are not acceptor substrates, D-γ-glutamyl-p-nitroanilide may be used in a convenient assay procedure that precludes autotranspeptidation (58,59). α-Substituted amino acids are not acceptors of the γ-glutamyl group, and, therefore, autotranspeptidation does not occur with α-methyl-L-glutamyl-α-aminobutyrate (21). The findings given in Fig. 2 thus indicate that it might be possible to achieve *in vivo* synthesis of the analogs of glutathione (e.g., those containing N-methylglutamyl or β-glutamyl residues), which would be inactive in transpeptidation. It would be of interest to learn whether such analogs might be active in reactions involving the sulfhydryl group. Of the several substrate analogs examined, only γ-(β-methyl)-glutamyl-L-α-aminobutyrate is a substrate of γ-glutamyl cyclotransferase. That glutamyl derivatives modified at the α-carbon atom (e.g., D-glutamyl and α-methylglutamyl) are not active seems to reflect the requirement in the cyclotransferase reaction of bringing the α-carbon atom quite close to the catalytic center of the active site. Requirements for size and alignment at the active site of this enzyme would be expected to be stringent. These studies have led to the finding of two good inhibitors of γ-glutamyl cyclotransferase, i.e., β-aminoglutaryl-L-α-aminobutyrate and D-γ-glutamyl-L-α-aminobutyrate. 5-Oxoprolinase exhibits activity towards piperidone-6-carboxylate and the 3- and 4-hydroxy derivatives of 5-oxoproline. As stated above, L-2-imidazolidone-4-carboxylate is a potent competitive inhibitor *in vitro;* this compound is also active *in vivo*.

The studies summarized in Fig. 2 suggest the feasibility of a number of *in vivo* approaches in which selective inhibition of individual reactions of the cycle might be achieved. There is evidence that the synthesis of glutathione is normally controlled by feedback inhibition of γ-glutamylcysteine synthetase by glutathione; this reaction is probably also affected by the tissue concentration of cysteine (43). Decreased tissue levels of glutathione occur after administration of D-glutamate (55), and this amino acid, as well as β-methylglutamate and γ-methylglutamate, would be expected to inhibit glutathione biosynthesis on the basis of the enzyme data. The findings also suggest that it might be possible to introduce *in vivo* a glutathione analog possessing a D-glutamyl or γ-methylglutamyl moiety; this could be accomplished by administering the appropriate carboxyl terminal cysteine dipeptide. The corresponding modified glutathiones would be expected to be substrates for the transpeptidase, but not of the cyclotransferase. Studies with D-γ-glutamyl or L-α-methylglutamyl-L-amino acids might therefore elucidate the extent to which such compounds are hydrolyzed *in vivo* by the action of γ-glutamyl transpeptidase.

It has recently been possible to examine the inhibition of γ-glutamyl cyclotransferase *in vivo* by giving the inhibitor (β-glutamyl-α-aminobutyrate (5,14b). In these studies, mice injected with β-glutamyl-α-aminobutyrate were found to have a moderate, but significant, depression of the steady-state concentration of 5-oxoproline in the kidney. Furthermore, administration of β-glutamyl-α-aminobutyrate prevented the accumulation of 5-oxoproline that occurs after the administration of large amounts of amino acids and that which occurs in the presence of the 5-oxoprolinase inhibitor, 2-imidazolidone-4-carboxylate. These observations provide strong support for the view that 5-oxoproline is formed *in vivo* by the action γ-glutamyl cyclotransferase. The findings with β-glutamyl-α-aminobutyrate, together with those cited above in which animals were treated with L-2-imidazolidone-4-carboxylate, are in accord with the conclusion that γ-glutamyl cyclotransferase and 5-oxoprolinase are, respectively, major *in vivo* catalysts for the formation and utilization of 5-oxoproline.

In considering the physiological function or functions of the γ-glutamyl cycle, it seems of significance that the first step in the degradation of glutathione is catalyzed by a membrane-bound enzyme, and that the reaction catalyzed by γ-glutamyl transpeptidase is greatly stimulated by certain amino acid and peptide acceptors.

The acceptor amino acids are not metabolized in the cycle, but are released unchanged. The other enzymes of the cycle are present in the cytosol. One must consider the possibility that the function of the transpeptidase is to regulate the intracellular concentration of glutathione. However, there is an effective feedback mechanism for the control of glutathione biosynthesis, namely, inhibition of γ-glutamylcysteine synthetase by glutathione (43). Furthermore, the transpeptidase is linked to the membrane, and a variety of studies have indicated that it is readily accessible to externally supplied substrates (33). Therefore, at least some of the transpeptidase appears to be on the outer surface of the membrane; however, one cannot exclude the possibility that it is a transmembrane enzyme. Its location is consistent with a role in the transport of compounds into or out of the cell, but other membrane-related functions, such as those involved in protection and structure, need to be considered.

Earlier studies on amino acid transport, based largely on the kinetics of this process, led to the proposal that the transport of amino acids involves a number of steps, such as the binding of the amino acid to a site on the cell membrane, carrier-mediated translocation, intracellular release of the amino acid from the carrier, and reactivation of the carrier in an energy-requiring process (18,41). The hypothesis that the γ-glutamyl cycle might function in transport followed from the recognition that the cycle has features previously postulated to be involved in amino acid transport. For example, the membrane-bound transpeptidase might mediate the binding of amino acid and its translocation. According to this idea, the enzyme interacts with extracellular amino acid and intracellular glutathione (or perhaps another γ-glutamyl donor derived from glutathione) to yield a γ-glutamyl amino acid which is translocated. (Whether the transpeptidase itself is involved in the translocation process is not yet clear.) Release of the amino acid from its γ-glutamyl carrier within the cell is catalyzed by γ-glutamyl cyclotransferase. The energy-requiring portion of the cycle involves resynthesis of the precursor of the γ-glutamyl carrier, i.e., glutathione. Various modifications of the γ-glutamyl cycle have been suggested; for example, glutamine might play a special role. γ-Glutamylglutamine might be readily formed by transpeptidation; γ-glutamylglutamine is an active γ-glutamyl donor, and the high concentration of glutamine in mammalian body fluids and tissues suggests that this amino acid might be formed. Indeed, there is evidence for the presence of γ-glutamylglutamine in a variety of tissues (20). A prominent role for glutamine in the cycle is also in accord with the relatively restricted specificity of γ-glutamyl cyclotransferase. The most active substrates for this enzyme are γ-glutamylglutamine and a variety of γ-glutamyl-γ-glutamyl amino acids (57).

One may also consider the possibility of a cycle that requires the cleavage of only 2 ATP molecules per turn; this might take place if the formation of 5-oxoproline were by-passed by hydrolysis (catalyzed by the transpeptidase) of the γ-glutamyl amino acid (33). Even a cycle involving cleavage of only one molecule of ATP is conceivable if there are successive transpeptidation reactions, one of which involves cysteine. A model for exchange diffusion has also been considered in which enzyme-γ-glutamyl amino acid complexes participate (33). The concept of the γ-glutamyl cycle thus seems to offer a number of possible pathways, based on the function of the γ-glutamyl moiety, for the translocation of amino acids across the cell membrane. Perhaps the transport of amino acids follows more than a single γ-glutamyl cycle pathway. The transpeptidase is most active toward glutamine, cystine, and several other neutral amino acids; these amino acids might be more effectively transported by the cycle than others. Aspartate and proline are poor substrates for the transpeptidase and are probably transported by other systems.

Various blocks or partial blocks of the γ-glutamyl cycle have been studied, and the observations made are consistent with the view that the γ-glutamyl cycle functions *in vivo* (33). Evidently, a block of glutathione synthetase, as found in 5-oxoprolinuria (32,68) does not stop the cycle because γ-glutamylcysteine is good substrate for the transpeptidase. On the other hand, a complete block of γ-glutamyl-cysteine synthetase would be expected to stop the cycle and to produce defects in amino acid transport if the transport hypothesis is correct. Therefore, it is interesting that the patients with a severe block of γ-glutamylcysteine synthetase exhibit aminoaciduria (32,45). A patient reported to have a marked deficiency of γ-glutamyl transpeptidase evidently shows only minor evidence for defective amino acid transport (33,53). It is notable that this patient has substantial glutathionemia and glutathionuria. It seems probable that the presence of glutathione in the urine of this patient reflects its presence in the blood plasma and leakage from the kidney. Perhaps one of the functions of the transpeptidase is to breakdown glutathione that enters the systemic circulation.

Another interpretation is that the genetic defect in this condition is associated with a defect that permits the slow leakage of glutathione from the cell. This patient may have a normal transport mechanism for glutathione, but in the absence of transpeptidase, glutathione appears unchanged outside the cell. Indeed, if the transpeptidase is located on the outer surface of the cell membrane, as seems to be the case, then it is probable that there must normally be transport of glutathione (or of its γ-glutamyl moiety) through the membrane to the enzyme. This follows from the observation that administration of high doses of amino acids to animals leads to a significant decrease in the concentration of glutathione (which is predominantly intracellular) and to a substantial increase in the formation of 5-oxoproline. Such an increase in 5-oxoproline formation is blocked by administration of a specific cyclotransferase inhibitor. Thus, the increased formation of 5-oxoproline and the decrease in intracellular glutathione that occurs after administration of amino acids must be associated with an increase in the intracellular concentration of substrate (γ-glutamyl amino acid) for the cyclotransferase. Such intracellular γ-glutamyl amino acid is presumably formed from intracellular glutathione and extracellular amino acid by the transpeptidase. There must also be transport of γ-glutamyl amino acid into the cell to account for the increase in 5-oxoproline formation. The collected findings support the γ-glutamyl cycle hypothesis for amino acid transport.

Although the presently available data are in general accord with the transport hypothesis, we need to know more about the various quantitative relationships and about the orientation of the transpeptidase in the cell membrane. It is not yet clear as to whether the transpeptidase itself plays a direct role in translocation or whether it functions to accept glutathione from within the cell and to form γ-glutamyl amino acids, which are then translocated by a separate mechanism.

It should be emphasized that we have not proposed that the γ-glutamyl cycle is the only transport system for amino acids. It may be active only at certain sites and only at certain stages of cellular development; it seems to be more active for certain amino acids than others. Nevertheless, although it has been generally thought that the transport of amino acids (and of a number of other compounds) is mediated by enzymes (or enzyme-like entities—hence, the term "permease" was proposed), the γ-glutamyl cycle hypothesis seems to be the only mechanism thus far suggested for amino acid translocation in which specific enzymes have been implicated.

Glutathione is used for detoxication reactions, some of which lead to the formation of mercapturic acids. It is generally believed the initial step in the formation of mercapturic acids involves a reaction of a foreign compound with the sulfhydryl group of glutathione within the cell to yield a glutathione adduct; this reaction may occur spontaneously or be catalyzed by glutathione-S-transferases. It is notable that the protein ligandin, a basic dimeric cytoplasmic protein that binds various electrophiles noncovalently and possibly some carcinogens covalently, exhibits glutathione-S-transferase activity (2). Thus, the ligandin system may function in the binding and transport of glutathione and of foreign compounds, and as a catalyst for coupling. The further metabolism of such S-substituted glutathione derivatives involves the removal of the γ-glutamyl group by a reaction that appears to be catalyzed by transpeptidase. This is followed by cleavage of the glycine moiety and acetylation (within the cell) of the amino group of the S-substituted cysteine moiety. Thus, it would appear that reactions of the γ-glutamyl cycle can be utilized in the formation of mercapturic acids and that the formation of mercapturic acids involves transport phenomena.

One must also consider various interrelationships between the γ-glutamyl cycle and other metabolic phenomena, which include the metabolism of glutamate, cysteine, and glycine. The cycle also functions to convert cysteine to an apparently less metabolically active tripeptide form. The intracellular concentration of cysteine is regulated at a level that is substantially lower than those of most of the other amino acids. The release of cysteine from glutathione takes place via the γ-glutamyl cycle, which could function in such a manner as to affect the rate of protein synthesis by providing cysteine, which may under some circumstances be rate limiting. Other aspects of the γ-glutamyl cycle have been considered in detail elsewhere (33).

In the course of studies on the conversion of glutamate to glutathione and to glutamine, we have examined in some detail the inhibition of glutamine synthetase and of γ-glutamylcysteine synthetase by methionine sulfoximine (Fig. 3). Investigation of the mechanisms of these enzymatic reactions indicate that enzyme-bound γ-glutamyl phosphate is an intermediate (28). Studies on glutamine synthetase showed that methionine sulfoximine is phosphorylated by ATP on the enzyme to

(a) Glutamate + ATP + Enzyme ⇌ Enzyme [γ-glutamyl-P; ADP]

(b) Enzyme [γ-Glu-P; ADP] + NH$_3$ ⇌ Enzyme + Glutamine + P$_i$ + ADP

(c) Enzyme [γ-Glu-P; ADP] + Cysteine ⇌ Enzyme + γ-Glu-cySH + P$_i$ + ADP

(d) Enzyme + Methionine sulfoximine + ATP ⟶ Enzyme [Methionine sulfoximine phosphate; ADP]

FIG. 3. Mechanism of the reactions catalyzed by glutamine synthetase—(a), (b) and γ-glutamylcysteine synthetase—(a), (c). Inhibition of both of these enzymes by methionine sulfoximine is described in (d).

form methionine sulfoximine phosphate, which binds tightly to the active site of the enzyme, resulting in irreversible inhibition (47). There is good evidence that methionine sulfoximine inhibits glutamine synthetase by serving as an inhibitory analog of the enzyme-bound tetrahedral intermediate or a transition state formed in the reaction catalyzed by this enzyme (14). Substantially the same type of inhibition by methionine sulfoximine occurs with γ-glutamylcysteine synthetase (44). It is interesting that only one of the four isomers of methionine sulfoximine (L-methionine-S-sulfoximine) inhibits glutamine synthetase and that only the same isomer inhibits γ-glutamylcysteine synthetase. A detailed consideration of the properties of the active sites of glutamine synthetase and γ-glutamylcysteine synthetase suggested the possibility of designing and synthesizing analogs of methionine sulfoximine that would selectively inhibit each of the synthetases. The full details of the reasoning and experimental work involved in these studies have been given elsewhere (14a,16). A number of methionine sulfoximine analogs were synthesized and examined not only for their effects on synthetases, but also for their convulsant activity. It has long been known that methionine sulfoximine induces convulsions in a number of species, and studies in our laboratory showed that of the four diastereoisomers of methionine sulfoximine only L-methionine-S-sulfoximine exhibits convulsant activity (48). Recent studies have led to the finding that α-ethylmethionine sulfoximine, which is a convulsant, inhibits glutamine synthetase effectively, but has no significant effect on γ-glutamylcysteine synthetase (16). Similarly, methionine sulfoximine analogs in which the S-methyl group is replaced with bulkier moieties (e.g., S-propyl homocysteine sulfoximine) do not inhibit glutamine synthetase appreciably, but markedly inhibit γ-glutamylcysteine synthetase (14a). Compounds of the latter type are very weak convulsants. The accumulated data support the hypothesis that the induction of convulsions is closely associated with the inhibition of glutamine synthetase rather than inhibition of glutathione synthesis. The structurally modified sulfoximine compounds that have been obtained thus make it possible to selectively inhibit two major pathways of glutamate metabolism, i.e., the one that leads to formation of glutamine (Fig. 1, reaction 4) and that leading to glutathione biosynthesis (Fig. 1, reaction 9). The availability of these one-enzyme inhibitors will hopefully make it possible to pursue in further detail the metabolic and physiological phenomena associated with these pathways of glutamate metabolism.

REFERENCES

1. Adamson, E. D., Szewczuk, A., and Connell, G. E. (1971): Purification and properties of γ-glutamylcyclotransferase from pig liver. *Can. J. Biochem.*, 49, 218:218–226.
2. Arias, I. M., and Jakoby, W. B., editors (1976): *Glutathione: Metabolism and Function*, Vol. 6. Kroc Foundation Series. Raven Press, New York.
3. Baruch, S., editor (May 1975): Symposium on Renal Metabolism. *Med. Clin. North Am.*, 59:3.
4. Batshaw, M., Brusilow, S., and Walser, M. (1975): Treatment of carbamyl phosphate synthetase deficiency with keto analogues of essential amino acids. *N. Engl. J. Med.*, 292:1085.
5. Bridges, R., and Griffith, O. W. (1978): γ-Glutamyl cyclotransferase-inhibition studies in vitro and in vivo using β-aminoglutaryl-L-α-aminobutyrate. *Fed. Proc.*, 37:388.
6. Buchanan, J. M. (1973): The amidotransferases. *Adv. Enzymol.*, 39:91–184.
7. Close, J. H. (1974): The use of amino acid precursors in nitrogen-accumulation diseases. *N. Engl. J. Med.*, 290:663.
8. Coles, N., Buckenberger, M. W., and Meister, A. (1962): Incorporation of dicarboxylic amino acids into soluble ribonucleic acid. *Biochemistry*, 1:317–322.
9. Cooper, A. J. L., and Meister, A. (1972): Isolation and properties of highly purified glutamine transaminase. *Biochemistry*, 11:661–671.
10. Cooper, A. J. L., and Meister, A. (1974): Isolation and properties of a new glutamine transaminase from rat kidney. *J. Biol. Chem.*, 249:2554–2561.
11. Cooper, A. J. L., and Meister, A. (1974): The glutamine transaminase-ω-amidase pathway. *CRC Crit. Rev. Biochem.*, 4:281–303.
12. Curtis, D. R., and Watkins, J. C. (1965): The pharmacology of amino acids related to γ-aminobutyric acid. *Pharmacol. Rev.*, 17:347.
13. Duffy, T. E., Cooper, A. J. L., and Meister, A. (1974): Identification of α-ketoglutaramate in rat liver, kidney, and brain. Relationships to glutamine transaminase and ω-amidase activities. *J. Biol. Chem.*, 249:7603–7606.
14. Gass, J. D., and Meister, A. (1970): Computer analysis of the active site of glutamine synthetase. *Biochemistry*, 9:1380–1390.
14a. Griffith, O. W., Anderson, M. E., and Meister, A. (1979): Inhibition of glutathione biosynthesis by prothionine sulfoximine (S-n-propyl-homocysteine sulfoximine), a selective inhibitor of γ-glutamylcysteine synthetase. *J. Biol. Chem.*, 254 (in press).
14b. Griffith, O. W., Bridges, R. J., and Meister, A. (1978): Evidence that the γ-glutamyl cycle functions in vivo using intracellular glutathione; effects of amino acids and selective inhibition of enzymes. *Proc. Natl. Acad. Sci. USA*, 75(11) (in press).
15. Griffith, O. W., and Meister, A. (1977): Selective inhibition of γ-glutamyl cycle enzymes by substrate analogs. *Proc. Natl. Acad. Sci., USA*, 74:3330–3334.
16. Griffith, O. W., and Meister, A. (1978): Differential inhibition of glutamine and γ-glutamylcysteine synthetases by α-alkyl analogs of methionine sulfoximine that induce convulsions. *J. Biol. Chem.*, 253:2333–2338.
17. Gross, M., Cooper, A. J. L., and Meister, A. (1976): On the utilization of L-glutamine by glutamate dehydrogenase. *Biochem. Biophys. Res. Commun.*, 70:373–380.
18. Heinz, E. (1972): Transport of amino acids by animal cells. In: *Metabolic Pathways*, Vol. 6, edited by L. E. Hokin, pp. 455–501. Academic Press.
19. Jones, M. E. (1965): Amino acid metabolism. *Ann. Rev. Biochem.*, 34:381–418.
20. Kanazawa, A., and Sano, I. (1967): The distribution of γ-L-glutamyl-L-glutamine in mammalian tissues. *J. Neurochem.*, 14:596–598.
21. Karkowski, A. M., Bergamini, M. V. W., and Orlowski, M. (1976): Kinetic studies of sheep kidney γ-glutamyl transpeptidase. *J. Biol. Chem.*, 251:4736–4743.
22. Lazzarini, R. A., and Mehler, A. H. (1964): Separation of specific glutamate- and glutamine-activating enzymes from *Escherichia coli*. *Biochemistry*, 10:1445–1449.
23. Levintow, L., and Meister, A. (1954): Reversibility of the enzymatic synthesis of glutamine. *J. Biol. Chem.*, 209:265–280.
24. Lowenstein, J. M. (1972): Ammonia production in muscle and other tissues: The purine nucleotide cycle. *Physiol. Rev.*, 52:382–414.
25. Maddrey, W. C., Chura, C. M., Coulter, A. W., and Walser, M. (1973): Effects of keto-analogues of essential amino acids in portal systemic encephalopathy. *Gastroenterology*, 65:559.
26. Meister, A. (1965): *Biochemistry of the Amino Acids*, 2nd ed., Academic Press, New York.
27. Meister, A. (1973): On the enzymology of amino acid transport. *Science*, 180:33–39.

28. Meister, A. (1974): Glutamine synthetase of mammals. In: *The Enzymes*, 3rd ed., edited by P. Boyer, pp. 699–754, Academic Press, New York.
29. Meister, A. (1974): Glutathione synthesis. In: *The Enzymes*, 3rd ed., edited by P. Boyer, pp. 671–697. Academic Press, New York.
30. Meister, A. (1975): Biochemistry of glutathione. In: *Metabolism of Sulfur Compounds, Metabolic Pathways*, 3rd ed., edited by D. M. Greenberg, pp. 101–188. Academic Press, New York.
31. Meister, A. (1975): Structure-function relationships in glutamine amidotransferases: Carbamyl phosphate synthetase. *PAABS Rev.*, 4:273–299.
32. Meister, A. (1977): 5-Oxoprolinuria (pyroglutamic aciduria) and other disorders of glutathione biosynthesis. In: *The Metabolic Basis of Inherited Diseases*, 4th ed., edited by J. B. Stanbury, J. B. Wyngaarden, and D. S. Frederickson, pp. 328–336. McGraw-Hill, New York.
33. Meister, A., and Tate, S. S. (1976): Glutathione and related γ-glutamyl compounds. Biosynthesis and utilization. *Annu. Rev. Biochem.*, 45:559–604.
34. Meister, A., Tate, S. S., and Ross, L. L. (1976): Membrane bound γ-glutamyl transpeptidase. In: *The Enzymes of Biological Membranes*, edited by A. Martinosi, pp. 315–347. Plenum Press, New York.
35. Meister, A., and Tice, S. V. (1950): Transamination from glutamine to α-keto acids. *J. Biol. Chem.*, 187:173–187.
36. Meister, A., Udenfriend, S., and Bessman, S. P. (1956): Diminished phenylketonuria in phenylpyruvic oligophrenia after administration of L-glutamine, L-glutamate, or L-asparagine. *J. Clin. Invest.*, 35:619–626.
37. Moldave, K., and Meister, A. (1957): Synthesis of phenylacetylglutamine by human tissue. *J. Biol. Chem.*, 229:463–476.
38. Orlowski, M., and Meister, A. (1970): The γ-glutamyl cycle: A possible transport system for amino acids. *Proc. Natl. Acad. Sci. USA*, 67:1248–1255.
39. Orlowski, M., and Meister, A. (1973): γ-Glutamyl cyclotransferase: Distribution, isozymic forms, and specificity. *J. Biol. Chem.*, 248:2836–2844.
40. Orlowski, M., Richman, P., and Meister, A. (1969): Isolation and properties of γ-L-glutamylcyclotransferase from human brain. *Biochemistry*, 8:1048–1055.
41. Pardee, A. B. (1968), Membrane transport proteins. *Science*, 162:632–637.
42. Ratner, S. (1973): Enzymes of arginine and urea synthesis. *Adv. Enzymol.*, 39:1–90.
43. Richman, P., and Meister, A. (1975): Regulation of γ-glutamyl-cysteine synthetase by nonallosteric feedback inhibition by glutathione. *J. Biol. Chem.*, 250:1422–1426.
44. Richman, P. G., Orlowski, M., and Meister, A. (1973): Inhibition of γ-glutamylcysteine synthetase by L-methionine-S-sulfoximine. *J. Biol. Chem.*, 248:6684–6690.
45. Richards, F. II, Cooper, M. R., Pearce, L. A., Cowan, R. J., and Spurr, C. L. (1974): Familial spinocerebellar degeneration, hemolytic anemia, and glutathione deficiency. *Arch. Intern. Med.*, 134:534–537.
46. Richards, P., Metcalfe-Gibson, A., Ward, E. E., Wrong, O., and Houghton, B. J. (1967): Utilisation of ammonia nitrogen for protein synthesis in man and the effects of protein restriction and ureamia. *Lancet*, 2:845–849.
47. Ronzio, R., and Meister, A. (1968): Phosphorylation of methionine sulfoximine by glutamine synthetase. *Proc. Natl. Acad. Sci. USA*, 59:164–170.
48. Rowe, W. B., and Meister, A. (1970): Identification of L-methionine-S-sulfoximine as the convulsant isomer of methionine sulfoximine. *Proc. Natl. Acad. Sci. USA*, 66:500–506.
49. Rudman, D. (1971): Capacity of human subjects to utilize keto analogues of valine and phenylalanine. *J. Clin. Invest.*, 50:90–96.
50. Sapir, D. G., Owen, O. E., Pozefsky, T., and Walser, M. (1974): Nitrogen sparing induced by a mixture of essential amino acids given chiefly as their keto-analogues during prolonged starvation in obese subjects. *J. Clin. Invest.*, 54:974–980.
51. Schloerb, P. R. (1966): Essential L-amino acid administration in uremia. *Am. J. Med. Sci.*, 252:650–659.
52. Schoenheimer, R. (1942): *The Dynamic State of Body Constituents*. Harvard University Press, Cambridge, Mass.
53. Schulman, J. D., Goodman, S. I., Mace, J. W., Patrick, A. D., Tietze, F., and Butler, E. J. (1975): Glutathionuria: Inborn error of metabolism due to tissue deficiency of γ-glutamyl transpeptidase. *Biochem. Biophys. Res. Commun.*, 65:68–74.
54. Sekura, R., and Meister, A. (1974): Glutathione turnover in the kidney: Considerations relating to the γ-glutamyl cycle and the transport of amino acids. *Proc. Natl. Acad. Sci. USA*, 71:2969–2972.

55. Sekura, R., Van Der Werf, P., and Meister, A. (1976): Mechanism and significance of the mammalian pathway for elimination of D-glutamate: Inhibition of glutathione synthesis by D-glutamate. *Biochem. Biophys. Res. Commun.*, 71:11–18.
56. Sherwin, C. P. (1917): Comparative metabolism of certain aromatic amino acids. *J. Biol. Chem.*, 31:307–310.
57. Taniguchi, N., and Meister, A. (1978): γ-Glutamyl cyclotransferase from rat kidney: Sulfhydryl groups and isolation of a stable form of the enzyme. *J. Biol. Chem.*, 253:1969–1978.
58. Thompson, G. A., and Meister, A. (1976): Hydrolysis and transfer reactions catalyzed by γ-glutamyl transpeptidase: Evidence for separate substrate sites and for high affinity of L-cystine. *Biochem. Biophys. Res. Commun.*, 71:32–36.
59. Thompson, G. A., and Meister, A. (1977): Interrelationships between the binding sites for amino acids, dipeptides, and γ-glutamyl donors in γ-glutamyl transpeptidase. *J. Biol. Chem.*, 252:6792–6797.
60. Van Der Werf, P., Griffith, O., and Meister, A. (1975): 5-Oxo-L-prolinase (L-pyroglutamate hydrolase): Purification and catalytic properties. *J. Biol. Chem.*, 250:6686–6692.
61. Van Der Werf, P., and Meister, A. (1975): The metabolic formation and utilization of 5-oxo-L-proline (L-pyroglutamate, L-pyrrolidone carboxylate). *Adv. Enzymol.*, 43:519–556.
62. Van Der Werf, P., Orlowski, M., and Meister, A. (1971): Enzymatic conversion of 5-oxo-L-proline (L-pyrrolidone carboxylate) to L-glutamate coupled with ATP cleavage to ADP: A reaction in the γ-glutamyl cycle. *Proc. Natl. Acad. Sci. USA*, 68:2982–2985.
63. Van Der Werf, P., Stephani, R. A., Orlowski, M., and Meister, A. (1973): Inhibition of 5-oxoprolinase by 2-imidazolidone-4-carboxylic acid. *Proc. Natl. Acad. Sci. USA*, 70:759–761.
64. Van Der Werf, P., Stephani, R. A., and Meister, A. (1974): Accumulation of 5-oxoproline in mouse tissues after inhibition of 5-oxoprolinase and administration of amino acids: evidence for function of the γ-glutamyl cycle. *Proc. Natl. Acad. Sci. USA*, 71:1026–1029.
65. Walser, M. (1975): Nutritional effects of nitrogen-free analogues of essential amino acids. *Life Sci.*, 17:1011.
66. Walser, M. (1975): Treatment of renal failure with keto acids. *Hosp. Pract.*, 10:59–66.
67. Walser, M., Coulter, A. W., Dighe, S., and Crantz, F. R. (1973): The effect of keto-analogues of essential amino acids in severe chronic uremia. *J. Clin. Invest.*, 52:678.
68. Wellner, V. P., Sekura, R., Meister, A., and Larsson, A. (1974): Glutathione synthetase deficiency, an inborn error of metabolism involving the γ-glutamyl cycle in patients with 5-oxoprolinuria (pyroglutamic aciduria). *Proc. Natl. Acad. Sci. USA*, 71:2505–2509.

Glutamic Acid: Advances in Biochemistry and Physiology, edited by L. J. Filer, Jr., et al.
Raven Press, New York © 1979.

Comparative Metabolism of Glutamate in the Mouse, Monkey, and Man

L. D. Stegink,* W. Ann Reynolds,† L. J. Filer, Jr.,* G. L. Baker,* T. T. Daabees,* and Roy M. Pitkin**

Departments of *Pediatrics and Biochemistry, and **Obstetrics and Gynecology, The University of Iowa College of Medicine, Iowa City, Iowa 52242; and †Department of Anatomy, University of Illinois at the Medical Center, Chicago, Illinois 60612

The mouse and monkey are often used as animal models to study the potential toxicity of glutamate salts for man. However, these species vary in apparent susceptibility to glutamate-induced neuronal necrosis. It is now generally accepted that the administration of large quantities of glutamate or aspartate to the neonatal rodent produces a variety of toxic effects (see review in ref. 27). However, even the rodent's susceptibility is partially strain dependent, some strains showing greater sensitivity than others to these amino acids (16). The potential for glutamate salts to produce toxic effects in the neonatal primate is controversial. Original reports associated the administration of large doses of glutamate salts with hypothalamic neuronal necrosis in the neonatal primate (23,25). However, at least four independent research groups have not been able to reproduce the lesion in the primate (1,2,17,26,32,36). These same research groups have no difficulty in inducing the rodent lesion.

Our research groups have focused on studies evaluating the effect of added glutamate salts in food systems on the plasma levels of amino acids. These data have been accumulated in an attempt to supplement the neuropathology data evaluating the safety of glutamate salts. Our premise has been the following: although the neonatal rodent is clearly susceptible to glutamate-induced neuronal necrosis (16,20), plasma glutamate levels must be substantially elevated prior to the occurrence of such lesions (33). In addition, the so-called blood-brain barrier of the species must be susceptible. Thus, our data indicate that plasma glutamate levels must be greater than six times normal in the infant mouse before neuronal necrosis is observed (33). Even the acutely sensitive neonatal mouse tolerates plasma glutamate plus aspartate levels under 60 μmoles/dl and develops neuronal necrosis only when plasma levels exceed this apparent threshold.

It has been suggested that the inability of research groups other than Olney's to induce lesions in the neonatal primate with glutamate reflected a failure to elevate plasma glutamate levels (25). However, this is not the case. Our combined research groups have studied animals in which plasma glutamate levels were grossly elevated without finding evidence of the hypothalamic lesion (32). These data clearly demonstrate that the neonatal primate and rodent differ in their susceptibility to glutamate-induced neuronal necrosis at equivalent plasma glutamate levels.

We have presented preliminary data (7,27) suggesting that glutamate metabolism might differ in the rodent and in the primate. These data were based on the radioactivity found in glutamate-derived metabolites after the administration of ^{14}C-labeled glutamate. The major metabolites in the neonatal primate were glucose and lactate, whereas the neonatal rodent accumulated α-ketoglutarate and acetoacetate as well. These studies led us to further comparisons of the absorption and metabolism of glutamate in mice, monkeys, and humans after ingestion of monosodium glutamate (MSG) either with meals or dissolved in water.

Our first experiment focused on plasma glutamate levels following the addition of MSG to meal systems. It had been suggested that MSG added to food might be absorbed more rapidly and metabolized less well than protein-bound glutamate, yielding markedly elevated plasma glutamate levels (21,22). To determine the appropriate experimental levels of MSG ingestion, we used the data of the Committee on GRAS List Survey—Phase III (9). The data in Table 1, taken from Appendix E of the Committee's report, show estimated daily intakes of MSG for individuals living in the United States. These data show an expected mean daily intake of 6.8 mg/kg body weight in the 12- to 23-month-old infant. This age group has the highest estimated daily intake of MSG in the United States. In this age group, 30 mg MSG/kg body weight represents the 90th percentile of total daily ingestion.

In our first study, we added MSG at a level of 34 mg/kg body weight to a high-protein meal and measured plasma amino acid levels with time (4). Normal adult subjects were fed a hamburger-milk shake meal providing 1 g protein/kg body weight. The composition of the meal is shown in Table 2. One group of six individuals ingested the meal alone, whereas a second group of six subjects ingested

TABLE 1. *Expected daily intake of MSG based on person-days*

	Total sample Intakes (mg/kg/day)			
		Percentile		
Age	Mean	90th	99th	99.9th
0–5 months	0.3	0	11	25
6–11 months	1.9	1.9	36	46
12–23 months	6.8	30	43	61
2–5 years	5.5	23	37	56
6–17 years	2.7	10	25	40
18+ years	1.5	7	12	19

From Committee on GRAS List Survey—Phase III, ref. 9.

TABLE 2. *Composition of the hamburger-milk shake test meal for a 70-kg adult*[a]

Component	Quantity (g)	Protein[b] (g)	Fat (g)	Carbohydrate (g)	Energy (kcal)
Hamburger	222	61	25.5	0	346
Bun	50	4.5	1.5	25.5	133
Milk	100	3.5	3.5	5	66
Ice cream	50	2	5	11	95
Total	442	71	35.5	72	640

[a] The quantity of the hamburger in each meal was varied with each individual so as to provide a uniform protein level of 1 g/kg body weight (4).
[b] Protein, 38% of total energy.

an identical meal to which MSG had been added at a level providing 34 mg/kg body weight. The plasma glutamate and aspartate levels in these subjects are shown in Fig. 1. These data demonstrate that the addition of MSG to the meal had no effect on plasma glutamate and aspartate levels beyond that of the meal itself.

The results from this study led us to evaluate the effects of higher MSG loads ingested with meals on plasma amino acid levels. The Acceptable Daily Intake (ADI) set for MSG by the WHO/FAO is 150 mg/kg body weight/day. We elected to study the effects of 100 and 150 mg MSG/kg added to a single meal on plasma glutamate levels. In addition, we wished to compare the effect of these dosages on mice, monkeys, and humans.

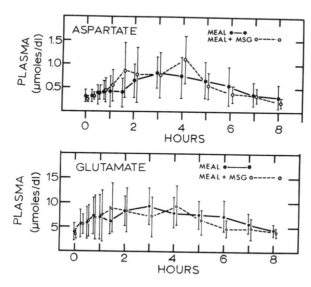

FIG. 1. Plasma glutamate and aspartate levels (mean ± SD) in normal adult subjects after ingestion of a high-protein meal (1 g protein/kg) with or without added MSG (34 mg/kg). (From Baker, et al., ref. 4.)

TABLE 3. Composition of the Sustagen[a] meal system used

Component	Quantity (g/kg)	Energy (kcal/kg)
Protein	0.40	1.6
Fat	0.059	0.53
Carbohydrate	1.12	4.48
Water	4.2	0
Total	5.78	6.61

[a] Mead-Johnson; formula also contains appropriate vitamins and minerals.

A Sustagen meal system was chosen for these studies. Sustagen (composition shown in Table 3) is a liquid, ready-to-feed solution that can be fed to mice, monkeys, and man. The Sustagen solution was fed to adult animals or humans at 4.2 ml/kg. This level provides 0.4 g protein, 1.12 g carbohydrate, and 1.6 kcal/kg body weight. The MSG was dissolved in Sustagen to give a 2.4% (w/v) solution when the dose was 100 mg/kg, and to give a 3.6% solution (w/v) when MSG was administered at 150 mg/kg body weight.

The data in Fig. 2 show mean plasma glutamate levels in adult mice (Webster Swiss albino strain) administered Sustagen meals providing 0, 100, or 150 mg MSG/kg body weight. Glutamate values at each time point represent plasma levels from pooled blood of five animals. The administration of the Sustagen meal alone resulted in a slight increase in plasma glutamate levels. The addition of MSG to the meal resulted in further, small, highly variable increases in plasma glutamate levels. Maximum levels of 15 μmoles/dl were observed at 1 hr after administration of the Sustagen meal providing 150 mg MSG/kg body weight.

The data in Fig. 3 show mean plasma glutamate levels in adult rhesus monkeys (*Macaca mulatta*) administered Sustagen meals providing MSG at 0, 100, or 150 mg/kg body weight. In each experiment, four adult animals were administered the Sustagen meals by stomach tube. Sequential blood samples were obtained from each animal after dosing as described previously (31,32). Plasma glutamate levels were essentially unchanged in the animals administered the Sustagen meal alone or

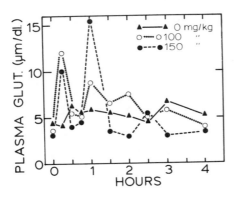

FIG. 2. Mean plasma glutamate (GLUT) levels in adult mice administered a Sustagen meal (0.4 g protein/kg) with and without added MSG (100 and 150 mg/kg body weight).

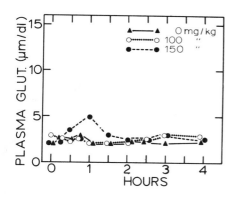

FIG. 3. Mean plasma glutamate (GLUT) levels in adult rhesus monkeys administered a Sustagen meal (0.4 g protein/kg) with and without added MSG (100 and 150 mg/kg body weight).

the Sustagen meal providing MSG at 100 mg/kg body weight. Ingestion of the Sustagen meal providing MSG at 150 mg/kg body weight resulted in a slight increase in glutamate levels at 1 hr after ingestion.

The data in Fig. 4 show plasma glutamate in normal human subjects fed the Sustagen meals (5). Six normal adult subjects (3 male, 3 female) were studied using a Latin Square Design (8) for the administration of the three meal systems studied (Sustagen alone, Sustagen providing 100 mg MSG/kg, and Sustagen providing 150 mg MSG/kg). The addition of MSG to the Sustagen meal produced slightly higher and broader plasma glutamate curves than observed after the ingestion of the Sustagen meal alone. However, peak plasma glutamate levels after ingestion of the meal providing 150 mg MSG/kg were no greater than those noted in subjects ingesting a high-protein meal (1 g protein/kg) without added glutamate (Fig. 1). These data demonstrate the excellent metabolism of MSG added to meals and show that added MSG is *not* absorbed preferentially to protein-bound glutamate as suggested by Olney (21,22).

The data in Fig. 5 compare plasma glutamate levels in adult mice, monkeys, and humans after the ingestion of the Sustagen meal providing MSG at a level of 150

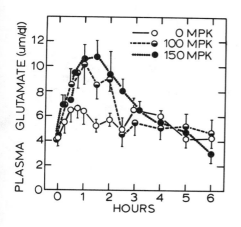

FIG. 4. Plasma glutamate levels (mean ± SEM) in six normal adult subjects administered a Sustagen meal (0.4 g protein/kg body weight) with and without added MSG (100 and 150 mg/kg body weight).

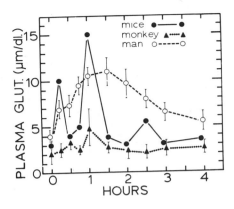

FIG. 5. Comparison of plasma glutamate (GLUT) levels in adult mice (mean) monkeys (mean ± SEM) and humans (mean ± SEM) administered a Sustagen meal providing 150 mg MSG/kg body weight.

mg/kg body weight. These data suggest that adult humans metabolize glutamate less rapidly than do adult mice or monkeys, since they show a broader curve. This was surprising, since we expected adult humans and monkeys to metabolize glutamate more rapidly than mice.

The MSG meal data led us to compare glutamate absorption-metabolism curves when MSG is administered to these three species in water. Adult mice, monkeys, and human subjects were administered MSG dissolved in water at levels providing 150 mg/kg body weight. The glutamate was dissolved to provide a 3.6% solution, and the solution was administered at 4.2 ml/kg body weight. The percentage of MSG in solution and the volume per kg administered were identical to those used in the Sustagen study. The data in Fig. 6 compare plasma glutamate levels in mice, monkeys, and human subjects after ingestion of MSG dissolved in water (150

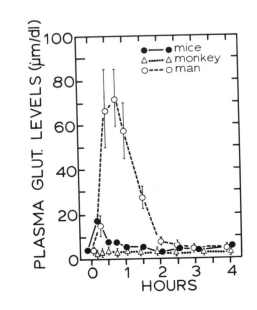

FIG. 6. Plasma glutamate (GLUT) levels in adult mice (mean), monkeys, and humans (mean ± SEM) administered 150 mg MSG/kg body weight dissolved in water.

mg/kg body weight). These data, like the Sustagen data, suggest that adult humans metabolize glutamate less rapidly than adult mice or monkeys. The levels seen in adult mice and monkeys were consistent with the levels observed in the neonatal pig given similar doses of MSG (28).

The data shown in Fig. 7 compare plasma glutamate levels in human subjects administered MSG at 150 mg/kg body weight either with the Sustagen meal or dissolved in water. These data demonstrate the modulating effect of food on MSG metabolism and absorption. The administration of the MSG in water produced much higher plasma glutamate levels than did the equivalent dose administered with a meal. Thus, MSG metabolism varies depending on the vehicle used to administer the dose. Most importantly, it becomes clear that MSG added to a meal is not preferentially absorbed.

The results obtained from mice and monkeys given MSG in water were puzzling when compared to the data obtained in man. Marked elevations in plasma glutamate levels were not obtained in either mice or monkeys given MSG at 100 or 150 mg/kg body weight. Failure to elevate plasma glutamate levels when MSG was administered with Sustagen was not surprising, since we had noted only small changes in plasma glutamate levels in humans under these conditions. However, the low plasma glutamate levels observed in monkeys and mice when MSG was administered in water was puzzling, since plasma glutamate levels had risen in humans under comparable conditions.

These results led us to consider possible causes for these low levels, especially the possibility that gastric emptying had failed to occur in these animals. To evaluate this possibility, loading studies in adult mice and monkeys were carried out using increasing doses of MSG dissolved in water. We wished to obtain a dose-

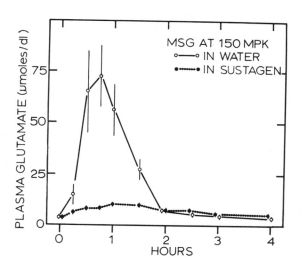

FIG. 7. Plasma glutamate levels in normal adult humans administered MSG at 150 mg/kg body weight either dissolved in water, or as part of a Sustagen meal. Data shown as mean ± SEM.

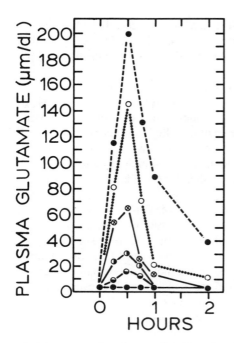

FIG. 8. Mean plasma glutamate levels in adult mice administered MSG dissolved in water at 0 (—●—), 150 (—○—), 250 (—◐—), 500 (—⊗—), 1,000 (···○···), and 2,000 (--●--) mg/kg body weight.

response curve for plasma glutamate levels with MSG dose. These data would allow us to determine if plasma glutamate levels measured in mice and monkeys were appropriate to the dose of MSG administered.

The data in Fig. 8 show mean plasma glutamate levels in adult mice administered MSG dissolved in water. The glutamate doses studied were 0, 250, 500, 1,000, 2,000, and 4,000 mg MSG/kg body weight. A 5% MSG solution was used for animals dosed at 250 and 500 mg/kg, a 10% solution was used for animals dosed at 1,000 mg/kg, a 20% solution was used for animals dosed at 2,000 mg/kg, and a 40% solution was used for animals dosed at 4,000 mg/kg. The data in Fig. 8 show a reasonable response of plasma glutamate levels to increasing doses of MSG and indicate that the plasma levels obtained at the 150 mg/kg dose are reasonable.

The data in Fig. 9 show similar results in adult monkeys administered MSG in water at doses of 150, 500, and 1,000 mg/kg body weight. In these studies, apparent delayed gastric emptying was observed in the two animals studied at 500 mg/kg dose. However, allowing for this shift, the overall response observed after MSG loading at 150 mg/kg is appropriate. The data shown in Figs. 8 and 9 support those data shown in Fig. 6 and indicate that adult humans metabolize glutamate less efficiently than either adult mice or monkeys.

Although peak plasma glutamate levels in adult mice and monkeys showed a reasonable response to increasing doses of MSG, the peak values observed were lower than expected, based on data available for the neonatal rodent and primate. This difference led us to speculate whether younger animals metabolize glutamate less well than adults.

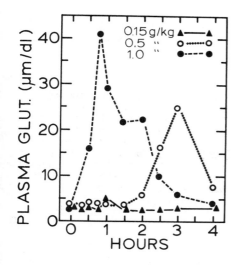

FIG. 9. Mean plasma glutamate (GLUT) levels in adult rhesus monkeys administered MSG dissolved in water at 150, 500, and 1,000 mg/kg body weight.

In 1974 we studied the plasma threshold levels of dicarboxylic amino acids required to produce neuronal necrosis after ingestion of protein hydrolysate solutions (33). Neonatal mice were injected with glutamate- and aspartate-containing solutions, and plasma amino acid levels were measured with time. As shown in Fig. 10, a series of dose-related curves of plasma glutamate and aspartate levels was

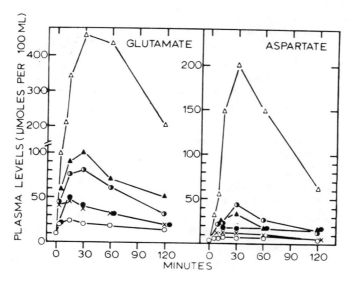

FIG. 10. Plasma glutamate and aspartate levels in 9- to 11-day-old mice injected with protein hydrolysate solutions containing glutamate and aspartate: Amigen at 20 μl/g (×), 50 μl/g (▲), or 100 μl/g (△), or Aminosol at 20 μl/g (○), 50 μl/g (●), or 100 μl/g (⊙). (From Stegink et al., ref. 33, with permission.)

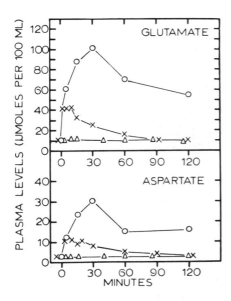

FIG. 11. Plasma glutamate and aspartate levels in 9- to 11-day-old mice following subcutaneous injection with either isotonic saline at 50 µl/g body weight (△) or glutamate and aspartate containing protein hydrolysate (Amigen) at 50 µl/g body weight (○). Plasma glutamate and aspartate levels in 25-day-old mice injected with Amigen at 50 µl/g body weight (×). (From Stegink et al., ref. 33, with permission.)

obtained. Comparison of these data to the degrees of neuronal necrosis observed by Olney et al. (24) in animals injected with these solutions indicated a plasma dicarboxylic amino acid threshold level of about 60 µmoles/dl was required for neuronal damage in the neonatal mouse.

Since the adult mouse is known to be more resistant to dicarboxylic amino acid-induced neuronal necrosis than the neonate (16,20), we compared plasma levels in infant and adult mice. The data in Fig. 11 show plasma glutamate and aspartate levels in 9- to 11-day-old and 25-day-old mice following subcutaneous injection with either isotonic saline or protein hydrolysate solutions at 50 µliters/gm body weight. Comparison of plasma glutamate and aspartate levels indicate that older animals metabolize injected glutamate and aspartate more rapidly than the younger animals.

These animals had been studied, however, after subcutaneous injection of glutamate rather than oral administration and with a glucose-protein hydrolysate solution rather than glutamate alone. To eliminate these differences, we studied neonatal mice that had received an orally administered solution of MSG in water. Nine- to ten-day-old mice were treated with 0, 250, 500, 1,000, and 2,000 mg MSG/kg body weight. The percentage of MSG in the solutions used was the same as that utilized for the adult animals shown in Fig. 8. The data in Fig. 12 show mean plasma glutamate levels in these animals. These data indicate much higher plasma glutamate levels are obtained in neonatal mice than in adult mice given equivalent oral doses of MSG in water.

These data indicate a plasma glutamate threshold for neuronal necrosis that is close to the one obtained in our studies with the protein hydrolysate (33). Reynold's data (*this volume*) indicate an absence of neuronal necrosis in neonatal mice given

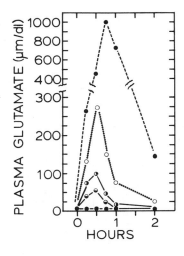

FIG. 12. Plasma glutamate levels in 9- to 11-day-old mice administered oral loads of glutamate dissolved in water: 0 (—●—), 250 (—⊙—), 500 (—○—), 1,000 (···○···), or 2,000 (--●--) mg/kg.

MSG at a dose of 250 mg/kg body weight. However, 22% of the neonatal mice administered MSG at 500 mg/kg show neuronal necrosis. The data in Fig. 12 show plasma levels of 40 μmoles/dl in animals given 250 mg/kg and levels of 80 to 100 μmoles/dl in animals given 500 mg/kg. This suggests a plasma glutamate threshold level of approximately 80 μmoles/dl, in good agreement with our value of 60 μmoles/dl obtained in the casein hydrolysate studies (33), where the osmotic load was high.

The peak plasma glutamate values obtained in these infant mice are similar to those reported by other investigators (Table 4). However, the peak plasma levels observed in our adult mice are significantly lower than values reported by other investigators (Table 5), although the values reported by others vary considerably, ranging from 208 to 344 μmoles/dl for an equivalent dose (6,15,18,19). These investigators (6,18,19) report little, or no, age-related effect on the metabolism of orally administered glutamate.

The available data confirm that the infant rodent metabolizes *injected* glutamate less well than the adult rodent. O'hara et al. (18,19) and Takasaki et al. (34) also report an age-related effect of subcutaneously injected glutamate, confirming our earlier report (33). O'hara et al. (18,19) reported peak plasma glutamate levels of 1,058 μmoles/dl in infant mice, 760 μmoles/dl in weanling mice, and 539 μmoles/dl in adult animals injected subcutaneously with MSG at a dose of 1 g/kg body weight. When the dose was given intraperitoneally rather than subcutaneously, this age-related effect was smaller. Our data are also consistent with data reported by Cresteil and Leroux (10) on the metabolism of injected glutamate by neonatal and adult rodents. They report a strong age-related effect in the conversion of injected ^{14}C-glutamate to $^{14}CO_2$. The rate of $^{14}CO_2$ production from injected glutamate increased from 130 nmoles/hr in the zero-hour neonatal rat to 2,000 nmoles/hr in the 21-day-old animal, indicative of an increased ability to metabolize glutamate.

TABLE 4. *Mean peak plasma glutamate levels in 7- to 11-day-old mice after oral administration of MSG dissolved in water*

Dose (mg/kg)	Plasma glutamate levels (μmoles/dl)		
	Iowa	Bizzi et al. (6)	O'hara et al. (18,19)
0	8	—	17
250	40	72	—
500	88	108	62
1,000	282	210	314
2,000	1,050	—	—

Since all laboratories report age-related effects on the metabolism of injected glutamate, the failure of other laboratories to note an age-related effect after oral administration was puzzling. These differences could reflect the differences in the strains of animals studied. In addition, it is possible that our mice had more residual food in their gut than those studied by other investigators. The simultaneous availability of food, particularly carbohydrate, has a marked effect on plasma glutamate levels after an oral glutamate load (30). In our studies, adult and infant mice were fasted from 2 to 4 hr prior to glutamate loading. However, adult and infant animals ingested diets of differing composition (mouse milk versus chow), and this difference could affect the quantity of food remaining in the gut at the time of glutamate loading. Such differences could account for the large variation in peak plasma glutamate levels reported by other investigators in adult mice after oral glutamate loads. For example, O'hara et al. (18,19) report higher levels than either Bizzi et al. (6) or James et al. (15) after equivalent glutamate loads (Table 5). However, O'hara et al. (18,19) fasted animals for 10 hr prior to dosing, whereas other groups apparently did not.

To test this point, adult mice were fasted for 24 hr prior to oral administration of MSG (1 g/kg body weight administered as a 10% solution). A comparison of plasma glutamate levels in adult mice fasted for 2 to 4 hr and those fasted for 24 hr is shown in Fig. 13. These data demonstrate that the length of fasting prior to dose has a

TABLE 5. *Mean peak plasma glutamate levels in infant, weanling, and adult mice after oral doses of MSG at 1 g/kg body weight using 10% solutions*

Age of animal	Peak plasma glutamate levels (μmoles/dl)			
	Iowa	Bizzi et al. (6)	O'hara et al. (18,19)	James et al. (15)
Infant	282	210	314	—
Weanling	—	—	219	—
Adult	145[a]	208[b]	344[c]	210[b]

[a] Animals fasted 2 to 3 hr.
[b] Animals not fasted.
[c] Animals fasted 10 hr prior to dosing.

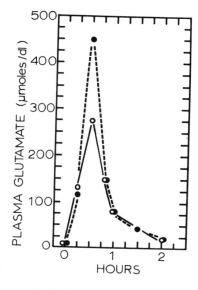

FIG. 13. Mean plasma glutamate levels in adult mice given 1 g MSG/kg body weight. Mice were fasted for 2 to 3 hr (—○—) or 24 hr (--●--) prior to dosing.

marked effect on peak plasma glutamate levels. The peak values observed in animals fasted for 24-hr were higher than those reported by O'hara et al. (18,19), where a 10-hr fast was used.

It is not certain whether the increased plasma glutamate levels noted with increasing length of fasting result entirely from a decrease in residual food in the gut. Intestinal enzymes have very short half-lives, and increased fasting will decrease the levels of intestinal enzymes needed to metabolize orally administered glutamate. However, these data support our contention of an age-related effect on glutamate metabolism in animals fasted for a standard length of time, such as 3 to 4 hr. The failure of O'Hara et al. (18,19) to observe an age-related effect with oral glutamate administration between neonatal, weanling, and adult mice probably reflects the differing fasting periods used. In their studies, adult animals were fasted for 10 hr, weanling animals for 2 hr, and infant mice were not fasted. Thus, the variables related to fasting could obscure an age effect. There are other reasons to believe that the age-related effect on glutamate metabolism observed in animals fasted 3 to 4 hr is real. Age-related effects are noted when glutamate is injected (18,19,33). Factors causing this age-related effect should also affect the metabolism of orally administered glutamate to some degree. Second, Wen and Gershoff (35) report that the levels of the two major glutamate transaminase enzymes present in rodent intestinal mucosa show a strong age-related effect. Enzyme levels are much lower in the neonatal rat than in postweanling animals. The intestinal transaminases play a major role in the metabolism of orally administered glutamate (27,30), and it is highly likely that differences in levels of mucosal glutamate transaminases would affect glutamate metabolism and clearance. Levels of these enzymes may decrease with increased fasting in adult animals, accounting for the increase in plasma glutamate levels when animals are fasted for 24 hr rather than 3 to 4 hr. Thus, the

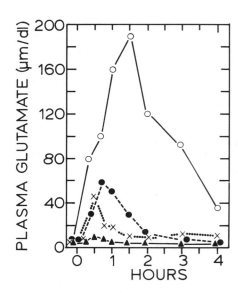

FIG. 14. Plasma glutamate levels in neonatal rhesus monkeys administered oral loads of MSG dissolved in water: 0 (—▲—), 250 (···x···), 500 (--●--), or 1,000 (—○—) mg/kg.

factors affecting glutamate metabolism and clearance are complex. The length of fasting prior to administration of the load, the simultaneous presence of food, and the age of the animal all affect peak plasma glutamate levels.

We also carried out experiments in neonatal rhesus monkeys to determine if similar age-related effects on glutamate metabolism were observed in the primate. The data in Fig. 14 show plasma glutamate levels in neonatal monkeys (1 to 2 weeks of age) given oral doses of MSG in water at 150, 250, 500, and 1,000 mg/kg body weight. Comparisons of these data with values obtained in adult monkeys (Fig. 9) indicate that the adult monkey metabolizes glutamate more rapidly than the neonate. The monkey, like man (30), may have a considerable individual-to-individual variation in the ability to metabolize glutamate. We have previously observed two neonatal monkeys who showed a marked inability to metabolize glutamate (32). However, the best interpretation of the data indicates an age-related increase in the ability of the rhesus monkey to metabolize glutamate.

The brains and retinas of the neonatal monkeys were concomitantly perfused with glutaraldehyde for future study after being prepared as thin (1-μ) plastic sections. No damage was found in the hypothalami (26,32) or the retinas (D. Apple, *personal communication*) of these newborn monkeys following MSG loads of 1 to 4 mg/g.

The role of the erythrocyte in the transport of glutamate in the blood is of considerable interest, since certain amino acids are known to be transported in the erythrocyte to a greater extent than in plasma under certain circumstances (3,11–14).

Our data differ from those reported by Bizzi et al. (6) concerning the ratio of erythrocyte to plasma glutamate in animals. They report this ratio to be about unity in both arterial and venous compartments, either in basal conditions or after oral administration of MSG. Our data (Table 6) indicate a considerable difference

TABLE 6. *Fasting plasma and erythrocyte glutamate levels in adult mice, monkeys, and humans*

Species	Plasma (μmoles/dl)	Erythrocyte (μmoles/dg)
Mice	6.8 ± 3.6[a]	21.0 ± 8.8
Monkeys	3.1 ± 0.4	59.9 ± 8.9
Humans	4.8 ± 1.6	18.9 ± 7.1

[a] Mean ± SD.

between plasma and erythrocyte levels of glutamate in fasting adult mice, monkeys, and man. This difference increases further if the erythrocyte levels are expressed as μmoles/100 ml water in the red cell, rather than as μmoles/100 g of cells. We do not see the simultaneous increase in both plasma and erythrocyte glutamate levels after glutamate loading as reported by Bizzi et al. (6). The data in Fig. 15 show plasma and erythrocyte glutamate levels in a single male adult human subject given MSG in water at 100 or 150 mg/kg body weight. Despite large increases in plasma glutamate levels, erythrocyte levels do not change. Similar results were obtained in adult mice.

The interaction between plasma glutamate and erythrocyte glutamate levels is complex. As shown in Table 6, the erythrocyte, like other tissues, maintains a considerably higher glutamate concentration within the cell than in plasma. Our data indicate that erythrocyte glutamate levels in the neonatal human and monkey are considerably higher than those in adult humans and monkeys. Our data suggest that the neonatal monkey erythrocyte may be more receptive to glutamate transfer from the plasma than the adult erythrocyte.

The data in Fig. 16 show plasma and erythrocyte glutamate levels in two neonatal monkeys administered oral doses of MSG in water. Considerable variation is noted between erythrocyte glutamate levels in these two animals. The data suggest that elevations in plasma glutamate that do not exceed the level of glutamate present in the erythrocyte do not affect erythrocyte levels. However, when plasma glutamate

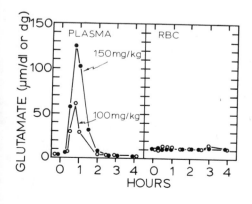

FIG. 15. Plasma (μmoles/dl) and erythrocyte (μmoles/dg) glutamate levels in a typical adult male subject ingesting MSG dissolved in water at 100 or 150 mg/kg body weight.

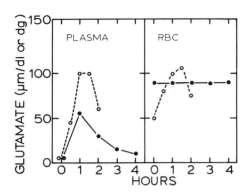

FIG. 16. Plasma (μmoles/dl) and erythrocyte (μmoles/dg) glutamate levels in two infant rhesus monkeys ingesting MSG dissolved in water: animal 1 (—●—) and animal 2 (--○--).

levels exceeded those normally present in the erythrocyte, erythrocyte glutamate levels also increased.

In summary, our data indicate that the administration of large doses of MSG with meals to humans, monkeys, and mice results in only small increases in plasma glutamate levels over those occurring with the meal alone. The data indicate that the added MSG is not absorbed preferentially to the protein-bound glutamate.

The administration of MSG dissolved in water results in higher plasma glutamate levels than when the equivalent dose is given with a meal. This effect is most obvious in humans.

Glutamate-loading studies in mice and monkeys indicate that the adult animal metabolizes glutamate more rapidly than the neonate, although many factors affect the rates of metabolism. The data suggest that adult humans metabolize glutamate less rapidly than do either adult mice or rhesus monkeys.

ACKNOWLEDGMENT

These studies were supported in part by a grant-in-aid from the International Glutamate Technical Committee.

REFERENCES

1. Abraham, R. W., Dougherty, W., Golberg, L., and Coulston, F. (1971): The response of hypothalamus to high doses of monosodium glutamate in mice and monkeys. Cytochemistry and ultrastructural study of lyosomal changes. *Exp. Mol. Pathol.*, 15:43–60.
2. Abraham, R. W., Swart, J., Golberg, L., and Coulston, F. (1975): Electron microscopic observations of hypothalami in neonatal rhesus monkeys (*Macaca mulatta*) after administration of monosodium L-glutamate. *Exp. Mol. Pathol.*, 23:203–213.
3. Aoki, T. T., Brennan, M. F., Muller, W. A., Moore, F. D., and Cahill, G. F., Jr. (1972): The effect of insulin on muscle glutamate uptake. Whole blood versus plasma glutamate analysis. *J. Clin. Invest.*, 51:2889–2894.
4. Baker, G. L., Filer, L. J., Jr., and Stegink, L. D. (1977): Plasma and erythrocyte amino acid levels in normal adults fed high protein meals. Effect of adding monosodium glutamate (MSG) or Aspartame. *Fed. Proc.*, 36:1154.
5. Baker, G. L., Filer, L. J., Jr., and Stegink, L. D. (1978): Plasma amino acid levels in normal adults after ingestion of high doses of monosodium glutamate (MSG) with a meal. *Fed. Proc.*, 37:752.

6. Bizzi, A., Veneroni, E., Salmona, M., and Garattini, S. (1977): Kinetics of monosodium glutamate in relation to its neurotoxicity. *Toxicol. Lett.*, 1:123–130.
7. Boaz, D. P., Stegink, L. D., Reynolds, W. A., Filer, L. J., Jr., Pitkin, R. M., and Brummel, M. C. (1974): Monosodium glutamate metabolism in the neonatal primate. *Fed. Proc.*, 33:651.
8. Cochran, W. G., and Cox, G. M. (1950): *Experimental Design*, p. 86. Wiley & Sons, New York.
9. Committee on GRAS List Survey—Phase III (1976): Estimating distributions of daily intake of monosodium glutamate (MSG), Appendix E. In: *Estimating Distribution of Daily Intake of Certain GRAS Substances*. Food and Nutrition Board, Division of Biological Sciences, Assembly of Life Sciences, National Research Council, National Academy of Sciences, Washington, D.C.
10. Cresteil, T., and Leroux, J-P. (1977): Early postnatal metabolism of amino acids in rat. *Pediatr. Res.*, 11:720–723.
11. Elwyn, D. H. (1966): Distribution of amino acids between plasma and red blood cells in the dog. *Fed. Proc.*, 25:854–861.
12. Elwyn, D. H., Launder, W. J., Parikh, H. C., and Wise, E. M., Jr. (1972): Roles of plasma and erythrocytes in interorgan transport of amino acids in dogs. *Am. J. Physiol.*, 222:1333–1342.
13. Elwyn, D. H., Parikh, H. C., and Shoemaker, W. C. (1968): Amino acid movements between gut, liver and periphery in unanesthetized dogs. *Am. J. Physiol.*, 215:1260–1275.
14. Felig, P., Wahren, J., and Raf, L. (1973): Evidence of interorgan amino acid transport by blood cells in humans. *Proc. Natl. Acad. Sci. USA*, 70:1775–1779.
15. James, R. W., Heywood, R., Worden, A. N., Garattini, S., and Salmona, M. (1978): The oral administration of MSG at varying concentrations to male mice. *Toxicol. Lett.*, 1:195–199.
16. Lemkey-Johnston, N., and Reynolds, W. A. (1974): Nature and extent of brain lesions in mice related to ingestion of monosodium glutamate. *J. Neuropathol. Exp. Neurol.*, 33:74–97.
17. Newman, A. J., Heywood, R., Plamer, A. K., Barry, D. H., Edwards, F. P., and Worden, A. N. (1973): The administration of monosodium-L-glutamate to neonatal and pregnant rhesus monkeys. *Toxicology*, 1:197–294.
18. O'hara, Y., Takasaki, Y., Iwata, S., and Sasaoka, M. (1978): Plasma glutamate levels in free feeding of monosodium glutamate and relationship between plasma glutamate and brain lesions in mice. *(in press)*.
19. O'hara, Y., Iwata, S., Ichimura, M., and Sasaoka, M. (1977): Effect of administration routes of monosodium glutamate on plasma glutamate levels in infant, weanling and adult mice. *J. Sci. Toxicol.*, 2:281–290.
20. Olney, J. W. (1969): Brain lesions, obesity and other disturbances in mice treated with monosodium glutamate. *Science*, 164:719–721.
21. Olney, J. W. (1975): Another view of Aspartame. In: *Sweeteners, Issues and Uncertainties*, pp. 189–195. Academy Forum, National Academy of Sciences, Washington, D.C.
22. Olney, J. W. (1975): L-Glutamic and L-aspartic acids—A question of hazard? *Food Cosmet. Toxicol.*, 13:595–600.
23. Olney, J. W., and Sharpe, L. G. (1969): Brain lesions in an infant rhesus monkey treated with monosodium glutamate. *Science*, 166:386–388.
24. Olney, J. W., Ho, O. L., and Rhee, V. (1973): Brain damaging potential of protein hydrolysates. *N. Engl. J. Med.*, 289:391–395.
25. Olney, J. W., Sharpe, L. G., and Feigin, R. D. (1972): Glutamate-induced brain damage in infant primates. *J. Neuropathol. Exp. Neurol.*, 31:464–488.
26. Reynolds, W. A., Lemkey-Johnston, N., Filer, L. J., Jr., and Pitkin, R. M. (1971): Monosodium glutamate: Absence of hypothalamic lesions after ingestion by newborn primates. *Science*, 172:1342–1344.
27. Stegink, L. D. (1976): Absorption, utilization and safety of aspartic acid. *J. Toxicol. Environ. Health*, 2:215–242.
28. Stegink, L. D., Filer, L. J., Jr., and Baker, G. L. (1972): Monosodium glutamate metabolism in the neonatal pig. Effect of load on plasma, brain, muscle and spinal fluid free amino acid levels. *J. Nutr.*, 103:1135–1145.
29. Stegink, L. D., Filer, L. J., Jr., and Baker, G. L. (1977): Effect of Aspartame and aspartate upon plasma and erythrocyte free amino acid levels in normal adult volunteers. *J. Nutr.*, 107:1837–1845.
30. Stegink, L. D., Filer, L. J., Jr., Baker, G. L., Mueller, S. M., and Wu-Rideout, M. Y-C. (1979): Factors effecting plasma glutamate levels in normal adult subjects *(this volume)*.
31. Stegink, L. D., Pitkin, R. M., Reynolds, W. A., Boaz, D. P., Filer, L. J., Jr., and Brummel, M. C. (1976): Placental transfer of glutamate and its metabolites in the primate. *Am. J. Obstet. Gynecol.*, 122:70–78.

32. Stegink, L. D., Reynolds, W. A., Filer, L. J., Jr., Pitkin, R. M., Boaz, D. P., and Brummel, M. C. (1975): Monosodium glutamate metabolism in the neonatal monkey. *Am. J. Physiol.*, 229:246–250.
33. Stegink, L. D., Shepherd, J. A., Brummel, M. C., and Murray, L. M. (1974): Toxicity of protein hydrolysate solutions. Correlation of glutamate dose and neuronal necrosis to plasma amino acid levels in young mice. *Toxicology,* 2:285–299.
34. Takasaki, Y., Matsuzawa, Y., Iwata, S., O'hara, Y., Yonetani, S., and Ichimura, M. (1979): Toxicological studies of monosodium L-glutamate in rodents: Relationship between routes of administration and neurotoxicity *(this volume)*.
35. Wen, C. P., and Gershoff, S. N. (1972): Effects of dietary vitamin B_6 on the utilization of monosodium glutamate by rats. *J. Nutr.,* 102:835–840.
36. Wen, C., Hayes, K. C., and Gershoff, S. M. (1973): Effects of dietary supplementation of monosodium glutamate on infant monkeys, weanling rats and suckling mice. *Am. J. Clin. Nutr.,* 26:803–813.

Glutamic Acid: Advances in Biochemistry and Physiology, edited by L. J. Filer, Jr., et al.
Raven Press, New York © 1979.

Glutamate Metabolism and Placental Transfer in Pregnancy

Roy M. Pitkin,* W. Ann Reynolds,** L. D. Stegink,[†] and L. J. Filer, Jr.[†]

Departments of *Obstetrics and Gynecology, and [†]Pediatrics and Biochemistry, The University of Iowa, Iowa City, Iowa 52242; and **Department of Anatomy, University of Illinois at the Medical Center, Chicago, Illinois 60612

For more than 50 years it has been recognized that amino acid levels in the fetus generally exceed those in the mother (5). The concentration of most amino acids in fetal plasma is approximately twice that of the mother (3), implying active transport by the placenta. Although this characteristic connotes certain obvious advantages with respect to facilitating fetal growth and development, it also carries a potential for adverse effects if the transplacental gradient is maintained in the face of elevated maternal levels of those amino acids known to be toxic in high concentration. Phenylalanine is a classic example of this phenomenon, for infants born to women with hyperphenylalanemia, though themselves genetically normal, frequently exhibit growth retardation and permanent brain damage as a consequence of their prolonged intrauterine exposure to high phenylalanine levels (2).

Glutamate represents an amino acid of special concern because of the demonstration of a neurotoxic effect of high doses in certain species. Studies of fetal toxicity of maternally administered glutamate have yielded conflicting results. Whereas Murakami and Inouye (6) reported brain lesions in the mouse fetus following maternal glutamate administration, no such effects were found by Lucas and Newhouse (4) in the mouse or by Newman et al. (7) in the monkey.

The question of fetal toxicity of maternally administered glutamate hinges in large part on the matter of placental transfer. Although the indirect study mentioned above has suggested that glutamate crosses the placenta, other more direct investigations indicate that it does not. Dierks-Ventling and associates (1) injected glutamate (1 g/kg) into the tail vein of pregnant rats and found no increase in fetal glutamate levels despite a 35-fold increase in the maternal concentration. In an *in vitro* study, Schneider and Dancis (9) measured uptake and release of 10 amino acids by human placental slices; acidic amino acids (glutamate and aspartate) and serine were taken up rapidly but released very poorly, consistent with poor to absent placental transfer, whereas the efflux system for other amino acids was much more efficient.

We have conducted studies of the bidirectional transfer of glutamate across the primate placenta, described in detail elsewhere (12). For comparative purposes, we have also investigated maternal-fetal transmission of aspartate (8), the other dicarboxylic amino acid normally present in plasma.

METHODS

Rhesus monkeys (*Macaca mulatta*) were studied during the last third of pregnancy. This species was utilized because it has a hemochorial placenta virtually identical morphologically, and in most instances functionally, with that of the human. The rhesus placenta is usually bipartite with each lobe connected by interplacental vessels that can be identified by transillumination of the exposed uterus, exposed by incising the myometrium, and catheterized with a T-tube. This preparation permits access to the fetal circulation under physiologic conditions; i.e., with the fetus *in utero* and the amniotic sac intact. Amniotic fluid samples can be obtained by intermittent transuterine puncture.

The animals were prepared by inserting venous catheters into the inferior vena cava (via a saphenous vein) and into an arm vein. Following tranquilization with phencyclidine hydrochloride and induction of inhalation anesthesia, the uterus was exposed through a midline celiotomy incision and an interplacental vein catheterized with a silastic T-tube.

Maternal-fetal glutamate transfer was examined in five animals in which monosodium L-glutamate (MSG) was infused into an antecubital vein over 1 hr and sequential samples were obtained from the maternal vena cava and the interplacental vessel at intervals during the infusion and for 3 hr thereafter. The dosage of glutamate administered ranged from 0.16 to 0.4 g/kg maternal weight, and each infusion included added 3-4-^{14}C-L-glutamate in amounts of 50 to 75 μCi.

Fetal-maternal glutamate transfer was examined in three animals. In one, 5 g/kg fetal weight (with 60 μc ^{14}C-glutamate) was infused into an interplacental artery over 1 hr and sequential maternal and fetal blood samples were obtained over 2 hr thereafter. In a second animal, glutamate (0.8 g/kg fetal weight with 10 μc ^{14}C-glutamate) was injected through the uterus and into the fetal chest wall. The third fetus received glutamate (1.5 g/kg fetal weight with 10 μc ^{14}C-glutamate) into the umbilical vein. In these two latter instances, maternal blood was sampled at intervals after administration, but only a single fetal sample was obtained at delivery 5 and 6 hr, respectively, after administration.

Maternal-fetal aspartate transfer was studied in a total of nine animals utilizing a protocol similar to that for glutamate studies, except that the infusing solution contained aspartate with added radioactive aspartic acid. Five animals received 0.1 g/kg and two each 0.2 and 0.4 g/kg.

All blood samples were centrifuged immediately and the plasma deproteinized with sulfosalicylic acid. Simultaneous radioactivity and amino acid analysis were done by a method (10) permitting the determination of amino acid levels as well as metabolically derived compounds (both ninhydrin positive and negative). Technicon NC 1 amino acid analyzers were employed.

RESULTS AND COMMENT

Maternal-Fetal Glutamate Transfer

Figure 1 illustrates chemical levels of glutamate in the five animals given maternal infusions of 0.15 to 0.4 g/kg over 1 hr. Loads of 0.15 to 0.22 g/kg raised maternal levels 10- to 20-fold (i.e., from a base line of 5 μmoles/dl to peak values of 50 to 100 μmoles/dl). Under these conditions, fetal plasma glutamate concentration did not change, implying a lack of placental transfer. At the highest dose (0.4 g/kg), the maternal glutamate level reached 280 μmol/dl (70 times base line), apparently promoting some transfer to the fetus and raising the fetal level to 44 μmoles/dl (10 times base line).

Figure 2 presents the radioactivity profile in a representative experiment in which 0.22 g/kg was infused, providing insights into glutamate metabolism in pregnant primates. During the infusion, approximately 75% of the total radioactivity in the maternal plasma represented glutamate itself with lesser, but nonetheless appreciable, quantities present in the two ninhydrin-negative compounds glucose and lactate. Aspartate, glutamine, and alanine were represented by small amounts of radioactivity. With termination of the infusion, glutamate fell rapidly, leaving glucose and lactate as the major sources of radioactivity in maternal plasma. Quite a different radioactivity profile was present in fetal plasma, in which glucose and lactate represented over 80% of the total counts, whereas glutamate and aspartate contained essentially no radioactivity.

Figure 3 illustrates the radioactivity profile in the animal infused at 0.4 g/kg, in

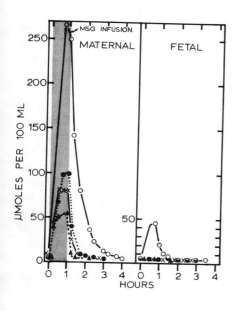

FIG. 1. Maternal and fetal plasma glutamate levels with maternal infusion of MSG at several dosage levels: ▲, 0.15; ×, 0.17 to 0.19 (mean of 2 animals); ●, 0.22; ○, 0.40 g/kg. (From Stegink et al., ref. 12, with permission.)

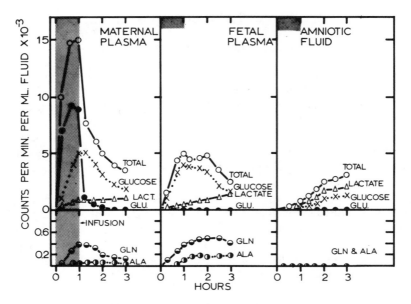

FIG. 2. Radioactivity profile of glutamate-derived metabolites in maternal and fetal plasma and amniotic fluid during and following maternal infusion of 0.22 g/kg MSG. (From Stegink et al., ref. 12, with permission.)

which it will be recalled that fetal plasma glutamate levels increased modestly in response to greatly elevated maternal levels. During the infusion, most of the maternal plasma radioactivity represented glutamate with smaller quantities present as glucose, lactate, aspartate, and glutamine. In the fetus, again almost all of the radioactivity was associated with glucose and lactate, although coincident with the maximal maternal level a small amount of radioactive glutamate was present in the fetal plasma.

From these results it is possible to construct a model describing glutamate metabolism in primate pregnancy. Most of the maternally administered glutamate circulates in the plasma as glutamate, raising maternal levels of this amino acid. However, despite plasma elevations up to 20-fold over base line, no transfer to the fetus occurs. With greater maternal loads, a limited amount is transferred to the fetus. From comparison of simultaneously determined maternal and fetal plasma levels, it appears that the "threshold" for transfer is a maternal concentration in the range of 250 to 300 μmoles/dl or 50 to 60 times the normal fasting value.

Approximately 20% of the infused glutamate is metabolized in the maternal compartment during the infusion. Glucose and lactate represent the principal metabolites, presumably resulting from the deamination of glutamate to α-ketoglutarate, which enters the tricarboxylic acid cycle and is then converted through pyruvate to glucose and lactate (11). Glucose and lactate cross the placenta readily, and their concentrations in the fetus generally mirror those in the mother.

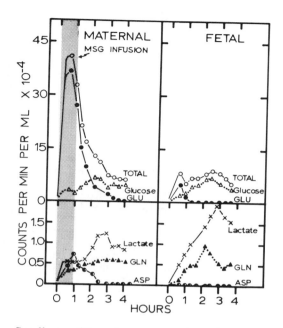

FIG. 3. Radioactivity profile of glutamate-derived metabolites in maternal and fetal plasma during and following maternal infusion of 0.4 g/kg MSG. (From Stegink et al., ref. 12, with permission.)

Small quantities of aspartate, glutamine, and alanine are also derived from glutamate. Aspartate appears to be handled in a manner similar to glutamate (i.e., it does not seem to cross the placenta), whereas glutamine and alanine reach the fetus in amounts appropriate to the normal transplacental gradient for these amino acids.

Fetal-Maternal Glutamate Transfer

In two experiments involving fetal glutamate administration by bolus injection of 1.5 and 2.4 g/kg (with added tracer radioactive glutamate), repeated analyses of maternal plasma failed to indicate any evidence of the transfer of any amino acid. However, increasing levels of radioactive glucose and lactate were found, reflecting fetal metabolism of glutamate to these two compounds, which then crossed to the maternal circulation.

Figure 4 illustrates the results of an experiment in which glutamate was infused into a fetus in an amount of 5 g/kg fetal weight (with added radioactive glutamate) over 1 hr. Fetal plasma glutamate levels rose to 2,000 μmoles/dl (400 times normal), and amniotic fluid values behaved similarly. Despite this extreme elevation, only a small amount of glutamate reached the maternal circulation. With cessation of the fetal infusion, fetal levels fell below 1,000 μmoles/dl, and there was a corresponding drop in the maternal level, implying that placental transfer had ceased.

These observations indicate that the placental "threshold" for glutamate is even higher with fetal-maternal than with maternal-fetal transfer. It probably lies above 1,000 μmoles/dl.

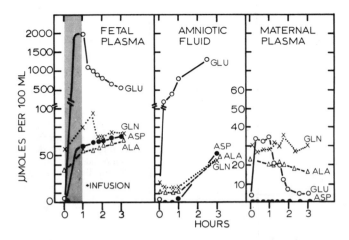

FIG. 4. Amino acid levels in fetal and maternal plasma and amniotic fluid during and following fetal infusion of MSG (5 g/kg fetal weight). (From Steglink et al., ref. 12, with permission.)

Maternal-Fetal Aspartate Transfer

Figure 5 summarizes the results of aspartate loading in pregnant monkeys. Infusion of 0.1 g/kg raised the maternal plasma levels approximately 150-fold to a mean value of 65 µmoles/dl without significantly altering fetal levels. Infusion of 0.4 g/kg produced maternal values more than 1,000 times base line, and under these

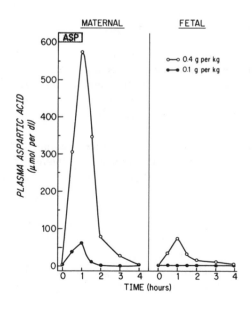

FIG. 5. Mean maternal and fetal plasma aspartate levels with maternal aspartate infusion at 0.1 g/kg (●) (5 experiments) and 0.4 g/kg (○) (2 experiments). (From Pitkin, ref. 8, with permission.)

conditions some aspartate was transferred to the fetus, but only in amounts sufficient to raise the fetal levels 200-fold.

Radioactivity studies confirmed the chemical measurements and indicated a metabolic pattern essentially identical to that of glutamate. During the infusion, most of the radioactivity in the maternal plasma involved aspartate itself with smaller quantities representing glucose and lactate, and only under conditions of very high maternal loading did labelled aspartate reach the fetal circulation.

Comparison of maternal and fetal blood levels indicates that the placental "threshold" for aspartate transfer occurs at approximately 100 μmoles/dl or 300 to 500 times base line. Below this level, no transfer to the fetus occurs, whereas above it relatively small quantities are transported.

SUMMARY

These studies have clarified the metabolism and distribution of the two dicarboxylic amino acids, glutamate and aspartate, in pregnant primates. The primary effect of maternal administration of either is to raise the circulating levels of the respective amino acid to an extent proportional to the dose. Approximately 20%, at least acutely, is metabolized, mainly to glucose and to a lesser extent to lactate. There is some interconversion between glutamate and aspartate, and small quantities of other amino acids are also formed.

The hemochorial placenta is virtually impermeable to glutamate and aspartate at maternal plasma levels less than 200 μmoles/dl (40 to 50 times fasting) and 100 μmoles/dl (300 to 500 times fasting), respectively. Above these threshold levels, some degree of transfer takes place. From a practical point of view, it should be emphasized that maternal concentrations of these magnitudes are never approached under physiologic conditions and can only be attained with intravenous infusion of large amounts.

REFERENCES

1. Dierks-Ventling, C., Cone, A. L., and Wapnir, R. A. (1971): Placental transfer of amino acids in the rat: I. L-Glutamic acid and L-glutamine. *Biol. Neonate*, 17:361–372.
2. Frankenberg, W. K., Dunlan, B. R., Coffelt, R. W., Koch, R., Coldwell, J. G., and Son C. D. (1968): Maternal phenylketonuria: Implications for growth and development. *J. Pediatr.*, 73:560–570.
3. Ghadimi, H., and Pecora, P. (1964): Free amino acids of cord plasma as compared with maternal plasma during pregnancy. *Pediatrics*, 30:500.
4. Lucas, D. R., and Newhouse, J. P. (1957): The toxic effect of sodium L-glutamate on the inner layers of the retina. *Arch. Ophthalmol.*, 58:193–201.
5. Morse, A. (1917): The amino acid nitrogen of the blood in cases of normal and complicated pregnancy and also in the newborn infant. *Bull. Johns Hopkins Hosp.*, 28:199.
6. Murakami, U., and Inouye, M. (1971): Brain lesions in the mouse fetus caused by maternal administration of monosodium glutamate. *Congenital Anomalies*, 11:171–177.
7. Newman, A. J., Heywood, R., Plamer, A. K., Barry, D. H., Edwards, F. P., and Worden, A. N. (1973): The administration of monosodium L-glutamate to neonatal and pregnant rhesus monkeys. *Toxicity*, 1:197–204.

8. Pitkin, R. M. (1977): Amino acids for the fetus. In: *Nutritional Impacts on Women*, edited by K. S. Moghissi and T. N. Evans, pp. 75–85. Harper & Row, Inc., Hagerstown, Md.
9. Schneider, H., and Dancis, J. (1974): Amino acid transport in human placental slices. *Am. J. Obstet. Gynecol.*, 120:1092–1098.
10. Stegink, L. D. (1971): Simultaneous measurement of radioactivity and amino acid composition of physiological fluids during amino acid toxicity studies. In: *Advances in Automated Analysis*, Vol. 1, edited by E. C. Barton, pp. 591–594. Thurman Associates, Miami, Fla.
11. Stegink, L. D., Brummel, M. C., Boaz, D. P., and Filer, L. J., Jr. (1973): Monosodium glutamate metabolism in the neonatal pig: Conversion of administered glutamate into other metabolites in vivo. *J. Nutr.*, 103:1146–1154.
12. Stegink, L. D., Pitkin, R. M., Reynolds, W. A., Filer, L. J., Jr., Boaz, D. P., and Brummel M. C. (1975): Placental transfer of glutamate and its metabolites in the primate. *Am. J. Obstet. Gynecol.*, 122:70–78.

Factors Influencing Dicarboxylic Amino Acid Content of Human Milk

G. L. Baker, L. J. Filer, Jr., and L. D. Stegink

Departments of Pediatrics and Biochemistry, The University of Iowa College of Medicine, Iowa City, Iowa 55242

Human milk is a major nutrient for infants throughout the world. In the United States, 50% of women breast feed their newborn infants at the time of hospital discharge. In developing countries, essentially all infants are breast fed. Glutamate is a major constituent of human milk, occurring in both the free and protein-bound forms. As such, it is a major source of nitrogen and energy for the growing infant. Free glutamate and aspartate are present in particularly large quantities in human milk.

Animal studies have raised questions about the safety of dicarboxylic amino acid ingestion by the lactating woman. Although human free plasma glutamate levels are low, milk and tissue levels are high (18). Thus, it has been questioned whether increased dietary ingestion of monosodium glutamate would elevate plasma glutamate levels, resulting in the movement of glutamate from plasma to human milk. To study this question, we gave lactating women loads of monosodium glutamate (MSG) or aspartate (as Aspartame—a dipeptide sweetener) and measured plasma and milk amino acid levels. In this chapter, we will review data previously reported on this subject (3,19) and present new information on the effects of dicarboxylic amino acid ingestion on plasma and human milk levels of glutamate and aspartate.

In 1972, we evaluated the effect of 6-g loads of MSG on plasma and human milk levels in lactating women (19). This load approximated a dose of 100 mg/kg body weight in these subjects. As shown in Table 1, the subjects studied had well-established lactation periods of 30- to 90-days duration. These subjects received the MSG load, contained in twelve 0.5-g capsules, at 0800 hr after an overnight fast. In four tests the MSG was given in conjunction with water; in nine tests the MSG was given with Slender (Carnation Products Co.), a liquid, ready-to-feed meal product; and in six tests a placebo (6 g lactose in the capsules) was given in water.

Milk samples were obtained at 0, 1, 2, 3, 4, 6, and 12 hr after administration of MSG or lactose. Blood samples were drawn at 0, 30, 60, 120, and 180 min after administration of MSG with water and at 0, 60, 90, 150, and 210 min after administration with Slender.

Mean plasma glutamate and aspartate levels in these subjects are shown in Fig. 1. Plasma glutamate and aspartate levels increased with time following administration

TABLE 1. Schedule of participation in loading tests

Subject	Days of lactation[a]		
	30	60	90
1	B	A	B
2	A	B	B
3	B	—	B
4	B	—	—
5	—	—	A
6	B	A	B
7	C	—	—
8	—	C	—
9	—	C	—
10	—	—	C

[a] A, MSG in water; B, MSG in Slender; C, lactose in water.

of MSG with both water and Slender. Peak plasma glutamate levels occurred earlier in subjects receiving MSG with water than in subjects receiving MSG with Slender. No change in plasma glutamate or aspartate levels was noted in subjects receiving the lactose placebo.

Figure 2 shows human milk glutamate and aspartate levels in these subjects. No significant differences between milk aspartate and glutamate levels were noted between subjects receiving MSG with water, MSG with Slender, or the lactose placebo. Concentrations of both amino acids in milk increased with time in all cases. It should be noted that the subjects were permitted to eat normally following the 4-hr milk sample.

Recently, we reviewed these older data in light of the more extensive data

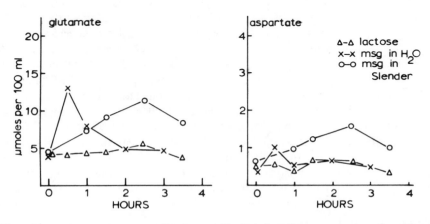

FIG. 1. Mean plasma glutamate and aspartate levels in lactating women given 6 g glutamate (MSG) in either water (×) or Slender (○), or given a lactose placebo (△). (From Stegink, ref. 18, with permission.)

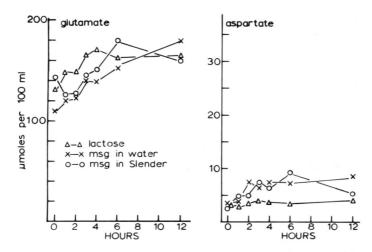

FIG. 2. Mean human milk free glutamate and aspartate levels in lactating women given 6 g glutamate (MSG) in either water (x) or Slender (o), or given a lactose placebo (△). (From Stegink, ref. 18, with permission.)

accumulated on the effects of glutamate loading on plasma amino acid levels (20). Two differences were apparent. First, plasma glutamate levels in subjects receiving MSG in capsules were lower than values obtained more recently (20), where an equivalent dose of MSG was dissolved in water. Second, in contrast to our earlier study (19), our recent data indicate that the addition of glutamate to a meal results in a much lower plasma glutamate level than that resulting from an equivalent dose of glutamate given in water (20).

Figure 3 shows the individual plasma glutamate levels in our original subjects receiving MSG in capsules with water. Considerable individual variation in plasma glutamate levels was noted. Plasma levels in some subjects hardly changed, whereas other subjects showed significant elevations. This variation raised a number of questions: (a) Do these differences arise because of individual variation in glutamate metabolism? (b) Do these differences reflect differing rates of capsule dissolution in the gut? (c) Were all subjects totally fasted? Some nursing notes from the metabolism unit evaluated retrospectively suggest that one or more subjects might have had toast and coffee before coming to the unit for study.

An important question to be resolved is whether our original conclusion that 6-g loads of MSG did not affect human milk glutamate levels was valid. Since two subjects given MSG in capsules with water did not show significantly increased plasma glutamate levels, we were concerned that these two subjects unduly influenced the statistical comparison of milk glutamate levels in subjects ingesting MSG with water with those receiving the lactose placebo. To evaluate this possibility, we studied two additional women with well-established lactation patterns. These two subjects received MSG dissolved in water (4.2 ml/kg body weight of a 2.4%

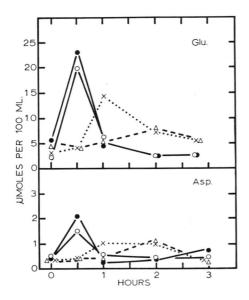

FIG. 3. Plasma glutamate (Glu) and aspartate (Asp) levels in individual subjects after administration of MSG with water. The symbols ○, ●, ×, and △ show values in individual subjects. (From Stegink et al., ref. 19, with permission.)

solution of MSG). This level provides a dose of 100 mg/kg body weight. Plasma and milk levels of free amino acids were determined as described in the original study (19).

The data in Table 2 compare plasma glutamate levels in these two subjects to values obtained in our previous study (19). Plasma glutamate values in these new subjects are significantly higher than values obtained previously (19), but are

TABLE 2. *Plasma glutamate levels in lactating women administered MSG in water at approximately 100 mg/kg body weight*

Time (min)	Current subjects[a]		Previous subjects[a,b] (mean ± SD)
	M.B.	M.M.	
0	6.5	3.2	3.9 ± 1.7
15	31.7	4.4	—
30	49.4	30.6	13.0 ± 10
45	28.6	33.1	—
60	22.6	19.7	7.5 ± 4.8
90	10.2	10.8	—
120	8.5	5.4	5.1 ± 2.7
150	6.2	3.9	—
180	7.7	5.0	4.8 ± 2.3
240	6.9	2.54	—

[a] In μmoles/dl.
[b] N = 4.
From Stegink et al., ref. 19.

TABLE 3. *Plasma glutamate levels in the four lactating women showing the highest plasma response to MSG ingestion in water at 100 mg/kg body weight*

Time (min)	Lactating subjects[a]				Mean ± SD for lactating subjects[a]	Mean ± SD for nonlactating subjects[a,b]
	M.S.	M.W.	M.B.	M.M.		
0	2.5	5.6	6.5	3.2	4.45 ± 1.90	2.70 ± 0.88
15	—	—	31.7	4.4	18.1 ± 19.3	14.7 ± 15.0
30	20.1	23.4	49.4	30.6	30.8 ± 13.1	47.1 ± 25.4
45	—	—	28.6	33.1	30.8 ± 3.11	50.1 ± 23.9
60	6.5	6.0	22.6	19.7	13.7 ± 8.68	24.9 ± 11.3
90	—	—	10.2	10.8	10.5 ± 0.42	8.19 ± 5.31
120	3.5	4.0	8.5	5.41	5.35 ± 2.24	4.28 ± 2.60
150	—	—	6.2	3.93	5.07 ± 1.61	3.72 ± 2.24
180	3.0	3.5	7.7	5.00	4.79 ± 2.10	4.19 ± 2.82
240	—	0	6.9	2.54	4.74 ± 3.11	3.05 ± 1.61

[a] In μmoles/dl.
[b] From Steginк et al., ref. 20.

comparable to those we have obtained in nonlactating subjects given an equivalent dose of MSG dissolved in water (20).

To determine the effect of elevated plasma levels on human milk levels, we evaluated the data obtained from the four subjects who had significant elevation of plasma glutamate levels and eliminated those two subjects showing little or no increase in plasma glutamate after loading. Table 3 lists the individual and mean plasma glutamate levels for these four subjects, along with mean values for nonlactating subjects administered an equivalent dose. Table 4 gives human milk glutamate levels in these four subjects compared with values in 10 subjects receiving the lactose placebo.

These data indicate that, although there was a significant increase in plasma glutamate levels, milk glutamate levels were not affected. The milk glutamate levels for these subjects are similar to those noted in subjects receiving the lactose placebo.

TABLE 4. *Milk glutamate levels in the four lactating subjects showing the highest plasma response to MSG ingestion at 100 mg/kg body weight*

Time after loading (hr)	Subjects[a]				Subjects given MSG[a] (mean ± SD)	Subjects given lactose (mean ± SD)[a,b]
	M.S.	M.W.	M.B.	M.M.		
0	69	158	65	117	102 ± 44	116 ± 29
1	82	145	62	108	99 ± 36	114 ± 39
2	140	150	82	78	113 ± 38	121 ± 27
3	117	155	112	129	128 ± 19	116 ± 27
4	126	165	103	112	126 ± 27	118 ± 30
12	121	166	128	82	122 ± 30	130 ± 37

[a] In μmoles/dl.
[b] From Baker et al., ref. 3; and Steginк et al., ref. 19.

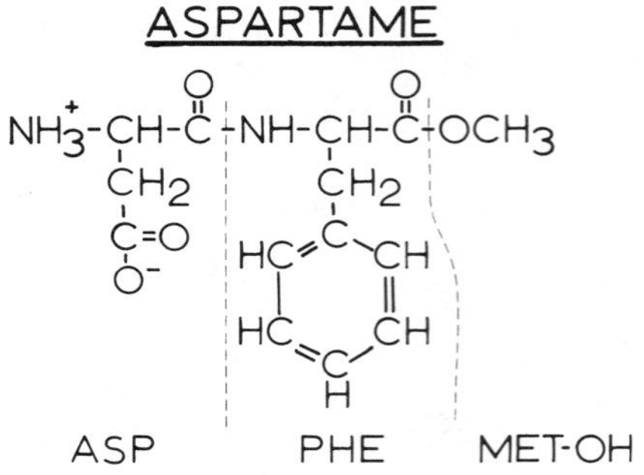

ASPARTAME

$$\overset{+}{NH_3}-CH-\overset{O}{\overset{\|}{C}}-NH-CH-\overset{O}{\overset{\|}{C}}-OCH_3$$

ASP　　　PHE　　　MET-OH

FIG. 4. Structure of Aspartame, L-aspartyl-L-phenylalanine methyl ester.

Thus, a load of MSG equivalent to two-thirds of the Acceptable Daily Intake (ADI) of glutamate in one dose produced no effect on milk glutamate levels.

These data suggest that a significant amount of glutamate in human milk is synthesized by the mammary gland. The data of Egan and Black (5) in the lactating cow indicate that most of administered ^{14}C-glutamate is converted to citrate by the mammary gland. Little lable was incorporated into milk free glutamate pools or milk protein.

Aspartate, like glutamate, is reported to exert neurotoxic effects in the neonatal rodent (13). However, in even the most sensitive animal species (the neonatal mouse), plasma aspartate and glutamate levels must be grossly elevated before neuronal necrosis is noted (23). To evaluate the possible effects of aspartate loading on human milk levels, we studied Aspartame (Searle Laboratories, Skokie, Ill.), a dipeptide containing aspartate. Aspartame is L-aspartyl-L-phenylalanyl-methyl ester (Fig. 4) and is 180 to 200 times sweeter than sucrose. It is hydrolyzed in the intestinal mucosa to its component amino acids and methanol. The latter are absorbed, metabolized, and cleared in a manner similar to that of aspartate, phenylalanine, and methanol arising from dietary protein and methylated polysaccarides.

Six healthy women with well-established lactation patterns were studied (3). The subjects were administered either Aspartame or a lactose load at a level of 50 mg/kg body weight. The order of administration was randomized in a crossover design. Aspartame or lactose was dissolved in 300 ml cold orange juice and administered to the subjects at 0800 hr after an overnight fast. Subjects were fasted for an additional 4 hr after administration of the test materials, but were allowed a normal diet after this time. Plasma and milk samples were collected and analyzed as described previously (19).

Table 5 shows plasma aspartate levels after Aspartame and lactose loading. No significant effect of either Aspartame or lactose loading was noted on plasma aspartate levels. Plasma phenylalanine levels increased significantly after Aspartame loading, reaching a mean peak value of 16.2 ± 4.9 μmoles/dl, but were not affected by lactose. The peak level of phenylalanine noted was only slightly higher than that observed postprandially in formula-fed infants (22).

TABLE 5. *Plasma aspartate and phenylalanine levels in lactating women given Aspartame or lactose at 50 mg/kg body weight*

Time (min)	Aspartate[a]		Phenylalanine[a]	
	Aspartame	Lactose	Aspartame	Lactose
0	0.42 ± 0.33	0.32 ± 0.20	4.61 ± 1.72	5.04 ± 1.13
15	0.42 ± 0.24	0.28 ± 0.07	8.34 ± 2.72	4.47 ± 0.81
30	0.54 ± 0.55	0.24 ± 0.09	14.5 ± 4.47	4.99 ± 0.98
45	0.31 ± 0.15	0.23 ± 0.05	16.2 ± 4.86	4.23 ± 0.95
60	0.34 ± 0.19	0.40 ± 0.40	14.2 ± 4.08	4.45 ± 0.55
90	0.30 ± 0.17	0.24 ± 0.14	15.7 ± 6.19	4.33 ± 0.83
120	0.30 ± 0.15	0.17 ± 0.05	12.8 ± 3.75	4.38 ± 0.93
180	0.17 ± 0.05	0.22 ± 0.03	8.07 ± 2.25	4.62 ± 0.86
240	0.23 ± 0.12	0.21 ± 0.03	6.42 ± 2.02	4.98 ± 1.18

[a] In μmoles/dl (mean ± SD).

Human milk levels in these subjects are shown in Table 6. Small, but statistically insignificant, increases in phenylalanine, aspartate, and glutamate levels were noted after Aspartame administration when compared to the same subjects after lactose loading. During the 4-hr fasting period after Aspartame loading, milk phenylalanine levels increased from 0.5 μmoles/dl to 2.3 μmoles/dl, whereas human milk aspartate levels increased from 2.3 μmoles/dl to about 4.8 μmoles/dl. The levels of these amino acids in milk samples collected, after the evening meal, 12 hr after the loading dose, were similar to those noted in milk samples collected 3 to 4 hr after loading.

EFFECT OF LENGTH OF LACTATION

Human milk glutamate levels vary considerably from individual to individual. We have evaluated the possibility that the duration of lactation might effect free amino acid levels. Free amino acid levels were measured in 225 milk samples obtained from 45 healthy women who had been lactating from 2 to 163 days. The effect of the duration of lactation on milk glutamate, aspartate, and taurine levels is shown in Table 7. Glutamate and aspartate levels were lower in colostrum (2 to 10 days lactation) than in mature milk. Glutamate levels were 47 μmoles/dl in colostrum and rose to a mean level of 128 μmoles/dl later in lactation. Similarly, aspartate levels increased from 2.7 to 5.2 μmoles/dl. In contrast, taurine levels decreased from 50 μmoles/dl in colostrum to 34 μmoles/dl in mature milk.

QUANTITY OF GLUTAMATE INGESTED BY THE TERM INFANT

These figures allow us to estimate the quantity of free and protein-bound glutamate and aspartate ingested by the normal 3.5-kg breast-fed term infant. As shown in Table 8, infants fed *ad libitum* with breast milk ingest a mean of 171 ml

TABLE 6. *Milk glutamate, aspartate, and phenylalanine levels in six lactating women given either Aspartame or lactose at 50 mg/kg body weight*

Time (hr)	Aspartate[a]		Glutamate[a]		Phenylalanine[a]	
	Aspartame	Lactose	Aspartame	Lactose	Aspartame	Lactose
0	2.25 ± 1.16	2.62 ± 0.82	109 ± 14	104 ± 24	0.48 ± 0.27	0.80 ± 0.35
1	2.82 ± 0.94	3.10 ± 0.54	106 ± 19	93 ± 31	2.06 ± 1.11	0.89 ± 0.39
2	4.67 ± 1.49	3.61 ± 0.85	122 ± 20	105 ± 22	2.29 ± 1.07	0.87 ± 0.32
3	4.53 ± 2.23	3.90 ± 1.46	128 ± 23	107 ± 32	2.08 ± 0.94	1.40 ± 1.85
4	4.82 ± 1.68	3.76 ± 1.84	120 ± 18	104 ± 29	1.99 ± 0.88	0.90 ± 0.53
12	5.59 ± 3.22	4.11 ± 2.50	155 ± 25	111 ± 23	1.19 ± 0.59	0.93 ± 0.59

[a] In µmoles/dl (mean ± SD).

TABLE 7. *Variation in human milk glutamate, aspartate, and taurine levels with duration of lactation*

Samples (N)	Days lactation Range	Days lactation Mean	Subjects (N)	Taurine[a]	Glutamate[a]	Aspartate[a]
16	2–10	6	13	50 ± 14	47 ± 34	2.7 ± 2
64	12–48	30	12	35 ± 11	138 ± 47	5.1 ± 3
36	49–69	56	5	35 ± 14	112 ± 36	4.4 ± 3
47	70–90	80	7	35 ± 16	144 ± 45	7.8 ± 13
62	> 90	110	8	29 ± 8	116 ± 28	3.7 ± 2

[a] In μmoles/dl (mean ± SD).

milk/kg/day (7). This level provides approximately 115 kcal/kg body weight. The free glutamate intake of this breast-fed infant is about 36 mg/kg/day, whereas the intake of protein-bound glutamate approximates 357 mg/kg/day. The breast-fed infant thus ingests a quantity of free glutamate equivalent to 46 mg/kg of MSG per day (36 mg × 1.28 to correct for the sodium ion and water of hydration in MSG).

The data in Table 9 show the expected daily per capita intake of MSG in the United States as estimated by the Committee on GRAS List Survey—Phase III (4). These data indicate that infants 12 to 23 months of age have the highest MSG intake of all groups, with a mean expected intake of 6.8 mg/kg/day. Since our calculations indicate that the breast-fed infant ingests 46 mg MSG/kg/day, such infants are at a level equivalent to the 99th percentile of expected daily intake.

TABLE 8. *Estimated glutamate and aspartate intake in the 3.5-kg breast-fed infant*

Variable	Mean		Range
Mean milk intake[a]	171	ml/kg/day	114–228
Free amino acid content[b]			
Glutamate	138	μmoles/dl (21 mg%)	
Aspartate	5	μmoles/dl (0.67 mg%)	
Total glutamate + aspartate[c]			
Glutamate (free and protein bound)	230	mg%	
Aspartate	116	mg%	
Total daily intake			
Free glutamate	36	mg/kg	24–48 mg/kg
Protein-bound glutamate	357		237–476
Total glutamate	393		262–524
Free aspartate	1.2	mg/kg	0.8–1.5 mg/kg
Protein-bound aspartate	197		131–262
Total aspartate	198		132–264

[a] From Fomon, ref. 7.
[b] See Table 7.
[c] Data from Macey et al., ref. 10 and Svanberg et al., ref. 24.

TABLE 9. Expected daily per capita intake of MSG[a]

		Total sample Intakes (mg/kg/day)		
		Percentile		
Age	Mean	90th	99th	99.9th
0–5 months	0.3	0	11	25
6–11 months	1.9	1.9	36	46
12–23 months	6.8	30	43	61
2–5 years	5.5	23	37	56
6–17 years	2.7	10	25	40
18+ years	1.5	7	12	19

[a] From Committee on GRAS List Survey—Phase III, ref. 4.

SPECIES VARIATION

It has been noted that the neonatal rodent is markedly sensitive to dicarboxylic amino acid-induced neuronal necrosis (9,12,18). The neonatal nonhuman primate does not appear to be sensitive to these amino acids, even at high ingestion levels (1,2,11,17,21,25), although this is a controversial finding (14,15). Table 10 compares milk glutamate and aspartate levels in a variety of animal species (8,16,19). These data indicate considerably higher glutamate levels in milk from human and nonhuman primates than in rodent milk. However, even the neonatal rodent ingests glutamate from its mother's milk. This glutamate must be rapidly metabolized, since it has no known effect on their growth and development. Similarly, data to be presented later in this symposium (6) indicate that the human infant rapidly metabolized the glutamate present in human milk.

SUMMARY

1. We have shown that the amino acid content of human milk varies with the duration of lactation. Levels of amino acids, such as glutamate and aspartate, increase with time, whereas others, such as taurine, decrease with time.

2. The breast-fed human in the U.S. ingests more glutamate on a kg per body weight basis, than at any other time during life.

3. Ingestion of MSG at 100 mg/kg increases the plasma levels of the lactating human but not milk levels.

4. Ingestion of MSG at 100 mg/kg body weight by lactating women has little or no effect on the glutamate intake of their infants.

ACKNOWLEDGMENTS

These studies have been supported in part by grants-in-aid from the Gerber Products Company, Searle Laboratories, and the International Glutamate Technical Committee.

TABLE 10. Milk glutamate and aspartate levels in various species[a]

Species	Rassin et al. (16)		Iowa		Ghadimi and Pecora (8)	
	Aspartate	Glutamate	Aspartate	Glutamate	Aspartate	Glutamate
Mouse	0.9					
Rat		5.7				
Colostrum	7.3 ± 1.2	15.1 ± 3.5	—	—	—	—
Mature	11.1 ± 1.2	11.1 ± 0.9	—	—	—	—
Guinea pig	1.6 ± 1.5	15.1 ± 4.1	—	—	—	—
Dog	0.2 ± 0.1	6.6 ± 0.7				
Cat	trace	17.7 ± 9.4	—	—	—	—
Cow						
Fresh	1.4 ± 0.2	6.8 ± 1.2	—	—	0.3	4.8
Middle	1.8 ± 0.1	12.8 ± 2.4			0.5	3.8
Rhesus monkey	5.8 ± 2.2	31.4 ± 7.0	—	—	—	—
Baboon	8.1 ± 1.3	43.9 ± 5.6				
Chimpanzee	18.5 ± 0.5	264 ± 7.5	—	—	—	—
Human						
Colostrum	5.1 ± 3.0	68.4 ± 12.9	2.7 ± 2.0	47 ± 34	22	76
Mature	4.2 ± 0.5	127 ± 8	5.2 ± 2.1	128 ± 20	7	42

[a] In μmoles/dl (mean ± SD), unless only one value available.

REFERENCES

1. Abraham, R. W., Dougherty, W., Golberg, L., and Coulston, F. (1971): The response of the hypothalamus to high doses of monosodium glutamate in mice and monkeys. Cytochemistry and ultrastructural study of lysosomal changes. *Exp. Mol. Pathol.*, 15:43–60.
2. Abraham, R., Swart, J., Golberg, L., and Coulston, F. (1975): Electron microscopic observations of hypothalami in neonatal rhesus monkeys (*Macaca mulatta*) after administration of monosodium L-glutamate. *Exp. Mol Pathol.*, 23:203–213.
3. Baker, G. L., Filer, Jr., L. J., and Stegink, L. D. (1976): Plasma, red cell and breast milk free amino acid levels in lactating women administered Aspartame or lactose at 50 mg per kg body weight. *J. Nutr.*, 106:xxxiii (July).
4. Committee on GRAS List Survey—Phase III (1976): Estimating distributions of daily intake of monosodium glutamate (MSG), Appendix E. In: *Estimating Distribution of Daily Intake of Certain GRAS Substances*. Food and Nutrition Board, Division of Biological Sciences, Assembly of Life Sciences, National Research Council, National Academy of Sciences, Washington, D.C.
5. Egan, A. R., and Black, A. L. (1968): Glutamic acid metabolism in the lactating dairy cow. *J. Nutr.*, 96:450–460.
6. Filer, L. J., Jr., Baker, G. L., and Stegink, L. D. (1979): Metabolism of free glutamate in clinical products fed infants (*this volume*).
7. Fomon, S. J. (1974): Voluntary food intake and its regulation. In: *Infant Nutrition*, 2nd ed., pp. 20–33. W. B. Saunders, Philadelphia.
8. Ghadimi, H., and Pecora, P. (1963): Free amino acids of different kinds of milk. *Am. J. Clin. Nutr.*, 13:75–81.
9. Lemkey-Johnston, N., and Reynolds, W. A. (1974): Nature and extent of brain lesions in mice related to ingestion of monosodium glutamate. *J. Neuropathol. Exp. Neurol.*, 33:74–97.
10. Macy, I. G., Kelly, H. J., and Sloan, R. E. (1953): *The Composition of Milks*, Publication 254. National Research Council, National Academy of Sciences, Washington, D.C.
11. Newman, A. J., Heywood, R., Plamer, A. K., Barry, D. H., Edwards, F. P., and Worden, A. N. (1973): The administration of monosodium L-glutamate to neonatal and pregnant rhesus monkeys. *Toxicology*, 1:197–204.
12. Olney, J. W. (1969): Brain lesions, obesity, and other disturbances in mice treated with monosodium glutamate. *Science*, 164:719–721.
13. Olney, J. W., and Ho, O-L. (1970): Brain damage in infant mice following oral intake of glutamate, aspartate or cysteine. *Nature*, 227:609–611.
14. Olney, J. W., and Sharpe, L. G. (1969): Brain lesions in an infant monkey treated with monosodium glutamate. *Science*, 166:386–388.
15. Olney, J. W., Sharpe, L. G., and Feigin, R. D. (1972): Glutamate-induced brain damage in infant primates. *J. Neuropathol. Expt. Neurol.*, 31:464–488.
16. Rassin, D. K., Sturman, J. A., and Gaull, G. E. (1978): Taurine and other free amino acids in milk of man and other mammals. *Early Human Dev.*, 2:1–13.
17. Reynolds, W. A., Lemkey-Johnston, N., Filer, L. J., Jr., and Pitkin, R. M. (1971): Monosodium glutamic: Absence of hypthalamic lesions after ingestion by newborn primates. *Science*, 172:1342–1344.
18. Stegink, L. D. (1976): Absorption, utilization, and safety of aspartic acid. *J. Toxicol. Environ. Health*, 2:215–242.
19. Stegink, L. D., Filer, L. J., Jr., and Baker, G. L. (1972): Monosodium glutamate: Effect on plasma and breast milk amino acid levels in lactating women. *Proc. Soc. Exp. Biol. Med.*, 140:836–841.
20. Stegink, L. D., Filer, L. J., Jr., Baker, G. L., Mueller, S. M., and Wu-Rideout, M. Y-C. (1978): Factors affecting plasma glutamate levels in normal adult subjects (*this volume*).
21. Stegink, L. D., Reynolds, W. A., Filer, L. J., Jr., Pitkin, R. M., Boaz, D. P., and Brummel, M. C. (1975): Monosodium glutamate metabolism in the neonatal monkey. *Am. J. Physiol.*, 229:246–250.
22. Stegink, L. D., Schmitt, J. L., Meyer, P. D., and Kain, P. (1971): Effects of DL-Methionine fortified diets on urinary and plasma methionine levels in young infants. *J. Pediatr.*, 79:648–659.
23. Stegink, L. D., Shepherd, J. A., Brummel, M. C., and Murray, L. M. (1974): Toxicity of protein hydrolysate solutions: Correlation of glutamate dose and neuronal necrosis to plasma amino acid levels in young mice. *Toxicology*, 2:285–299.
24. Svanberg, U., Gebre-Medhin, M., Ljungqvist, B., and Olsson, M. (1977): Breast milk composition

in Ethiopian and Swedish mothers. III. Amino acids and other nitrogenous substances. *Am. J. Clin. Nutr.*, 30:499–507.
25. Wen, C., Hayes, K. C., and Gershoff, S. M. (1973): Effects of dietary supplementation of monosodium glutamate on infant monkeys, weanling rats and suckling mice. *Am. J. Clin. Nutr.*, 26:803–813.

Glutamic Acid: Advances in Biochemistry and Physiology, edited by L. J. Filer, Jr., et al. Raven Press, New York © 1979.

Regulation of Amino Acid Availability to Brain: Selective Control Mechanisms for Glutamate

William M. Pardridge

Department of Medicine, Division of Endocrinology and Metabolism, UCLA School of Medicine, Los Angeles, California 90024

The transport of circulating amino acids into brain is of much importance to brain function since many pathways of cerebral amino acid metabolism are influenced by precursor availability (29). The rate of synthesis of several putative neurotransmitters (serotonin, catecholamines, histamine, or carnosine) is affected by the level in brain cells of precursor neutral amino acids (tryptophan, tyrosine, histidine) (6,9,43,47). In addition, when brain levels of essential neutral or basic amino acids fall to very low levels, e.g., due to a hyperaminoacidemia, cerebral protein synthesis may become substrate limited. [Protein synthesis in the CNS (35) proceeds at rates independent of precursor amino acid supply at normal brain levels of amino acids.] Unlike the essential neutral or basic amino acids, the acidic amino acids, glutamate and aspartate, can be synthesized in brain cells at rates commensurate with the metabolic demands for these compounds. Consequently, the rate of transport of the acidic amino acids from blood to brain is much lower than for the neutral or basic amino acids (22). Despite the relative independence of the brain on the circulating acidic amino acids, the mechanisms controlling the flux into the CNS of glutamate and aspartate are of much interest; these compounds are putative excitatory neurotransmitters (16) and are neurotoxic when plasma levels are elevated by the administration of large doses (25). This chapter will review the mechanisms regulating the transport of amino acids from the blood into the brain, and emphasis will be placed on the fundamental differences between the factors regulating the brain uptake of glutamate and aspartate versus the neutral and basic amino acids. The three factors (29) controlling the rate of amino acid transport from the blood into the brain are (a) cerebral blood flow, (b) plasma concentration, and (c) the blood-brain barrier (BBB) permeability.

CEREBRAL BLOOD FLOW

The rate of influx of a substrate into brain will be proportional to the rate of cerebral blood flow as long as the BBB permeability constant (PS in ml/min/g) is within an order of magnitude of the rate of flow (F in ml/min/g). The relationship (4) between PS and F is given by the fractional extraction (E) of the unidirectional influx of the substrate into brain (see Appendix). When E is greater than approximately 15% (4), the influx will vary with changes in flow. Although at normal plasma levels the E value for phenylalanine or leucine is approximately 15% (22), the E value for glutamate or aspartate is generally not higher than 2 or 3% (22); therefore, cerebral blood flow does not normally influence the brain uptake of glutamate or aspartate. However, there are several circumventricular organs (CVO) of the brain that lack a BBB (46). Conceivably, the regional E value for glutamate influx into the CVOs is greater than 15%. Therefore, conditions that increase F, e.g., seizures (which are induced by high levels of glutamate), may accelerate the rate of brain uptake of circulating glutamate within the CVOs on the basis of increased cerebral blood flow (39).

PLASMA CONCENTRATIONS

As long as the K_m (half-saturation constant) of BBB amino acid transport is greater than or equal to the plasma level, then the rate of brain uptake of amino acid will be proportional to the plasma concentration. Postprandial plasma amino acid levels are a function of (a) the dietary amino acid composition and (b) the transport of amino acids across the splanchnic (gut and liver) barriers. Unlike the large neutral or basic amino acids, physiologic doses of glutamate or aspartate are readily metabolized to alanine via transamination by gut epithelial cells (41). However, pharmacologic doses of glutamate or aspartate, or even doses found in a 50% casein diet (41), exceed the capacity of the gut transamination sites, and the acidic amino acids gain access to the portal circulation.

Amino acids in the portal plasma must clear the hepatocyte bed before entering the systemic circulation. Neutral amino acids are transported into liver cells by specific transport systems of very high capacity (high V_{max}) and low affinity (high K_m). Although more quantitative studies are needed, preliminary investigations indicate the K_m of neutral amino acid transport into liver cells *in vivo* is greater than 10 mM (31). Therefore, the neutral amino acid transport systems are probably never saturated during the absorption of even pharmacologic doses of amino acids. The very high capacity of the liver neutral amino acid transport system may explain why the oral administration of cysteine (a neutral amino acid that is oxidized by tissues to form a neurotoxic sulfonic amino acid, cysteic acid) is ineffective in producing brain lesions, whereas doses of this amino acid administered parenterally are effective (25).

Acidic amino acids are commonly believed to enter liver cells poorly. This misconception is due largely to a study of glutamate transport into liver slices (11); glutamate was shown to rapidly equilibrate with liver cells (half-time less than 1 min), but to reach a tissue/medium distribution ratio of only 0.5. Although the short half-time of equilibration suggests glutamate readily enters liver cells, the authors concluded acidic amino acids penetrate liver cells poorly based on the lack of complete equilibration with liver water (11). However, the latter function may be ATP-dependent and Krebs has since shown liver ATP is rapidly depleted, within minutes of preparing the liver slices (14). Data obtained *in vivo* indicate acidic amino acids readily penetrate the liver cell membrane via a class-specific transport system (31). The transport data for glutamate is shown in Fig. 1 and was obtained with a tissue sampling-single injection technique using a ^{3}H-water internal standard

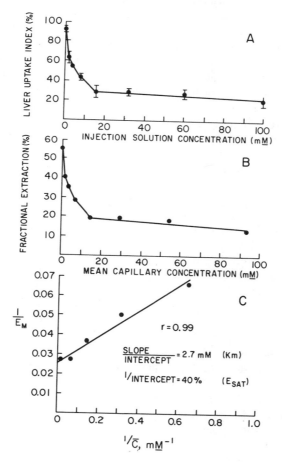

FIG. 1. The transport of ^{14}C-L-glutamate vs a ^{3}H-water reference into rat liver *in vivo* (31). **A:** The LUI for ^{14}C-L-glutamate is plotted against the portal injection solution concentration (C_P) of unlabeled glutamate. Data are means ± SEM (n = 4 to 6). **B:** The fractional extraction (F) of the unidirectional influx of glutamate into liver is plotted against the mean capillary concentration ($\overline{C}$) of glutamate; E = (LUI) (E_{HOH}), where E_{HOH} = 0.64 is the fractional extraction of influx into liver of the ^{3}H-water reference at 18 sec following portal injection (30); $\overline{C}$ is estimated from C_P (see Appendix). **C:** A double-reciprocal plot of $1/E_m$ vs $1/\overline{c}$, where $E_m = E_O - E_S$; E_O and E_S are the fractional extraction of glutamate influx at a tracer concentration and at an inhibiting concentration, respectively. The K_m of glutamate transport (2.7 mM) is equal to the slope/intercept ratio; the fractional extraction of saturable transport (E_{sat} = 40%) is equal to the reciprocal of the intercept. Since E = 55%, the fractional extraction of nonsaturable uptake is 15%. (See Appendix for details of the kinetic analysis.)

in anesthetized rats. A complete kinetic analysis of the original data (31) was not possible due to lack of information on the hepatic clearance of the ^{3}H-water reference, as well as portal blood flow. This information is now available (30) and permits computation of the K_m, V_{max}, and K_d (constant of nonsaturable transport) of glutamate and aspartate transport into liver cells. As shown in Table 1, the K_m of aspartate and glutamate transport is 1.9 mM and 2.7 mM, respectively. Since normal portal plasma levels of the acidic amino acids, up to 0.3 mM for glutamate (41), are less than 10% of the transport system K_m, the liver cell membrane is never saturated by physiologic doses of acidic amino acids. However, since peripheral glutamate levels reach 2.5 mM (18) at 30 min following an oral dose of 2 g/kg of glutamate, portal concentrations of glutamate under these conditions are probably above the K_m of the transport system (Table 1). Therefore, toxic doses of glutamate (1 to 4 g/kg p.o.) achieve portal amino acid levels that approach the maximal capacity of the liver to take up glutamate. Since the capacity of the splanchnic barriers to clear toxic doses of glutamate is somewhat limited, the major factor protecting brain cells from toxic levels of glutamate is the BBB.

BLOOD-BRAIN BARRIER

The BBB [and the blood-retina barrier (BRB)] are the products of a unique capillary structure. The brain capillary endothelial cells are fused together by tight junctions, which convert the brain capillary wall into an epithelial barrier (5,44). The BBB segregates the cerebral and systematic extracellular fluids and is effectively a plasma membrane (with regional specializations) for the entire brain. Due to the presence of the BBB, circulating compounds enter brain via either (a) lipid mediation (lipid-soluble compounds, e.g., drugs) or (b) carrier mediation (water-soluble compounds, e.g., metabolic substrates). The rate at which metabolic substrates penetrate the BBB is a function of the kinetic characteristics (K_m, V_{max}, K_d) of the specific carrier systems (Table 2) that transport the respective substrates (34). Although it has been known for some time that the neutral and basic amino acids readily penetrate the BBB via class-specific transport systems (22), Oldendorf has recently documented the presence of an acidic amino acid carrier for glutamate and aspartate (23). The BBB glutamate carrier is of particular interest in that the failure

TABLE 1. *Kinetics of glutamate and aspartate transport into liver cells* in vivo[a]

Amino acid	K_m (mM)	V_{max} (μmoles/min/g)	K_D (ml/min/g)
Glutamate	2.7	1.3	0.15
Aspartate	1.9	1.2	0.16

[a] Calculated (see Appendix) from previously reported data (30,31). The K_d represents nonsaturable uptake by both the intracellular and extracellular (sucrose) spaces of liver. Since the K_d for sucrose is 0.08 to 0.14 ml/min/g, more than 90% of the nonsaturable uptake of glutamate or aspartate is due to distribution in the extracellular space.

TABLE 2. The BBB transport systems[a]

Transport system	Representative substrate	K_m (mM)	V_{max} (nmoles/min/g)	K_d (ml/min/g)
Hexose	Glucose	9	1,600	0.023
Monocarboxylic acid	Lactate	1.9	120	0.028
Neutral amino acid	Phenylalanine	0.12	30	0.018
Basic amino acid	Lysine	0.10	6	0.007
Amine	Choline	0.22	6	0.003
Purine	Adenine	0.027	1	0.006
Nucleoside	Adenosine	0.018	0.7	0.001
Acidic amino acid	Glutamate	0.04	0.4	0.002

[a] From Pardridge and Oldendorf, ref. 34; except values for glutamate calculated from data of Oldendorf and Szabo, ref. 23. The K_d represents nonsaturable transport into both the intracellular and extracellular spaces of the brain. In the case of choline, adenine, adenosine, and glutamate, the component of K_d due to extracellular uptake has been subtracted (^{113m}indium-EDTA was used as an extracellular space reference), and the K_d reported here represents only nonsaturable transport into brain cells. The K_d for glucose, lactate, phenylalanine, and lysine includes the extracellular component and, therefore, overestimates the K_d of nonsaturable transport into brain cells by approximately 0.006 ml/min/g. The latter value is based on the fractional extraction, 0.01, of ^{113m}indium-EDTA uptake by brain.

to raise whole brain glutamate levels after systemic administration has been attributed to the impermeability of the BBB to glutamate (15).

The neutral and basic amino acids penetrate the BBB via their respective transport systems. The K_m and V_{max} for each amino acid are listed in Table 3, as well as their respective plasma concentrations for the rat. Two important points in regard to BBB amino acid transport are to be emphasized: (a) the close approximation of the transport K_m and plasma levels and (b) the bidirectional nature of amino acid transport across the BBB. As shown in Table 3, the K_m of BBB transport of the neutral and basic amino acids approximates the plasma level (29). Since the effect of competition on transport rates is directly related to the ratio of plasma level to K_m (28), it can be seen that when the transport K_m exceeds the plasma level by 10-fold or more, the effect of competition among similar amino acids will be negligible. Since the K_m of neutral amino acid transport in mammalian erythrocytes, renal tubule, liver cells, gut epithelia, and probably skeletal muscle is in the 1- to 10-mM range or greater (29), competition for transport under physiologic conditions does not occur in most tissues. However, amino acid competition does normally occur in brain under day-to-day conditions, e.g., the elevation of brain tryptophan after a carbohydrate meal is due to the insulin-mediated hypoaminoacidemia (9). The fundamental basis for the sensitivity of the CNS to amino acid competition (both physiologic and pathologic, e.g., the hyperaminoacidemias) is the close approximation of the BBB transport K_m and plasma amino acid levels (28,29).

The second important point about BBB transport of the neutral and basic amino

TABLE 3. Kinetics of the BBB amino acid transport systems

Amino acid	Plasma level[a] (mM)	K_m[b] (mM)	V_{max} (nmoles/min/g)
Neutral amino acids			
Phenylalanine	0.05	0.12	30
Leucine	0.10	0.15	33
Tyrosine	0.09	0.16	46
Tryptophan	0.10	0.19	33
Methionine	0.04	0.19	33
Histidine	0.05	0.28	38
Isoleucine	0.07	0.33	57
Valine	0.14	0.63	49
Threonine	0.19	0.73	37
Cycloleucine	—	0.75	55
Basic amino acids			
Arginine	0.10	0.09	9
Lysine	0.30	0.10	6
Ornithine	0.09	0.23	11

[a] From Pratt, ref. 40.
[b] From Pardridge and Oldendorf, ref. 33.

acids is that the rate of net uptake (i.e., influx-efflux, determined by multiplying the arteriovenous difference × cerebral blood flow) is much less than the rate of unidirectional influx (determined by either single-injection or constant-infusion isotope dilution techniques). As shown in Fig. 2, the rate of unidirectional influx of the large neutral or basic amino acids is on the order of 1 to 5 nmoles/min/g (28,40). This value approximates the rate at which essential amino acids are incorporated into proteins in the adult rat brain *in vivo* (17). However, the rate of *net* amino acid utilization by brain is small compared to rates of influx. [For example, aromatic amino acids are converted to neurotransmitters at rates (20) less than 50 pmoles/min/g, which are too low to be detected by arteriovenous differences.] In fact, the rate of net uptake of amino acids is not statistically different from zero for most neutral or basic amino acids (8). One exception to this rule is the branched-chain amino acids, leucine and isoleucine, for which positive net uptakes up to 40% of the rate of influx are observed (8). The branched-chain amino acids may be serving as precursors to sterol synthesis in the brain (45).

The relationships between (a) the K_m to plasma level and (b) the rate of influx to net uptake across the BBB for glutamate contrasts with the mechanisms mediating the transport of the basic or neutral amino acids. Although the K_m of glutamate transport across the BBB has not been reported, estimates can be made from data reported by Oldendorf and Szabo (23); linear transformation of their data according to previously reported methods (28) indicates the K_m and V_{max} values for glutamate transport are approximately 0.04 mM and 0.4 nmole/min/g, respectively (Table 2). Since the normal plasma glutamate level, 0.15 mM (40), is nearly fourfold the glutamate K_m, the glutamate carrier is virtually saturated by physiologic plasma

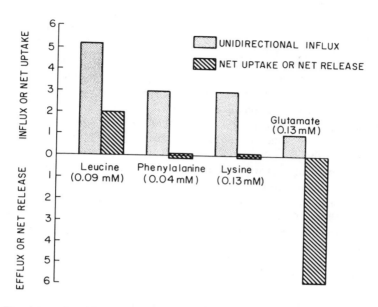

FIG. 2. The rates of unidirectional influx and net uptake for four amino acids—leucine (branched-chain neutral), phenylalanine (aromatic neutral), lysine (basic), and glutamate (acidic)—are shown for a given plasma concentration. Rates of leucine and phenylalanine influx (28) and lysine and glutamate influx (40) are shown for the barbiturate-anesthetized rat. Rates of net amino acid uptake (or output) are calculated from arteriovenous differences across the isolated perfused dog brain (8). The latter preparation provides the most accurate measure of arteriovenous differences across the brain because there is no extracerebral contamination of the venous drainage. Extracerebral contamination is particularly likely in studies with small animals, such as the rat (see ref. 12). The rate of net glutamate efflux from brain shown here is calculated from arteriovenous differences of plasma amino acid concentrations. If analyses of whole blood, i.e., plasma plus erythrocytes, are made, net glutamate release from brain is as high as 10 nmoles/min/g (8), indicating a considerable fraction of glutamate that is transported out of brain is rapidly taken up by circulating red cells.

levels of the amino acid. Therefore, brain glutamate does not rise or fall in parallel with changes in plasma levels, as is the case with the neutral or basic amino acids.

Although the rate of influx of the neutral or basic amino acids is much greater than the respective rates of net uptake, the reverse is true for glutamate. While the rate of glutamate influx into brain is 1.0 nmoles/min/g (40), the rate of *net* flux of glutamate across the BBB is 5.9 nmoles/min/g in favor of production (8). Therefore, glutamate *leaves* the brain and does so faster than any of the amino acids. Since glutamate influx into brain is 1 nmoles/min/g (40) and the net release of glutamate by brain is 6 nmoles/min/g (8), the rate of unidirectional efflux of glutamate, 7 nmoles/min/g, is sevenfold greater than the rate of influx. If the BBB glutamate carrier were symmetric, i.e., equal V_{max}/K_m ratio on both the blood and the brain sides of the BBB, as is the neutral amino acid carrier (33), then the glutamate concentration in the brain interstitial space would have to be over 1 mM to account for the sevenfold disparity between influx and efflux rates. It is unlikely that the brain interstitial space contains 1 mM glutamate, a putative excitatory transmitter (16), particularly since the CSF level of glutamate is less than 10 μM (36). Therefore, it seems probable that the sevenfold disparity between glutamate influx and efflux across the BBB is due to an asymmetry of the BBB glutamate carrier, i.e., the V_{max}/K_m ratio of the glutamate carrier on the *brain* side of the BBB is several-fold greater than the V_{max}/K_m ratio on the *blood* side of the BBB. Such a condition defines the glutamate carrier as an *active efflux* system, actively transporting glutamate from the brain interstitium to the blood against a concentration gradient.

In addition to the saturable route of glutamate influx into brain, there is also a nonsaturable route of transport, as is the case with the seven other classes of compounds that penetrate the BBB via carrier systems (Table 2). Although the nonsaturable mechanism is generally ascribed to free diffusion, it is unlikely that any polar metabolic substrate "freely" diffuses through the BBB. The nonsaturable systems are probably very low affinity ($K_m \geq$ 100 mM) carrier systems. Since the nonsaturable mechanism of glutamate transport across the BBB is characterized by a K_d of 0.002 ml/min/g (Table 2), the rate of glutamate influx via this mechanism is only 0.3 nmoles/min/g at a plasma level of 0.15 mM. However, when plasma glutamate is raised to 40 mM by the parenteral administration of 2 g/kg glutamate (37), the rate of glutamate influx into the brain might be as high as 80 nmoles/min/g (plasma level × K_d). The latter value for glutamate influx exceeds the V_{max} of the neutral amino acid carrier. The observation that brain glutamate is not increased with the administration of toxic doses of glutamate (37), despite the presence of the nonsaturable route of glutamate transport through the BBB, indicates the capacity of the BBB active efflux system for glutamate is relatively high.

Certain regions of brain, however, take up glutamate under conditions of very high plasma levels (37). These regions are the CVO, i.e., periventricular areas that lack a BBB (Table 4). The capillaries in these regions lack tight junctions and have large interendothelial pores and active pinocytosis (5). The neuronal necrosis that occurs when large doses of glutamate (0.5 to 4 g/kg) are administered is typically in areas contiguous with a CVO (26), e.g., the arcuate nucleus (near the median eminence) or the preoptic area [near the organum vasculosum of the lamina terminales (OVLT)]. The selective vulnerability of regions such as the arcuate nucleus to toxic doses of glutamate may be due to one or both of two possible mechanisms.

TABLE 4. *The CVOs of the brain*[a]

CVO	Contiguous brain region
Median eminence	Arcuate nucleus, hypothalamus
OVLT	Preoptic area, hypothalamus
Subfornical organ	Roof of third ventricle
Subcommissural organ	Dorsum of third ventricle
Area postrema	Base of fourth ventricle

[a] These periventricular areas of brain lack a BBB; other areas of brain that do not have a BBB are the choroid plexus, the pineal gland, and the neurohypophysis (see ref. 46).

First, these areas may be exposed to high levels of glutamate simply because of radial diffusion of glutamate from the extracellular space of the CVO to contiguous brain areas. A second possible mechanism is the retrograde axoplasmic flow of glutamate from the nerve ending in the CVO, e.g., the median eminence, to the body of neurosecretory cells within the BBB proper, e.g., the arcuate nucleus. The retrograde axoplasmic flow mechanism explains why certain neurosecretory regions of the hypothalamus, e.g., the supraoptic, paraventricular, and arcuate nuclei, selectively take up horseradish peroxidase (a protein marker that does not cross the BBB) at 8 to 12 hr after the intravenous injection of the protein (5).

When very high doses of glutamate are administered (2 to 4 g/kg), even non-CVO areas of brain may be affected (25). This signifies a breakdown of the BBB and may be due to the seizure activity induced by large doses of glutamate (21). Convulsions cause a breakdown of the BBB (38) due to the acute hypertensive response (e.g., chordotomy prevents the hypertension of seizures and also the breakdown of the BBB).

One non-CVO area of brain that appears to be affected by toxic doses of glutamate, via a mechanism that is apparently not related to convlusive activity, is the retina (25). The retina is protected from the systemic circulation by the BRB, which is formed by tight junctions (44) between capillary endothelial cells analogous to the brain capillary. Retinal capillaries mediate the active efflux of many amino acids from the vitreous humor to the retinal capillary lumen (3). In this way, a concentration gradient of amino acids is maintained from the posterior chamber, where amino acids are actively taken up across the ciliary body, through the vitreous to the retina, where amino acids are actively removed across retinal capillaries (3). Conceivably, the selective vulnerability of the retina to high plasma levels of glutamate is due to a relative decrease in the activity of the BRB active efflux system for acidic amino acids or to a relatively high activity of the active influx system for glutamate transport across the ciliary body.

In addition to the selective vulnerability of the retina and certain regions of brain (CVOs) to toxic doses of glutamate, there are also developmental and species selectivities. Younger animals (less than 10 days old) are more susceptible to the amino acid as compared to older animals (25). This increased sensitivity is often attributed to an "immature or leaky BBB" in the young animal. The historical background leading to the present-day misconception of a leaky BBB in the young has recently been reviewed (42). Actually, the brain endothelial tight junctions are formed in the first trimester of human fetal life (19); the BBB to proteins is complete by the 15th day of fetal life in the rat (24). An anatomically intact BBB in the newborn, however, does not rule out developmental modulations in specific carrier systems (34). For example, the active efflux system for organic acids (e.g., 5-HIAA) does not develop in the rat until the period between 5 to 30 days (2). Conceivably, a low activity of the BBB active efflux system for glutamate in the young animals may explain their greater sensitivity to toxic doses of glutamate. In addition to developmental differences, the active efflux system for organic acids demonstrates a marked species specificity, e.g., this system is probenecid sensitive (presumably due to a lower transport K_m) in rats and mice, but is probenecid insensitive in rabbits and rhesus monkeys (13). The greater sensitivity of rodents to toxic doses of glutamate, as compared to the rhesus monkey (10), may be due to a lower capacity of the rodent BBB active efflux system for glutamate.

In conclusion, comparison of rates of glutamate influx and net release in the brain suggests an active efflux system for this amino acid exists within the BBB. However, the evidence for regional, developmental, and species selectivity of this transport system is at present only indirect, and future quantitative studies will be needed to adequately characterize the BBB transport of the acidic amino acids.

APPENDIX

The liver uptake index (LUI) is a ratio of the fractional extraction (E) of the unidirectional influx of a ^{14}C-labeled compound (at tracer concentrations) relative to the fractional extraction of a ^{3}H-water reference (E_{HOH}) at 18 sec following a single portal vein injection (31), i.e.,

$$\text{LUI} = \frac{E}{E_{HOH}} \quad (1)$$

Given E_{HOH} (0.64 at 18 sec), the LUI may be converted to E (30). E represents the sum of both saturable (E_{sat}) and nonsaturable (E_{ns}) routes of transport; E_{sat} may be computed from a double-reciprocal plot (Fig. 1).

Extraction values may be converted to permeability-surface constants (PS) with the use of Crone's equation (7),

$$PS = (F) \ln \frac{1}{1-E} \quad (2)$$

which has been shown to be applicable to liver transport studies (1); F is the rate of portal blood flow (0.93 ml/min/g) in the barbiturate-anesthetized, hepatic artery-ligated rat (30). The PS value is related to Michaelis-Menten kinetics (32) according to

$$PS = \frac{(V_{max}}{K_m)} + K_d \quad (3)$$

where $PS_{sat} = V_{max}/K_m$ and $PS_{ns} = K_d$. Since PS_{sat} and K_m may be calculated from the data in Fig. 1, V_{max} may be estimated. It should be emphasized that an accurate estimate of K_d is essential in accurately calculating the transport K_m and V_{max}, particularly when K_d is large, as in liver transport studies. The method described here calculates K_d by extrapolation from data over the entire curve, which provides the best estimate of K_d. The analysis used here for liver transport studies is analogous to that used for BBB transport (28).

One modification of the above analysis that must be made for liver, but is not necessary for most studies in brain, is the estimation of the mean capillary concentration ($\overline{C}$) from the arterial or portal plasma concentration (C_P). Assuming the capillary concentration falls logarithmically as the bolus traverses the capillary bed (27), then

$$\overline{C} = \frac{(E)}{-\ln(1-E)} (C_P) \qquad (4)$$

When E is small, $\overline{C}$ approximates C_P; however, when E is large, C_P overestimates $\overline{C}$; e.g., if $E = 25\%$, $\overline{C}$ is 87% of C_P; but when $E = 0.75$, $\overline{C}$ is only 54% of C_P (33).

ACKNOWLEDGMENTS

Studies in the author's laboratory are supported by a grant from the National Science Foundation and a Clinical Investigator Award from the National Institutes of Health.

REFERENCES

1. Bass, L., Keiding, S., Winkler, K., and Tygstrup, N. (1976): Enzymatic elimination of substrates flowing through the intact liver. *J. Theor. Biol.*, 61:393–409.
2. Bass, N. H., and Lundborg, P. (1973): Postnatal development of mechanisms for the elimination of organic acids from the brain and cerebrospinal fluid system of the rat: rapid efflux of ^{3}H paraaminohippuric acid following intrathecal infusion. *Brain Res.*, 56:285–298.
3. Bito, L. Z. (1977): The physiology and pathophysiology of intraocular fluids. *Exp. Eye Res.* (Suppl.), 25:273–289.
4. Bradbury, M. W. B., Patlak, C. S., and Oldendorf, W. H. (1975): Analysis of brain uptake and loss of radiotracers after intracarotid injection. *Am. J. Physiol.*, 229:1110–1115.
5. Brightman, M. W. (1977): Morphology of blood-brain interfaces. *Exp. Eye Res.* (Suppl.), 25:1–25.
6. Chung-Hwang, E., Khurana, H., and Fisher, H. (1976): The effect of dietary histidine level on the carnosine concentration of rat olfactory bulbs. *J. Neurochem.*, 26:1087–1091.
7. Crone, C. (1963): The permeability of capillaries in various organs as determined by use of the "indicator diffusion" method. *Acta Physiol. Scand.*, 58:292–305.
8. Drewes, L. R., Conway, W. P., and Gilboe, D. D. (1977): Net amino acid transport between plasma and erythrocytes and perfused dog brain. *Am. J. Physiol.*, 233:E320–E325.
9. Fernstrom, J. D., and Wurtman, R. J. (1971): Brain serotonin content: increase following injection of carbohydrate diet. *Science*, 174:1023–1025.
10. Goldberg, L., Abraham, R., and Coulston, F. (1974): When is glutamate neurotoxic? *N. Engl. J. Med.*, 290:1326–1327.
11. Hems, R., Stubbs, M., and Krebs, H. A. (1968): Restricted permeability of rat liver for glutamate and succinate. *Biochem. J.*, 107:807–815.
12. Hertz, M. M., and Bolwig, T. G. (1976): Blood-brain barrier studies in the rat: An indicator dilution technique with tracer sodium as an internal standard for estimation of extracerebral contamination. *Brain Res.*, 107:333–343.
13. Kessler, J. A., Fernstermacher, J. D., and Patlak, C. S. (1976): Homovanillic acid transport by the spinal cord. *Neurology*, 26:434–440.
14. Krebs, H. A., Cornell, N. W., Lund, P., and Hems, R. (1974): Isolated liver cells as experimental material. In: *Regulation of Hepatic Metabolism*, edited by F. Lundquist and N. Tygstrup, pp. 726–750. Munksgaard, Copenhagen.
15. Himwich, W. A., Petersen, J. C., and Allen, M. L. (1957): Hematoencephalic exchange as a function of age. *Neurology*, 7:705–710.

16. Krnjevic, K. (1970): Glutamate and γ-aminobutyric acid in brain. *Nature*, 228:119–124.
17. Lajtha, A. (1974): Amino acid transport in the brain in vivo and in vitro. In: *Aromatic Amino Acids in the Brain* edited by G. E. W. Wolstenholme and D. W. Fitzsimons, pp. 25–41. Ciba Foundation Symposium 22, Elsevier, London.
18. McLaughlan, J. M., Noel, F. J., Botting, H. G., and Knipfel, J. E. (1970): Blood and brain levels of glutamic acid in young rats given monosodium glutamate. *Nutr. Rep. Int.*, 1:131–138.
19. Møllgard, K., and Saunders, N. R. (1975): Complex tight junctions of epithelial and of endothelial cells in early foetal brain. *J. Neurocytol.*, 4:453–468.
20. Neff, N. H., Spano, P. F., Groppetti, A., Wang, C. T., and Costa, E. (1971): A simple procedure for calculating the synthesis rate of norepinephrine, dopamine, and serotonin in rat brain. *J. Pharmacol. Exp. Ther.*, 176:701–710.
21. Nemeroff, C. B., and Crisley, F. D. (1975): Monosodium L-glutamate-induced convulsions: Temporary alteration in blood-brain barrier permeability to plasma proteins. *Environ. Physiol. Biochem.*, 5:389–395.
22. Oldendorf, W. M. (1971): Brain uptake of radiolabeled amino acids, amines, and hexoses after arterial injection. *Am. J. Physiol.*, 221:1629–1639.
23. Oldendorf, W. H., and Szabo, J. (1976): Amino acid assignment to one of three blood-brain barrier amino acid carriers. *Am. J. Physiol.*, 230:94–98.
24. Olsson, Y., Klatzo, I., Sourander, P., and Steinwall, O. (1968): Blood-brain barrier to albumin in embryonic new born and adult rats. *Acta Neuropathol.*, 10:117–122.
25. Olney, J. W. (1976): Brain damage and oral intake of certain amino acids. *Adv. Exp. Biol. Med.*, 69:497–506.
26. Olney, J. W., Rhee, V., and de Gubareff, T. (1977): Neurotoxic effects of glutamate on mouse area postrema. *Brain Res.*, 120:151–157.
27. Pappenheimer, J. R., and Setchell, B. P. (1973): Cerebral glucose transport and oxygen consumption in sheep and rabbits. *J. Physiol.*, 233:529–551.
28. Pardridge, W. M. (1977): Kinetics of competitive inhibition of neutral amino acid transport across the blood-brain barrier. *J. Neurochem.*, 28:103–108.
29. Pardridge, W. M. (1977): Regulation of amino acid availability to the brain. In: *Nutrition and the Brain*, Vol. 1, edited by R. J. Wurtman and J. J. Wurtman, pp. 141–204. Raven Press, New York.
30. Pardridge, W. M. (1977): Unidirectional influx of glutamine and other neutral amino acids into liver of fed and fasted rat in vivo. *Am. J. Physiol.*, 232:E492–E496.
31. Pardridge, W. M., and Jefferson, L. S. (1975): Liver uptake of amino acids and carbohydrates during a single circulatory passage. *Am. J. Physiol.*, 228:1155–1161.
32. Pardridge, W. M., and Oldendorf, W. H. (1975): Kinetics of blood-brain barrier transport of hexoses. *Biochim. Biophys. Acta*, 382:377–392.
33. Pardridge, W. M., and Oldendorf, W. H. (1975): Kinetic analysis of blood-brain barrier transport of amino acids. *Biochim. Biophys. Acta*, 401:128–136.
34. Pardridge, W. M., and Oldendorf, W. H. (1977): Transport of metabolic substrates through the blood-brain barrier. *J. Neurochem.*, 28:5–12.
35. Parks, J. M., Ames, III, A., and Nesbett, F. B. (1976): Protein synthesis in central nervous tissue: Studies on retina in vitro. *J. Neurochem.*, 27:987–997.
36. Perry, T. L., Hansen, S., and Kennedy, J. (1975): CSF amino acids and plasma-CSF amino acid ratios in adults. *J. Neurochem.*, 24:587–589.
37. Perez, V. J., and Olney, J. W. (1972): Accumulation of glutamic acid in the arcuate nucleus of the hypothalamus of the infant mouse following subcutaneous administration of monosodium glutamate. *J. Neurochem.*, 19:1772–1782.
38. Petito, C. K., Schaefer, J. A., and Plum, F. (1977): Ultrastructural characteristics of the brain and blood-brain barrier in experimental seizures. *Brain Res.*, 127:251–267.
39. Plum, F., and Duffy, T. E. (1975): The couple between cerebral metabolism and blood flow during seizures. In: *Brain Work: The Coupling of Function, Metabolism and Blood Flow in the Brain*, edited by D. H. Ingvar and N. A. Lassen, pp. 197–214. Munksgaard, Copenhagen.
40. Pratt, O. E. (1976): The transport of metabolizable substances into the living brain. *Adv. Exp. Biol. Med.*, 69:55–75.
41. Remesy, C., Demigne, C., and Aufrere, J. (1978): Interorgan relationships between glucose, lactate and amino acids in rats fed on high-carbohydrate or high-protein diets. *Biochem. J.*, 170:321–329.
42. Saunders, N. R. (1977): Ontogeny of the blood-brain barrier. *Exp. Eye Res. (Suppl.)*, 25:523–550.
43. Schwartz, J. C., Lampart, C., and Rose, C. (1972): Histamine formation in rat brain in vivo: Effects of histidine loads. *J. Neurochem.*, 19:801–810.

44. Shiose, Y. (1970): Electron microscopic studies on blood-retinal and blood-aqueous barriers. *Jpn. J. Ophthalmol.*, 14:73–87.
45. Smith, M. E. (1974): Labelling of lipids by radioactive amino acids in the central nervous system. *J. Neurochem.*, 23:435–438.
46. Weindl, A. (1973): Neuroendocrine aspects of circumventricular organs. In: *Frontiers in Neuroendocrinology*, edited by W. F. Ganong and L. Martin, pp. 3–32. Oxford University Press, New York.
47. Wurtman, R. J., Larin, F., Mostafapour, S., and Fernstrom, J. D. (1974): Brain catechol synthesis: Control by brain tyrosine concentration. *Science*, 185:183–184.

Glutamic Acid: Advances in Biochemistry and Physiology, edited by L. J. Filer, Jr., et al. Raven Press, New York © 1979.

Biochemical Aspects of the Neurotransmitter Function of Glutamate

R. P. Shank* and M. H. Aprison**

*Department of Physiology, Temple University School of Medicine, Philadelphia, Pennsylvania 19140; and **Institute of Psychiatric Research and Departments of Psychiatry and Biochemistry, Indiana University Medical Center, Indianapolis, Indiana 46101*

The metabolism of glutamate in CNS tissues has been studied extensively over the past 40 years. Prior to the late 1960s, most studies focused on the central role of glutamate in nitrogen metabolism (35,37), the involvement of this amino acid in energy metabolism (6), its metabolic relationship to GABA (4), and its metabolic compartmentation (6). Since the appearance of neurophysiological (10,22) and neurochemical (13,20,21) evidence that glutamate may function as an excitatory neurotransmitter, interest has developed in the metabolism of glutamate as it relates specifically to this putative synaptic function (1,5,7,12,29,31,32,36). Unfortunately, investigations into the metabolic processes that underlie the neurotransmitter function of glutamate are hindered because of the complexity of glutamate metabolism and the complex morphology of CNS tissues.

At the present time our knowledge is not sufficient to draw many definite conclusions regarding the biochemical aspects of the neurotransmitter function of glutamate. Despite this limited amount of definitive information, a model has emerged that outlines a series of the biochemical events that appear to be associated with the transmitter function (Fig. 1). This model was developed from experimental observations reported by a number of investigators. Notable contributors include Berl and his colleagues (6), Van den Berg and Garfinkel and their colleagues (36), Balázs et al. (1), Quastel and Benjamin (5,29), Bradford and his colleagues (7), Shank and Baxter (32), Shank and Aprison (31), and, more recently, Hamberger et al. (15).

BIOCHEMICAL MODEL OF THE NEUROTRANSMITTER ROLE OF GLUTAMATE

According to this model, the neurotransmitter pool of glutamate resides within the synaptic vesicles of the nerve terminals of those neurons utilizing glutamate as a neurotransmitter (hereafter such neurons will be referred to as glutamate neurons). The glutamate molecules within these vesicles are presumed to be derived from

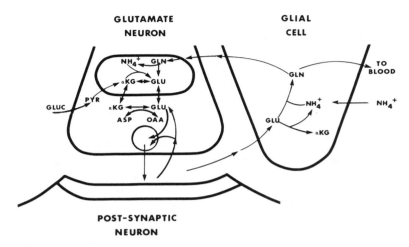

FIG. 1. A model illustrating some of the presumed biochemical events that occur at a glutamate synapse.

cytoplasmic and mitochondrial pools. Since glutamate in these latter pools serves several other cellular functions, it is unlikely that the transmitter pool exists as a distinct metabolic entity, but instead is part of a common metabolic pool.

The transmitter pool of glutamate is thought to have its metabolic origin in both glucose and glutamine. Each molecule of glutamine may serve as a source of amino nitrogen for two molecules of glutamate. This model implies that the carbon portion of glutamate is supplied by glutamine and α-ketoglutarate in equal amounts. However, the metabolism of glutamate is probably far too complex for such a simple stoichiometric relationship to exist.

The mechanism by which glutamate is released from nerve terminals is not yet established, although the favored view is that excitation-secretion coupling in glutamate neurons is mediated by a Ca^{2+}-dependent exocytotic process. Present evidence indicates that the release of glutamate is at least partially dependent on extracellular Ca^{2+} (27). However, glutamate can also be released to some extent in the absence of extracellular Ca^{2+}, indicating either that Ca^{2+} released from intracellular storage sites can mediate exocytosis or that glutamate can be released by a carrier mechanism.

Subsequent to being released from the presynaptic neuron, the molecules of glutamate must be rapidly cleared out of the synaptic cleft. Present evidence leaves little doubt that this inactivation process occurs by a combination of two mechanisms: (a) transport back into the nerve terminal and (b) diffusion out of the cleft followed by uptake into glial cells. Much of the glutamate taken up by glial cells is apparently converted to glutamine, which is subsequently released into the extracellular fluid. This glutamine is then available to be taken up into the nerve terminal in order to replenish the neurotransmitter pool. This model therefore suggests that a metabolic cycle exists between glutamate neurons and glial cells in which there is a *net* flow of glutamate from neurons to glia that is compensated by a *net* flow of glutamine in the reverse direction. The model does not necessarily imply that there is a stoichiometric relationship between the fluxes of these compounds.

In this review we will focus on three key elements of this model: (a) glucose and glutamine as metabolic precursors of the neurotransmitter pool of glutamate, (b) uptake as the mechanism of transmitter inactivation, and (c) the role that glial cells play in supporting the neurotransmitter function.

GLUCOSE: PRECURSOR OF THE NEUROTRANSMITTER POOL OF GLUTAMATE

It is experimentally well established that glucose is the principal substrate of energy metabolism in CNS tissues and that much of the glucose carbon passes through glutamate before being oxidized to CO_2. Since energy metabolism is especially vigorous in the CNS, the synthesis of glutamate from precursors derived from glucose is quite rapid. There are at least two metabolic pathways that can account for this role of glutamate as an intermediate in energy metabolism. Both of these pathways can be thought of as attachments to the normal citric acid cycle (Fig. 2). One involves the synthesis of GABA and is frequently referred to as the "GABA

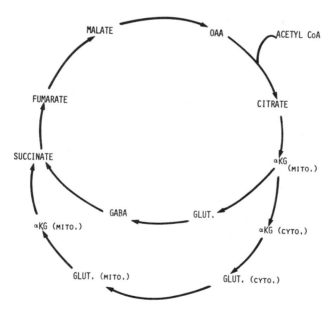

FIG. 2. Depiction of the role of glutamate as an intermediate in energy metabolism. In CNS tissues much of the glucose carbon is incorporated into glutamate prior to being oxidized to CO_2. Therefore, glutamate may be regarded as an intermediate in an expanded citric acid cycle. There are at least two functions served by this metabolic role of glutamate. One is the synthesis of GABA (*inner loop*) in those neurons utilizing GABA as a neurotransmitter. The second function is to aid the transfer of reducing equivalents from the cytoplasm into mitochondria by serving as a component of the malate-aspartate shuttle (see ref. 8). The involvement of glutamate in the malate-aspartate shuttle is depicted by the outer loop.

shunt." This pathway is probably restricted to neurons utilizing GABA as a neurotransmitter. Present evidence indicates that about 10% of the carbon oxidized in the CNS passes through the GABA shunt (1). The second pathway reflects the involvement of glutamate in the malate-aspartate shuttle, which serves as a mechanism by which reducing equivalents are transferred from the cytosol into mitochondria (8).

Although these pathways result in a rapid turnover of glutamate, they are not necessarily useful for restoring the transmitter pool. For every molecule of glutamate formed via these pathways, another one must be converted to the next intermediate (GABA or α-ketoglutarate). Otherwise, the content of each of the intermediates in the pathways would eventually be depleted unless a mechanism were available to replenish the carbon drained off through the loss of glutamate. One possible mechanism by which the content of the intermediates in these pathways can be replenished is the carboxylation of pyruvate to form oxaloacetate.

Because glucose carbon is incorporated into glutamate so extensively, it is likely that a very high percentage of the glutamate molecules within the transmitter pool have glucose as their metabolic origin. However, the actual value that glutamate derived from the citric acid cycle has in restoring the transmitter pool is more likely to be reflected in the rate at which pyruvate is carboxylated to form oxaloacetate. Although it is well established that CNS tissues have the capacity to convert pyruvate to oxaloacetate (28), the extent to which this occurs in the terminals of glutamate neurons is yet to be determined.

GLUTAMINE: PRECURSOR OF THE NEUROTRANSMITTER POOL OF GLUTAMATE

In absolute terms, the rate at which glutamate is formed from glutamine is probably much less than the rate at which it is formed from glucose. However, in terms of the *net* synthesis of glutamate needed to replace the molecules lost through transmitter release, the conversion of glutamine to glutamate can conceivably equal or exceed that from glucose.

Studies designed to provide evidence regarding the possible role of glutamine as a precursor for the neurotransmitter pool of glutamate have been reported recently by several investigators (7,15,29,31). In our own experiments we have examined the uptake and metabolism of ^{14}C-glutamine by the isolated toad-brain preparation and by a synaptosomal preparation obtained from the rat brain. When the isolated toad brain was incubated in a medium containing 0.2 mM glutamine, there was no net uptake of this amino acid; indeed, there was a marked net efflux of glutamine from the brain (Fig. 3). Such an efflux is to be expected since glutamine is probably the major end product of nitrogen metabolism in the CNS (26,36). Despite the net efflux of glutamine there was an accumulation of ^{14}C-glutamine in the tissue that could reflect a net uptake of glutamine into some cellular compartment, such as the terminals of glutamate neurons. Much of the ^{14}C-glutamine taken up was metabolized, as evidenced by an accumulation of label in glutamate (Fig. 3),

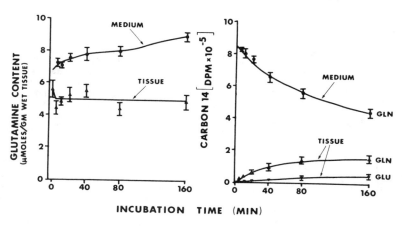

FIG. 3. The time course of change in glutamine content (**left**) and radioactivity (**right**) in isolated toad-brain tissue and the medium when the tissue was incubated in a medium initially containing 0.2 mM L-glutamine and 0.4 μCi (U-^{14}C)L-glutamine. Hemisected toad brains (average mass of approximately 50 mg) were incubated in 2 ml oxygenated bicarbonate-buffered medium at 23° C. The values for the content of glutamine in the medium were obtained by dividing the amount (μmoles) of glutamine in the medium by the tissue mass. Each data point is the mean ± SEM of 6 brain hemisections. In addition to the accumulation of ^{14}C in glutamine and glutamate in the tissue, a considerable amount also accumulated in GABA and aspartate, and presumably much of the ^{14}C was eventually expelled as CO_2. (From Shank and Aprison, ref. 31.)

GABA, and aspartate (31). When the specific radioactivity of glutamate was plotted relative to that of glutamine in the tissue, an unusual metabolic relationship was revealed. The value of this ratio (S.A.[1] glutamate/S.A. glutamine) rose very quickly for a brief period (5 min or less), but increased slowly thereafter (Fig. 4). When the toad-brain tissue was incubated in a standard bicarbonate buffered medium, the average specific activity of glutamate was 7% of that for glutamine after a 5-min incubation period, but after a 160-min incubation period, the specific activity of glutamate had risen to only 15% of that for glutamine. In one set of experiments, the toad-brain tissue was incubated in a medium in which sucrose was substituted for NaCl, and K^+ was made 29 mM. This incubation condition should promote the release of glutamate from nerve terminals and greatly reduce the reuptake of glutamate into either neurons or glial cells. This condition markedly increased the amount of glutamine rapidly converted to glutamate, but did not markedly effect the slope during which the relative specific activity of glutamate rose slowly (Fig. 4).

Since the metabolism of glutamine and glutamate in CNS tissues is quite complex, it is likely that any single explanation for the biphasic rise in the relative specific radioactivity of glutamate shown in Fig. 4 will be an oversimplification. One reasonable explanation for the rapid initial rise would be that a small portion (~10%) of the glutamine taken up by the tissue is converted to glutamate before

---

[1] S.A., specific radioactivity.

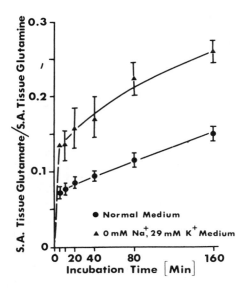

FIG. 4. The change in the specific radioactivity of glutamate in the toad-brain tissue is shown relative to the specific radioactivity of glutamine in the tissue. The data for the normal medium relate to the experiments described in Fig. 3. The data for the 0 mM Na^+, 29 mM K^+ medium were obtained when the toad-brain tissue was incubated in a medium in which sucrose was substituted for NaCl and $KHCO_3$ was substituted for $NaHCO_3$. This incubation medium was used in order to enhance the release of glutamate and block the uptake. (From Shank and Aprison, ref. 31.)

mixing with the bulk of the endogenous glutamine. This could happen in either of two ways. Some of the glutamine could be converted to glutamate during transit across the cell membrane. In this situation the glutamine rapidly metabolized to glutamate would mix with very little of the endogenous pool of glutamine before being converted to glutamate. A biochemical system that could both transport glutamine across the cell membrane and deamidate it to glutamate is the γ-glutamyl cycle described by Meister and colleagues (26). A second way in which glutamine could be metabolized to glutamate without mixing with the bulk of the endogenous glutamine is that the exogenous glutamine could be transported into a compartment that contains little glutamine but is rich in glutaminase activity.

Because the specific radioactivity of glutamate remained much lower than that of glutamine, it is likely that most (~90%) of the exogenous glutamine was transported into a tissue compartment where it was fully or partially mixed with a large pool of endogenous glutamine. Furthermore, the glutamine in this pool must be metabolized to glutamate quite slowly.

The marked increase in the relative amount of glutamine rapidly converted to glutamate when the toad-brain tissue was incubated in the low Na^+-high K^+ medium suggests, but certainly does not establish, that the terminals of glutamate neurons constitute one site where glutamine is rapidly converted to glutamate. Isolated synaptosomes are known to be rich in glutaminase activity (7), and have the capacity to metabolize considerable amounts of glutamine to glutamate (Table 1) (3,7). In addition, the content of glutamine in synaptosomes is relatively low (Table 1) (7). These observations further suggest, but unfortunately do not establish, that the terminals of glutamate neurons represent one site where glutamine is rapidly converted to glutamate.

TABLE 1. Metabolism of (U-^{14}C) glutamine by synaptosomes from rat brain[a]

Amino acid	Content (nmoles/mg protein)	Total radioactivity (dpm/mg protein)	R.S.A.[b]
Glutamine	3.55	14,600	1.00
Glutamate	14.55	29,100	0.49
Aspartate	7.78	8,010	0.31
GABA	4.44	7,360	0.20

[a] Synaptosomes were obtained from the whole brain of rats by a procedure similar to that of Gray and Whittaker (14). The synaptosomes were incubated for 15 min at 37° C in an oxygenated bicarbonate-buffered medium. After incubation, a tissue pellet was obtained by centrifugation. The tissue was homogenized in 80% ethanol, and the content and radioactivity in each amino acid was determined by the method of Shank and Aprison (30).
[b] R.S.A., specific radioactivity relative to glutamine which is arbitrarily assigned a value of 1.0. From Shank and Aprison, *unpublished*.

NEURONAL AND GLIAL UPTAKE OF GLUTAMATE

Until the early 1970s, studies pertaining to the uptake of glutamate were primarily concerned with the transport of glutamate across the blood-brain barrier (BBB) and the role of transport mechanisms in regulating the intracellular concentration of glutamate. These studies demonstrated that the flux of glutamate between the blood and CNS parenchyma is severely restricted by the BBB and that CNS tissues can accumulate large amounts of glutamate even when there is a large intra- to extracellular concentration ratio (23).

In recent years, most uptake studies have been concerned with the role that transport mechanisms serve in supporting the neurotransmitter function of glutamate. These studies leave little doubt that transport systems present in both neuronal and glial cell membranes are responsible for inactivating glutamate subsequent to release from nerve terminals. These transport systems presumably also serve to ensure that the extracellular steady-state concentration of glutamate is maintained below levels that can cause neuronal excitation. A matter that has yet to be resolved is the relative amount of glutamate taken back into the nerve terminal and returned to the transmitter pool.

Initial uptake studies first demonstrated that CNS slices (2,16), synaptosomes (6), and glial preparations (17) take up glutamate by two transport systems with widely different affinity constants. One of the transport systems exhibits a relatively high affinity for glutamate ($K_m \sim 20$ μM) and was Na$^+$ dependent. The other system has a low affinity for glutamate, but the V_{max} is greater. More recently, a third transport system has been identified. This system has a K_m value of ~ 2 μM and has been observed in granule cells and glial cells obtained from the cerebellum of mice (9), and in CNS tissue slices (11). Data obtained with cerebellar granule cells and glial-enriched preparations suggest that although all three systems may be instrumental in inactivating glutamate (Table 2), the highest affinity system is particularly

TABLE 2. *Relative uptake capacity of glutamate transport systems during synaptic resting and active states*[a]

Transport system	K_m (μM)	Relative V_{max}	Relative uptake activity	
			Resting state[b]	Synaptic activation[c]
Very high affinity	2	1	100	200
High affinity	20	2	36	390
Low affinity	500	5	4	670

[a] Data based on values obtained for granule cell and glial-enriched preparations obtained from the cerebellum of 10-day-old mice.
[b] Assumed concentration of glutamate in synaptic cleft is 0.002 mM.
[c] Assumed concentration of glutamate in synaptic cleft is 1.0 mM.

instrumental in keeping the extracellular concentration at very low levels during synaptic resting periods.

That CNS tissues do indeed have the ability to rapidly remove glutamate from extracellular fluids has been shown by net uptake studies using the isolated toad brain. In these studies, whole or hemisected toad brains were incubated in a bicarbonate-buffered physiological solution containing 0.04 or 0.1 mM L-glutamate (33). During incubation there was a rapid net uptake of glutamate, and eventually a steady-state extracellular concentration of less than 0.002 mM was attained (Fig. 5). When no glutamate was initially present in the medium, there was a net efflux into

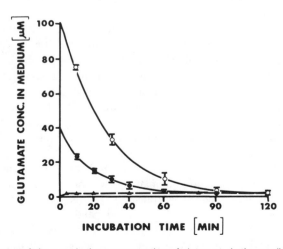

FIG. 5. Time course of changes in the concentration of glutamate in the medium in which isolated toad brains were incubated. Each toad brain (~60 mg wet tissue) was incubated in 2 ml oxygenated bicarbonate-buffered medium initially containing glutamate concentrations of 100 (○), 40 (●), or 0 μmoles/liter (△). Each point is the mean of 2 to 7 experiments. The vertical lines represent the range (where $N = 2$) or SEM of the data. In some instances, the SEM values were smaller than the size of the symbol. (From Shank et al., ref. 33; and Shank and Aprison, ref. 31.)

the medium until a similar steady-state level was achieved. Therefore, on the basis of these results, it would appear that the CNS regulates extracellular glutamate at a level between 0.001 and 0.002 mM. Physiological studies indicated that 0.002 mM is the concentration at which glutamate just begins to cause membrane depolarization (19). This extracellular concentration (0.002 mM) is about 0.05% of the intracellular concentration of glutamate; thus, CNS tissues have a remarkable capacity to remove this amino acid from the extracellular fluid.

METABOLISM OF GLUTAMATE TAKEN UP BY CNS TISSUES

Exogenous glutamate taken up by the isolated toad brain does not equilibrate with the endogenous glutamate before being metabolized. Most of the exogenous glutamate is rapidly metabolized to glutamine, and much of the glutamine so formed is subsequently released without being further metabolized (Fig. 6). Mammalian brain preparations have also been shown to selectively metabolize exogenous glutamate to glutamine (6). Glutamine synthetase is now known to be located predominantly, if not exclusively, within neuroglia (25,34); consequently, it is reasonable to conclude that most of the exogenous glutamate was taken into glial cells. This does not necessarily mean that glutamate released from nerve terminals is selectively taken

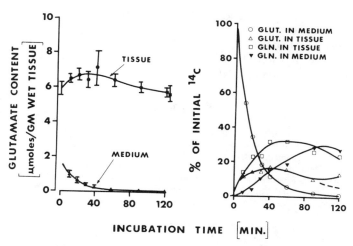

FIG. 6. Uptake and metabolism of glutamate by the isolated toad brain. **Left:** The change in the content of glutamate in the tissue and medium. **Right:** The change in the distribution of radioactivity in glutamate and glutamine in the tissue and medium are plotted. Each toad brain (~60 mg wet mass) was incubated in a bicarbonate-buffered medium (2 ml) initially containing 0.04 mM L-glutamate and 0.02 μCi of (U-^{14}C)L-glutamate. The content of glutamate (μmoles/g wet tissue) in the medium was obtained by multiplying the actual concentration of glutamate in the medium by the ratio of the medium volume (in ml) to the tissue mass (in g). The data are the mean ± SEM of 3 or 4 experiments. The last data point for the content of glutamate in the tissue represents the mean content of glutamate when the tissue was incubated 120 min without glutamate initially present in the medium. (From Shank and Baxter, ref. 32, and *unpublished*; and Shank et al., ref. 33.)

up by glia. The exogenous glutamate is likely to have a greater probability of being exposed to glial uptake sites than neuronal uptake sites, whereas the reverse is the case for endogenous glutamate released from nerve terminals.

CONCLUSION

At the present time, our model of the biochemical processes associated with the neurotransmitter function of glutamate serves more as a working hypothesis than an established and well-understood series of biochemical events. It does appear reasonably certain that uptake into the presynaptic terminal and surrounding glial cells functions as the mechanism by which glutamate is inactivated after being released into the synaptic cleft. However, we still have much to learn about the transport systems that mediate this uptake. It is likely that glutamine and α-ketoglutarate serve as the major precursors of the transmitter pool of glutamate; however, virtually nothing is known about the mechanisms that regulate the synthesis of glutamate from these precursors. With regard to the role that glial cells serve, we as yet have no appreciation of the quantitative significance of their involvement in the inactivation of glutamate or in supplying glutamine for the restoration of the transmitter pool.

ACKNOWLEDGMENTS

The authors' research was supported in part by Research Grant MH 03225-16,17 from NIMH, and by Biomedical Research Support Grant No. RR05417 from the Division of Research Sources, NIH.

REFERENCES

1. Balázs, R., Patel, A. J., and Richter, D. (1973): Metabolic compartments in the brain: Their properties and relation to morphologic structures. In: *Metabolic Compartmentation in the Brain*, edited by R. Balázs and J. E. Cremer, pp. 167–184. Macmillan, London.
2. Balcar, V. J., and Johnston, G. A. R. (1972): The structural specificity of the high affinity uptake of L-glutamate and L-aspartate by rat brain slices. *J. Neurochem.*, 19:2657–2666.
3. Baldessarini, R. J., and Yorke, C. (1974): Uptake and release of possible false transmitter amino acids by rat brain tissue. *J. Neurochem.*, 23:839–848.
4. Baxter, C. F. (1970): The nature of GABA. In: *Handbook of Neurochemistry*, Vol. 3, edited by A. Lajtha, pp. 289–353. Plenum Press, New York.
5. Benjamin, A. M., and Quastel, H. H. (1974): Fate of glutamate in the brain. *J. Neurochem.* 23:457–464.
6. Berl, S., and Clarke, D. D. (1969): Metabolic compartmentation of glutamate in the CNS. In: *Handbook of Neurochemistry*, Vol. 2, edited by A. Lajtha, pp. 447–472. Plenum Press, New York.
7. Bradford, H. F., and Ward, H. K. (1976): On glutaminase activity in mammalian synaptosomes. *Brain Res.*, 110:115–125.
8. Brand, M. D., and Chappell, J. B. (1974): Glutamate and aspartate transport in rat brain mitochondria. *Biochem. J.*, 140:205–210.
9. Campbell, G. LeM., and Shank, R. P. (1978): Glutamate and GABA uptake by cerebellar granule and glial cell enriched populations. *Brain Res.*, 153:618–622.
10. Curtis, D. R., Phillis, J. W., and Watkins, J. C. (1960): The chemical excitation of spinal neurones by certain acidic amino acids. *J. Physiol. (Lond.)*, 150:656–682.

11. Fagg, G. E., Jones, I. M., and Jordan, C. C. (1978): Multi-component accumulation of L-^{14}C glutamate by rat spinal cord slices. *Neurosci. Lett. (in press).*
12. Graham, L. T. Jr., and Aprison, M. H. (1969): Distribution of some enzymes associated with the metabolism of glutamate, aspartate, γ-aminobutyrate and glutamine in cat spinal cord. *J. Neurochem.*, 16:559–566.
13. Graham, L. T. Jr., Shank, R. P., Werman, R., and Aprison, M. H. (1967): Distribution of some synaptic transmitter suspects in cat spinal cord: Glutamic acid, aspartic acid, γ-aminobutyric acid, glycine and gluamine. *J. Neurochem.*, 14:465–472.
14. Gray, E. G., and Whittaker, V. P. (1962): The isolation of nerve endings from brain: An electron-microscopic study of cell fragments derived by homogenization and centrifugation. *J. Anat. (Lond.),* 96:79–87.
15. Hamberger, A. C., Chiang, G. H., Nylen, E. S., Scheef, S. W., and Cotman, C. W. (1978): Glutamate as a CNS transmitter I: Evaluation of glucose and glutamine as precursors for the synthesis of preferentially released glutamate. *J. Neurochem. (in press).*
16. Hammerschlag, R., and Weinreich, D. (1972): Glutamic acid and primary afferent transmission. In: *Advances in Biochemical Psychopharmacology, Vol. 6: Studies of Neurotransmitters at the Synaptic Level,* edited by E. Costa, L. L. Iverson, and R. Paoletti, pp. 165–180. Raven Press, New York.
17. Henn, F. A., Goldstein, M. N., and Hamberger, A. (1974): Uptake of the neurotransmitter candidate glutamate by glia. *Nature.*, 249:663–664.
18. Hills, A. G., Reid, E. L., and Kerr, W. D. (1967): Circulatory transport of L-glutamine in fasted mammals: Cellular sources of urine ammonia. *Am. J. Physiol.*, 223:1470–1476.
19. Hösli, L., Andres, P. F., and Hösli, E. (1976): Ionic mechanisms associated with the depolarization by glutamate and aspartate on human and rat spinal neurons in tissue culture. *Pfluegers Arch.*, 363:43–48.
20. Johnson, J., and Aprison, M. H. (1970): The distribution of glutamic acid, a transmitter candidate, and other amino acids in the dorsal sensory neuron of the cat. *Brain Res.*, 24:285–292.
21. Johnson, J., and Aprison, M. H. (1971): The distribution of glutamate and total free amino acids in thirteen specific regions of the cat central nervous system. *Brain Res.*, 26:141–148.
22. Krnjevic, K., and Phillis, J. W. (1963): Iontophoretic studies of neurones in the mammalian cerebral cortex. *J. Physiol. (Lond.),* 165:274–304.
23. Lajtha, A. (1968): Transport as a control mechanism of cerebral metabolite levels. In: *Progress in Brain Research, Vol. 29: Brain Barrier Systems,* edited by A. Lajtha and D. Ford, pp. 201–218. Elsevier, Amsterdam.
24. Logan, W. J., and Snyder, S. H. (1972): High affinity uptake systems for glycine, glutamic and aspartic acids in synaptosomes of rat central nervous system. *Brain Res.*, 42:413–431.
25. Martinez-Hernandez, A., Bell, K. P., and Norenberg, M. D. (1977): Glutamine synthetase: Glial localization in brain. *Science,* 195:1356–1358.
26. Meister, A. (1973): On the enzymology of amino acid transport. *Science,* 180:30–39.
27. Nadler, J. V., White, W. F., Vaca, K. W., Redburn, D. A., and Cotman, C. W. (1977): Characterization of putative amino acid transmitter release from slices of rat dentate gyrus. *J. Neurochem.*, 29:279–290.
28. Patel, M. S. (1974): The relative significance of CO_2-fixing enzymes in the metabolism of rat brain. *J. Neurochem.*, 22:717–724.
29. Quastel, J. H. (1974): Amino acids and the brain. *Biochem. Soc. Trans.*, 2:725–744.
30. Shank, R. P., and Aprison, M. H. (1970): Method of multiple analyses of concentration and specific radioactivity of amino acids in nervous tissue extracts. *Anal. Biochem.*, 35:136–145.
31. Shank, R. P., and Aprison, M. H. (1977): Glutamine uptake and metabolism by the isolated toad brain: Evidence pertaining to its proposed role as a transmitter precursor. *J. Neurochem.*, 28:1189–1196.
32. Shank, R. P., and Baxter, C. F. (1975): Uptake and metabolism of glutamate by isolated toad brains containing different levels of endogenous amino acids. *J. Neurochem.*, 24:641–646.
33. Shank, R. P., Whiten, J. T., and Baxter, C. F. (1973): Glutamate uptake by the isolated toad brain. *Science,* 181:860–862.
34. Utley, J. D. (1964): Glutamine synthetase, glutamotransferase, and glutaminase in neurons and non-neural tissue in the medical geniculate body of the cat. *Biochem. Pharmacol.*, 13:1383–1392.
35. Van den Berg, C. J. (1970): Glutamate and glutamine. In: *Handbook of Neurochemistry,* Vol. 3, edited by A. Lajtha, pp. 355–379. Plenum Press, New York.
36. Van den Berg, C. J., Reignierse, G. L. A., Blochuis, G. G. C., Kron, M. C., Rhonda, G., Clarke,

D. D., and Garfinkel, D. (1976): A model of glutamate metabolism in brain: A biochemical analysis of a heterogeneous structure. In: *Metabolic Compartmentation and Neurotransmission-Relation to Brain Structure and Function,* edited by S. Berl, D. D. Clarke, and S. Schneider, pp. 515–544. Plenum Press, New York.
37. Weil-Malherbe, H. (1950): Significance of glutamic acid for the metabolism of nervous tissue. *Physiol. Rev.,* 30:549–568.

Glutamic Acid: Advances in Biochemistry and Physiology, edited by L. J. Filer, Jr., et al. Raven Press, New York © 1979.

Glutamic Acid as a Transmitter Precursor and as a Transmitter

E. Costa, A. Guidotti, F. Moroni, and E. Peralta

Laboratory of Preclinical Pharmacology, National Institute of Mental Health, Saint Elizabeths Hospital, Washington, D.C. 20032

Glutamate is one of the most active neuroexcitatory substances present in the CNS of vertebrates, where it may function as an important synaptic transmitter (12). In brain, glutamate also functions as the precursor for gamma-aminobutyric acid (GABA), a very important inhibitory transmitter. The quantity of glutamic acid decarboxylase (GAD), the enzyme that converts glutamate into GABA, does not appear to be rate limiting although it is mainly located in presynaptic GABAergic terminals. If the enzyme were ubiquitous, local applications of the excitatory transmitter glutamate would trigger an increased formation of GABA, and if the postsynaptic cells possessed specific GABAergic receptors to GABA, a biphasic response may ensue. This consideration brings up the question of whether GAD is a regulatory step for GABA and/or glutamate steady state. Recent reports have indicated (17) that GAD is only partially saturated by its cofactor, pyridoxal-5-phosphate (pyridoxal-P) in the intact brain and that there is a rapid postmortem increase in the degree of saturation of the enzyme by pyridoxal-P. It appears that in the absence of glutamate, pyridoxal-P is tightly bound to GAD, but glutamate, the substrate of GAD, promotes dissociation of the cofactor from the enzyme (16). This GAD inactivation promoted by glutamate appears to be very slow at low concentrations of glutamate and increases rapidly as glutamate reaches saturating concentrations. Pyridoxal-P is tightly bound to GAD in the absence of glutamate but this cofactor is loosely bound when this substrate is present. Thus, since GAD utilizes glutamate to form GABA, the pyridoxal-P will tend to stay on the enzyme; therefore, a continuous function of the enzyme in these conditions requires a constant supply of pyridoxal-P. The question then is: What are the regulatory biochemical events that maintain at steady state the glutamate pool that serves as a substrate for GAD? Is GAD regulation by the supply of pyridoxal-P the mechanism that nerve impulses utilize to increase the rate of GABA formation in the presence of an increased rate of GABAergic neuronal activity?

In cortex, hippocampus, cerebellum, and other brain nuclei, both glutamatergic (12) and GABAergic (7) neurons have been described. Lesion studies have shown that in striatum a degeneration of specific glutamatergic or GABAergic tracts causes little or no change in the tissue content of glutamate or GABA, respectively (5,13).

Thus, in certain brain nuclei, the amount of glutamate and GABA involved in neuronal transmission is not a preponderent part of the total brain content. Moreover, this finding suggests that a change in GABA and glutamate content cannot be a ubiquitous biochemical index of the changes in activity of glutamatergic and GABAergic neurons; probably, the turnover rate of GABA and glutamate are better indices. Biochemical measurements directed to ascertain the dynamic state of the glutamate functioning as a transmitter must differentiate this pool from the other neuronal pool where glutamate functions as a GABA precursor. Both pools must in turn be differentiated from the neuronal and glial cell pool where glutamate is a product of intermediary metabolism. If the precursors of glutamate were different in these various pools, one would have gained an important advantage in using glutamate turnover rate to estimate participation of glutamatergic neurons in brain function. Based on metabolic considerations, Quastel (24) suggested that glutamine might function in replenishing the transmitter pools of glutamate and GABA. Undoubtedly, blood glutamine is an excellent source of C and N for GABA or glutamate because intra- and extracellular levels of glutamine are high (32) and it can readily pass from blood to brain (20,22). In brain, glutamine could originate from glial cells (29) where CNS glutamine synthetase appears to be located (31). The glutamine taken up by axons could be metabolized by glutaminase (4) and by glutamate decarboxylase to generate GABA (26). Having none of the excitatory or inhibitory properties of glutamate and GABA, glutamine would be ideally suited for the function of precursor of glutamate and GABA. Thus, with glutamine present in the cerebral spinal fluid at concentrations around 500 μM, a considerable concentration gradient would drive glutamine into nerve terminals where it would be converted into glutamate and thereby either contribute to the energy metabolism or in particular neurons contribute to the glutamate pool functioning as a transmitter (4). The evidence that at least certain types of terminals convert glutamine into GABA was provided by studies with brain synaptosomes showing that ^{14}C-glutamine is rapidly metabolized to glutamate and GABA (4). According to current knowledge (33), there is a net flux of glutamate and GABA from neurons to glial cells caused by the impulse-mediated release of these amino acids from nerve terminals. Probably, glutamine formed in glial cells is taken back into the neurons for restoration of the transmitter pools of glutamate and GABA. This theoretical model has not yet been verified by *in vivo* studies; however, some *in vitro* experiments in which the brain structure was not disrupted were performed with hemisections of toad brain (28). These experiments supported a role of glutamine in replenishing neuronal pools of GABA and glutamate by showing increased uptake and conversion of radioactive glutamine to glutamate and GABA following K$^+$-induced depolarization (28). However, these results failed to indicate the extent to which glutamine contributes to maintain the steady state of the glutamate in the various brain areas where glutamate functions both as a transmitter and as a precursor of GABA. Unfortunately, experiments on glutamine uptake and synthesis using synaptosomes are not useful for extrapolations to "*in vivo*" conditions. One important drawback concerning the use of synaptosomes is their low content of glutamate and GABA

(25) caused by the lack of Na$^+$ during synaptosomal preparation. In the absence of Na$^+$, the amino acid carrier cannot function in maintaining amino acid steady state and transports glutamate and GABA out of the cell rather than keeping them inside.

In order to evaluate glutamatergic function, it appears necessary to resort to *in vivo* measurements of glutamate turnover; therefore, it is important to consider the neuroanatomy of the glutamatergic pathways and select one pathway that can be manipulated with facility and is located in a brain structure of known neurochemical characteristics.

NEUROANATOMY OF THE GLUTAMATERGIC SYSTEM

Glutamate is the most abundant amino acid in the adult CNS, with the highest concentration in n. accumbens, cerebral cortex, and cerebellum (Table 1). When the tissues are ranked according to the concentration of glutamate present, it appears evident that the amount of glutamate present bears no relationship with that of GABA (Table 1). Although GABA content may change from area to area by as much as sixfold (compare cerebellum and substantia nigra contents), the glutamate content changes by less than twofold and the glutamine content practically fails to change (Table 1). An interpretation of these data could be that in various brain areas the content of glutamine, a compound that readily crosses the blood-brain barrier, is in equilibrium with the blood; in contrast, glutamate and GABA contents vary in function of abundance of specific cells that store these amino acids. As a corollary of these considerations, we can conclude that it is difficult to assume that the glutamine content plays a regulatory role for the rate of glutamate synthesis because glutamate content in various brain areas appears to be independent from blood glutamine.

Support for the idea that glutamate may be an excitatory neurotransmitter in mammalian brain is provided by a specific Na$^+$-dependent high-affinity uptake system in brain synaptosomes (15) and slices (14). Using biochemical and elec-

TABLE 1. *GABA, glutamate, and glutamine content in various areas of the rat brain*

	nmoles/mg protein ± SEM		
	Glutamate	GABA	Glutamine
N. accumbens	125 ± 3.4	47 ± 2.2	—
Cortex	120 ± 6.0	15 ± 2.0	49 ± 4.9
Cerebellum	110 ± 3.0	15 ± 0.5	58 ± 6
Globus pallidus	95 ± 2.8	89 ± 5.5	—
Diencephalon	95 ± 5.0	22 ± 0.4	51 ± 7.2
Hippocampus	93 ± 4.0	21 ± 2.0	66 ± 2.7
N. caudatus	81 ± 4.0	20 ± 3.0	72 ± 7
Substantia nigra	79 ± 2.8	95 ± 4.4	—

Each value is the average of at least four assays. Sprague-Dawley rats weighing 150 g were microwaved and the brain processed for mass fragmentographic assay as described previously (2) or in the chapter by T. Giacometti (*this volume*).

trophysiological indices, the presence of glutamatergic neurons was identified: (a) in the lateral olfactory tract (4,10); (b) in the afferents from the entorhinal cortex to the hippocampal formation via the fibers of the perforant pathway (3,21,33a); (c) in the pathways connecting the frontal cortex with striatum (13); (d) in the cerebellar parallel fibers originating from the granule cells (27); and (e) in the septum, the septal-hippocampal cholinergic pathway is regulated by a glutamatergic feed back loop that links the hippocampus to the septum (J. T. Coyle, *personal communication;* and this laboratory, *unpublished*).

In selecting an appropriate model from these five pathways to study biochemically the regulation of the turnover of transmitter glutamate, we gave certain weight to the availability of additional knowledge on transmitter neurochemistry and ordinary neuroanatomical information of the selected brain structure. Thus, we have selected the striatum because in this area we can separate extrinsic afferent pathways

TABLE 2. *Effect of striatal injections of kainic acid and cortical ablation on the neurotransmitter profile of rat striatum*

Neurochemical parameters	Cortical ablation	Kainic acid
	Percent controls	
GABAergic neurons		
GAD	99	42^a
GABA content	100	35^a
GABA turnover	45^a	—
^{3}H-GABA binding	70^a	—
GABA uptake	94	46^a
Cholinergic neurons[b]		
CAT	83	49^a
ACh content	95	30^a
ACh turnover	64^a	30^a
Dopaminergic neurons		
TH	108	97
DA content	102	98
HVA	105	110
^{3}H-halop-binding	68^a	64^a
DA-sensitive adenylcyclase	95	5^a
Glutamatergic neurons		
Glutamate content	87^a	97
Glutamate uptake	61^a	98
Enkephalinergic neurons		
Enkephalin content	92	45^a
^{3}H-naloxone binding	70^a	60^a

The different parameters were determined 21 days after cortical ablation or 15 days after intrastriatal injection of 1 µg of kainic acid, according to the following methods: GABA, glutamate content, and GABA turnover rate (18), GABA binding (8), GABA and glutamine uptake (11), ACh (acetylcholine) content and turnover (35), TH (36), DA and HVA (6), enkephalin (34), and naloxone binding (23). The other data were from Schwarcz et al. (30).
[a] The difference from controls was significant, $p < 0.05$.
[b] From Moroni et al., ref. 19.

from intrinsic interneurons by comparing neurochemical parameters in rats with cortical deafferentation and with striatal lesion caused by local injection of kainic acid (5). The results of Table 2 show that striatal deafferentation reduces GABA and ACh turnover but fails to change CAT, GABA uptake, or the content of ACh and GABA, suggesting that the reduction in the turnover of ACh and GABA is due to a decrease of an afferent stimulation caused by decortication. Probably, decortication eliminates glutamatergic afferents among a number of other corticostriatal connections. This suggestion is confirmed by the data in Table 2, which shows that cortical deafferentation reduces glutamate content and uptake. Both parameters fail to change following kainic acid treatment, which destroys striatal interneurons (5) but not afferent axons (Table 2). Cortical ablation fails to change HVA, TH, or DA content, adding significance to the changes in the parameters that are indices of glutamergic activity that were changed by decortication (Table 2). In contrast, the data of Table 2 show that kainic acid injections that destroy a large number of intrinsic striatal neurons reduce the indices of enkephalinergic, GABAergic, and cholinergic activities, indicating that these three types of neurons reside within the striatum and, presumably, receive the corticostriatal innervation. Finally, the data of Table 2 also indicate that the corticostriatal afferent fibers that include glutamatergic axons may contain in their membranes GABA, enkephalin, and opiate receptors. The neuroanatomical inferences that can be drawn from the data of Table 2 and from other data from this laboratory are illustrated in the working model of Fig. 1.

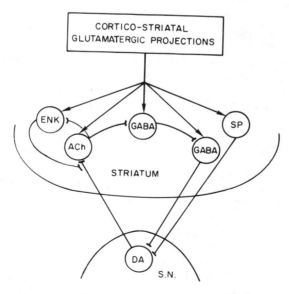

FIG. 1. Schematic representation of interaction of various striatal transmitters as revealed by the data of Table 2. S.N., substantia nigra; DA, dopamine; S.P., substance P; ENK, enkephalin; ACh, acetylcholine.

MASS FRAGMENTOGRAPHIC STUDIES OF THE *IN VIVO* RELATIONSHIPS BETWEEN GLUTAMINE AND GLUTAMATE

In order to ascertain whether or not glutamate biosynthesis depends on a conversion of glutamine into glutamate, male Sprague-Dawley rats were infused intraventricularly with 400 nmoles of 1,2,3,3,D_4-glutamine/min in a volume of 2 μl/min for various time periods. The rats (approximately 140 g in weight) were killed by exposing their heads for 2.5 sec to a focused high-intensity microwave beam as described by Guidotti et al. (9). The brains were removed from the skull and the parietal cortex, cerebellum, caudate, hippocampus, and diencephalon were carefully dissected and frozen on Dry Ice. Each brain area was subsequently homogenized in 1 ml of 1 N-acetic acid containing 300 nmoles of glutamine-D_4, 200 nmoles of glutamate-D_5, and 50 nmoles of N-aminoisovaleric acid (AVA), which were used as internal standards for glutamine, glutamate, and GABA, respectively. The whole homogenate was centrifuged and the supernatant was loaded onto a small column containing 0.5-ml Dowex 50 $\times$ 8, 200–400 mesh in the acid form. The columns were washed previously with 1 ml 2N-hydrochloric acid, which was followed by water rinses until the eluate had a neutral pH.

The amino acids were eluted from the resin with 1 ml of 3N NH_4OH and aliquots (100 to 200 μl) of the eluate were transferred to glass vials and evaporated to dryness with a stream of nitrogen. To the dry residue, 50 μl of 1,1,1,3,3,3-hexafluoroisopropanol (HFIP) (Pierce Co., Rockford, Ill.) and 100 μl pentafluoropropionic anhydride (PFPA) (Pierce Co.) were added. The vials were sealed, heated for 1 hr at 60° C and stored at 4° C overnight. Just before the mass fragmentographic analysis, the reaction mixture was evaporated to dryness. The residue was dissolved in 15 to 50 μl of ethylacetate and aliquots of 1 to 3 μl were injected into the gas chromatograph-mass spectrometer (LKB 9000). The separation was made on a 2 m $\times$ 3 mm i.d. silanized glass column packed with 3% OVI on Gas chrom Q 100–200 mesh (Applied Science Lab., State College, Pa.), maintained at a temperature of 150° C. The temperature of the flash heater was 200° C and the ion source was kept at 250° C. The electron energy and the trap current were set at 70 eV and 60 A, respectively. The following ions were recorded with the m.i.d.: 230 and 234 for glutamine and glutamine-D_4; 230 and 235 for glutamate and glutamate-D_5; 399 for GABA and 413 for AVA.

Polyethylene cannulae were implanted into the lateral ventricle of rats and D_4-glutamine was infused for various periods of time. The brains from animals infused with glutamine-D_4 were homogenized in a solution containing only AVA as an internal standard. The next steps of the analytical procedure were identical to those used for the measurements of the levels of amino acids, but when the samples were injected into the gas chromatograph-mass spectrometer in order to monitor the percent of incorporation of the deuterium into the endogenous glutamine, glutamate, and GABA pools, the following fragments were recorded: 230 and 234 for glutamine and glutamine-D_4; 250 and 254 for glutamate and glutamate-D_4; 399 and 403 for GABA and GABA-D_4, respectively.

The data reported in Table 3 show the enrichment of deuterated glutamine at

TABLE 3. *Relative enrichment of the D_4-variant into glutamine and glutamate during various periods of intraventricular infusion[a] with glutamine-D_4*

	Glutamate-D_4 (nmoles/mg prot.)	Glutamate-D_4/Glutamate-$D_0 \cdot 100$			Glutamine-D_4/Glutamine $D_0 \cdot 100$			Glutamine-D_0 (nmoles/mg prot.)
		8'	10'	15'	8'	10'	15'	
Caudate	81 ± 4.0	0.26 ± 0.09	0.58 ± 0.15	0.59 ± 0.19	12 ± 4.5	12 ± 8	21 ± 6.6	72 ± 7
Hippocampus	93 ± 4.0	0.34 ± 0.22	1.23 ± 0.29	2.0 ± 0.5	24 ± 6.5	60 ± 13	88 ± 19	66 ± 2.7
Cerebellum	110 ± 6.0	0	0	0	6.9 ± 2.1	4.2 ± 2	7.9 ± 28	58 ± 6
Cortex	120 ± 6.0	0.65 ± 22	1.2 ± 0.34	1.1 ± 0.33	37 ± 9	54 ± 14	55 ± 5	49 ± 4.9

[a] 400 nmoles/min at 2 μl/min.

various times during the intraventricular infusion of this deuterated compound. It appears that cortex and hippocampus contain a greater amount of deuterated glutamine than caudate and cerebellum. Since these areas contain approximately the same amount of glutamine, this difference can be explained by the pattern of the location of the uptake mechanism in various brain regions or by differences in the turnover rate of glutamine in various brain areas. Since, as discussed earlier, there is no blood-brain barrier for glutamine, regional differences in the uptake mechanism may not be operative in explaining the distribution pattern of D_4-glutamine of Table 3. We have therefore invoked a faster turnover rate in cerebellum and caudate than in hippocampus and cortex to explain the lower amount of D_4-glutamine in this area. We are, however, aware that other factors may be operative. Our interest in developing a method to measure glutamate turnover rate from the conversion of glutamine into glutamate was reduced by the data shown in Table 3. Glutamine in hippocampus and cortex has a specific activity higher than that in caudate and cerebellum. Also, the specific activity of the glutamate formed in the hippocampus and cortex is several fold greater than that formed in caudate or cerebellum. Thus, it appears that not only is the conversion of glutamine to glutamate minimal but it follows first-order kinetics. On the basis of these results, we should conclude that the role of glutamine in the formation of glutamate should be minimal. However, it could be argued that this conversion, which we consider small, reflects almost exclusively the turnover of that pool of glutamate that functions as a transmitter.

To test this possibility we plan to compare, in rats, the conversion of deuterated glutamine in glutamate in the striatum ipsilateral to that contralateral to a monolateral ablation of cortex. Since we failed to detect labeling in the GABA pool of striatum, hippocampus, and cortex, we are also interested in pursuing further the possibility that blood-borne glutamine may be a suitable precursor to use in measuring the turnover of glutamate in the glutamate pool that functions as a transmitter.

BIOCHEMICAL STUDIES OF GLUTAMATE AS A GABA PRECURSOR

The development of a very sensitive method for the simultaneous quantitation of glutamic acid and GABA content by mass fragmentography (1) has made it possible to measure the turnover rate of GABA in small brain structures. With this method, the carboxylic groups of glutamate and GABA are esterified with HFIP, whereas the amino groups are acylated with PFPA. To obtain an estimation of the GABA utilization, the changes with time in the ^{13}C enrichment of GABA and glutamate during constant rate of infusion of ^{13}C-glucose are monitored (2). Similar to the methods to measure the turnover rate of catecholamines (6) or acetylcholine (35) also the method to measure GABA turnover is based on a certain number of assumptions and it involves a number of approximations concerning the precursor pool. The method tacitly implies that the conversion of glucose into the various glutamate pools present in the brain areas where we measure GABA turnover proceeds at the same rate constant. Naturally, this is never completely true because glutamate functions not only as a metabolite of intermediary metabolism, but also as a precursor of GABA and a transmitter in its own right. We know that the rate of transmitter biosynthesis is never constant and depends on transsynaptic regulation. Hence, in certain brain areas such as the cerebellum and hippocampus, where various glutamate pathways exist, it has been proved impossible to measure GABA turnover using ^{13}C-glucose, because the specific activity of GABA fails to relate to that of glutamate as it is known for a product and precursor relationship. In practice, if one minimizes the error due to the lack of linearity caused by feedback, one can measure the turnover rate of GABA in globus pallidus, n. caudatus, n. accumbens, and substantia nigra but not in hippocampus, cerebellar cortex, and deep cerebellar nuclei (2).

In the brain nuclei shown in Table 4, the value of the turnover rate of GABA is completely unrelated to the GABA content or to the glutamate content but relates to the efflux rate of ^{13}C-GABA (k_{GABA}), which is a function of the rate in which the specific activity of glutamate and GABA reach equilibrium. It is important to note from Table 3 that the lack of glutamatergic fibers reaching the striatum (Fig. 1) is associated with a decrease of GABA and ACh turnover rate (Table 2).

CONCLUSIONS

Evidence was reviewed suggesting that glutamine may serve as a precursor of toad brain glutamate, and probably as a precursor of that pool of glutamate that in this preparation functions as a transmitter. However, it is difficult to extrapolate from these experiments to *in vivo* experiments in the mammalian brain. A number of appropriate brain models were discussed to study the regulation of glutamate pool functioning as a transmitter. The validity of caudate as such a model was documented by a number of neurochemical data available. These experiments show that glutamatergic afferents to striatum stimulate the GABA and ACh metabolism presumably because glutamate functions as an excitatory transmitter. Using deuterated glutamine injected intraventricularly it is possible to label glutamate in hippocampus, cortex, and caudatus, but this labeling does not appear to have the

TABLE 4. *GABA turnover rate in various brain nuclei*

Nucleus	GABA (nmoles/mg prot.)	Glutamate (nmoles/mg prot.)	k_{GABA}/hr	TR_{GABA} (nmoles/mg prot/hr)
Substantia nigra	95 ± 4.4	79 ± 2.8	3.1 ± 0.50	290
Globus pallidus	89 ± 5.5	95 ± 2.8	4.0 ± 0.30	360
N. caudatus	19 ± 0.86	81 ± 4	18 ± 1.5	340
N. accumbens	47 ± 2.2	125 ± 3.4	5.1 ± 0.70	240

From Bertilsson et al., ref. 2, and Miller et al., ref. 16, with permission.

properties typical for a precursor–product relationship between glutamine and glutamate. However, interestingly enough, we could not detect labeling of GABA. Since the striatal GABA that is not labeled with D_4-glutamine is labeled with ^{13}C-glucose, which also labels glutamate, it is concluded that ^{13}C-glucose (and not D_4-glutamine) is a better precursor to study glutamate as a precursor, whereas D_4-glutamine warrants further study as a possible precursor of glutamate that functions as a transmitter in cortex, hippocampus, and caudatus.

REFERENCES

1. Bertilsson, L., and Costa, E. (1976): Mass fragmentographic quantitation of glutamic acid and gamma aminobutyric acid in cerebellar nuclei and sympathetic ganglion of rats. *J. Chromatogr.*, 118:395–402, 1976.
2. Bertilsson, L., Mao, C. C., and Costa, E. (1977): Application of principles of steady state kinetics to the estimation of gamma-aminobutyric acid turnover rate in nuclei of rat brain. *J. Pharmacol. Exp. Ther.*, 200:277–284.
3. Bradford, H. F., and Richards, C. D. (1976): Specific release of endogenous glutamate from piriform cortex stimulated in vitro. *Brain Res.*, 105:168–172.
4. Bradford, H. F., and Ward, H. K. (1976): On glutaminase activity in mammalian synaptosomes. *Brain Res.*, 110:115–125.
5. Coyle, J. T., Molliver, M. E., and Kuhar, M. J. (1978): In situ injection of kainic acid: A new method for selectively lesioning neuronal cell bodies while sparing axons of passage. *J. Comp. Neurol.*, 180:301–308.
6. Costa, E., Green, A. R., Koslow, S. H., LeFevre, H. F., Revuelta, A., and Wang, C. (1972): Dopamine and norepinephrine in noradrenergic axons: A study in vivo of their precursor product relationship by mass fragmentography and radiochemistry. *Pharmacol. Rev.*, 24:167–190.
7. Curtis, D. R., and Johnston, G. A. R. (1974): Amino acid transmitters in the mammalian central nervous system. *Ergebn. Physiol.*, 69:97–188.
8. Enna, S. J., and Snyder, S. H. (1977): Influence of ions, enzymes and detergents on gamma-aminobutyric acid receptor binding in synaptic membranes of rat brain. *Mol. Pharmacol.*, 13:442–453.
9. Guidotti, A., Cheney, D. L., Trabucchi, M., Doteuchi, M., Wang, C. T., and Hawkins, P. (1974): Focussed microwave radiation: A technique to minimize postmortem changes of cyclic nucleotides, dopa and choline and to preserve brain morphology. *Neuropharmacology*, 13:1115–1122.
10. Harvey, J. A., Scholfield, C. N., Graham, L. T., Jr., and Aprison, M. H. (1975): Putative transmitters in denervated olfactory cortex. *J. Neurochem.*, 24:445–449.
11. Iversen, L. L., and Kelly, J. S. (1975): Uptake and metabolism of γ-aminobutyric acid by neurons and glial cells. *Biochem. Pharmacol.*, 24:933–938.
12. Johnson, J. L. (1972): Glutamic acid as a synaptic transmitter in the nervous system: A review. *Brain Res.*, 37:1–19.
13. Kim, J.-S., Hassler, R., Haug, P., and Paik, K.-S. (1977): Effect of frontal cortex ablation on striatal glutamic acid level in rat. *Brain Res.*, 132:370–374.
14. Kuhar, M. J., and Snyder, S. H. (1970): The subcellular distribution of free ^3H-glutamic acid in rat cerebral cortical slices. *J. Pharmacol. Exp. Ther.*, 171:141–152.
15. Logan, W. J., and Snyder, S. H. (1972): High affinity uptake systems for glycine, glutamic and aspartic acids in synaptosomes of rat central nervous tissues. *Brain Res.*, 42:413–431.
16. Miller, L. P., Martin, D. L., Mazumder, A., and Walters, J. R. (1978): Studies on the regulation of GABA synthesis: Substrate promoted dissociation of pyridoxal-5-phosphate from GAD. *J. Neurochem.*, 30:361–369.
17. Miller, L. P., Walters, J. R., and Martin, D. L. (1977): Postmortem changes implicate adenine nucleotides and pyridoxal-5'-phosphate in regulation of brain glutamate decarboxylase. *Nature (Lond.)*, 266:847–848.
18. Moroni, F., Cheney, D. L., Peralta, E., and Costa, E. (1978): Opiate receptor agonists as modulators of GABA turnover in the n. caudatus, globus pallidus and substantia nigra of the rat. *J. Pharmacol. Exp. Ther.* (in press).

19. Moroni, F., Cheney, D. L., and Costa, E. (1978): Turnover rate of acetylcholine in brain nuclei of rats injected intraventricularly and intraseptally with alpha and beta-endorphin. *Neuropharmacology*, 17:191–196.
20. Moroni, F., Cheney, D. L., and Costa, E. (1977): Beta-endorphin inhibition of acetylcholine turnover rate in nuclei of rat brain. *Nature (Lond.)*, 267:267–268.
21. Nadler, J. V., Vaca, K. M., White, W. F., Lynch, G. S., and Cotman, C. W. (1976): Aspartate and glutamate as possible transmitters of excitatory hippocampal afferents. *Nature (Lond.)*, 260:538–540.
22. Oldendorf, W. H. (1971): Brain uptake of radiolabeled amino acids, amines, and hexoses after arterial injection. *Am. J. Physiol.*, 221:1629–1639.
23. Pert, C. B., Pasternak, G., and Snyder, S. H. (1977): Opiate agonists and antagonists discriminated by receptor binding in brain. *Science*, 182:1359–1361.
24. Quastel, J. H. (1974): Amino acids and the brain. *Biochem. Soc. Trans.*, 2:765–780.
25. Rassin, D. K. (1972): Amino acids as putative transmitters: Failure to bind to synaptic vesicles of guinea pig cerebral cortex. *J. Neurochem.*, 19:139–148.
26. Roberts, E. (1974): Gamma-aminobutyric acid and nervous system function—a perspective. *Biochem. Pharmacol.*, 23:2737–2749.
27. Roffler-Tarlov, S., and Sidman, R. L. (1978): Concentrations of glutamic acid in cerebellar cortex and deep nuclei of normal mice and weaver, staggerer and nervous mutants. *Brain Res.*, 142:269–283.
28. Shank, R. P., and Aprison, M. H. (1977): Glutamine uptake and metabolism by the isolated toad brain: Evidence pertaining to its proposed role as a transmitter precursor. *J. Neurochem.*, 28:1189–1196.
29. Shank, R. P., and Baxter, C. F. (1975): Uptake and metabolism of glutamate by isolated toad brains containing different levels of endogenous amino acids. *J. Neurochem.*, 24:641–646.
30. Schwarcz, R., Creese, I., Coyle, J. T., and Snyder, S. H. (1978): Dopamine receptors localized on cerebral cortical afferents to rat corpus striatum. *Nature (Lond.)*, 271:766–768.
31. Utley, J. D. (1964): Glutamine synthetase, glutamotransferase, and glutaminase in neurons and nonneural tissue in the medial geniculate body of the cat. *Biochem. Pharmacol.*, 13:1383–1392.
32. Van Den Berg, C. J. (1970): Glutamate and glutamine. In: *Handbook of Neurochemistry*, Vol. 3, edited by A. Laytha, pp. 355–379. Plenum Press, New York.
33. Van Den Berg, C. J., Reijnierse, G. L. A., Blockuis, G. G. D., Kroon, M. C., Ronda, G., Clarke, D. D., and Garfinkel, D. (1976): In: *Metabolic Compartmentation and Neurotransmission—Relation to Brain Structure and Function*, edited by S. Berl, D. D. Clarke, and D. Schneider, pp. 515–544. Plenum Press, New York.
33a. White, W. F., Nadler, J. V., Hamberger, A., Cotman, C. W., and Cummins, J. T. (1977): Glutamate as transmitter of hippocampal perforant path. *Nature (Lond.)*, 270:356–357.
34. Yang, H.-Y. T., Hong, J. S., Fratta, W., and Costa, E. (1978): Rat brain enkephalins: Distribution and biosynthesis. In: *Advances in Biochemical Psychopharmacology*, Vol. 18, edited by E. Costa and M. Trabucchi, pp. 149–160. Raven Press, New York.
35. Zsilla, G., Racagni, G., Cheney, D. L., and Costa, E. (1977): Constant rate infusion of deuterated phosphorylcholine to measure the effects of morphine on acetylcholine turnover rate in specific nuclei of rat brain. *Neuropharmacology*, 16:25–30.
36. Zivkovic, B., Guidotti, A., and Costa, E. (1974): Effects of neuroleptics on striatal tyrosine hydroxylase: Changes in affinity for the pteridine cofactor. *Mol. Pharmacol.*, 10:727–753.

Glutamic Acid: Advances in Biochemistry and Physiology, edited by L. J. Filer, Jr., et al.
Raven Press, New York © 1979.

Problems in the Evaluation of Glutamate as a Central Nervous System Transmitter

D. R. Curtis

Department of Pharmacology, John Curtin School of Medical Research, Australian National University, Canberra City, ACT 2601 Australia

There is now an extensive literature concerned with the excitation of neurones in the mammalian CNS by L-aspartate, L-glutamate, and by straight-chain and cyclic analogs of these naturally occurring acidic amino acids.

When the depolarization of single neurones by microelectrophoretic aspartate and glutamate was first reported (21), a role for these amino acids as excitatory transmitters was considered unlikely, despite the presence of substantial amounts of each within the brain. The major reason was the very close similarity in the time courses of recovery from excitation by D- and L-enantiomorphs of these amino acids (see also ref. 11), which was considered to exclude extracellular enzymic degradation as a means of inactivating synaptically released aspartate or glutamate. It is now clear, however, that extracellular enzymic modification, as occurs at cholinergic synapses, is of little or no significance at central synapses utilizing amino acids as transmitters (18). Under *in vitro* conditions, both low- and high-affinity sodium-dependent transport processes have been described for excitatory and inhibitory amino acids (66), and the similar time courses of recovery from excitation by microelectrophoretic D- and L-aspartate, D- and L-glutamate, and L-homocysteate most likely result from the sharing by these amino acids of a low-affinity, high V_{max} uptake mechanism (2,11).

Another reservation regarding a transmitter function for the excitant amino acids was the large variety of neurones that were excited by aspartate and glutamate, but this may well reflect the widespread occurrence of "aspartergic" and "glutamergic" neurones within the mammalian CNS, rather than the nonphysiological, "nonspecific," nature of the excitation. Since both aspartate and glutamate are possibly excitatory transmitters, it has been convenient to classify the appropriate postsynaptic receptors as "aspartate-" and "glutamate-preferring" (47). Both amino acid molecules are flexible; if that of glutamate interacts with the glutamate receptor in an extended conformation, interaction with the aspartate receptor would be possible in a folded conformation. Aspartate, however, having a shorter chain length, would be unlikely to interact with such a glutamate receptor, and aspartate and N-methyl-D-aspartate are possibly highly selective for aspartate-preferring sites.

On the other hand, kainate, a conformationally restricted analog of glutamate, may be more selective for a glutamate-preferring receptor.

In this brief review I wish to outline the investigations of the mode of action of excitant amino acids that may be relevant to their toxic effects on neurones, the use of agonists and antagonists to study aspartergic and glutamergic pathways *in vivo*, and neurochemical evidence for such pathways in certain regions of the brain.

DEPOLARIZATION OF NEURONES BY AMINO ACIDS

There are major technical difficulties in investigating the ionic mechanism of the synaptic excitation of central neurones *in vivo*, largely because of the somatic and dendritic location of excitatory synapses. The same investigational problems confound attempts to compare the ionic mechanism of amino acid depolarization with that induced synaptically. Although it is probable that not all excitatory transmitters have an identical postsynaptic action (see ref. 52), at those synapses which have been most thoroughly investigated there appears to be an increase in the membrane permeability to sodium and potassium ions.

The reversible depolarization of spinal, cortical, and caudate neurones by aspartate and glutamate is accompanied by an increase in membrane conductance, and the measured "reversal" potential is not inconsistent with an increased membrane permeability to both sodium and potassium ions (4,14,21,53,63,93). The involvement of chloride ions seems unlikely, and, since tetrodotoxin does not block the depolarization, the sodium permeability increase differs from that associated with action potentials (17,92). The participation of sodium ions is also apparent from the observation that the depolarization of cultured human and rat spinal neurones by aspartate and glutamate can be abolished reversibly by replacement of extracellular sodium with choline (41). The enhancement of the excitatory effects of L-aspartate and L-glutamate on spinal neurones by *p*-mercuriphenylsulfonate, an inhibitor of amino acid uptake, suggests that the depolarization is not, however, generated by a carrier-linked transport of sodium and amino acids (18).

Although the failure to excite neurones by a number of agents that are more powerful calcium chelators than glutamate suggests that the excitation is not produced merely by a lowering of the extracellular calcium ion concentration (20), the interaction between excitant amino acid molecules and external membrane receptors may initiate a sodium ion permeability increase by displacing calcium ions from critical membrane sites (22).

Thus, the ionic mechanism of the depolarization of neurones by amino acids *in vivo* may well be identical to that produced by some synaptically released transmitters. With cultured mouse spinal neurones, however, the discrepancy between the reversal potentials for synaptic and glutamate depolarizations suggests different ionic mechanisms and that this amino acid is not a naturally occurring synaptic transmitter (75).

A number of recent findings suggest that the excitatory effect of acidic amino acids is not as simple as originally proposed and that several other factors have to be taken into consideration in addition to the interaction with membrane receptors and the initiation of a change in membrane permeability. These include ion movements linked with amino acid uptake and the disturbances of intra- and extracellular ion concentrations produced by prolonged activation of receptors and excitation of neurones.

Uptake may be very important in determining the effectiveness of an amino acid as an excitant. The only available technique for comparing the potencies of amino acid excitants *in vivo* is electrophoretic administration into the vicinity of single neurones. Assuming a similarity of transport number (13), the potencies are expressed relative to glutamate by comparing equally effective electrophoretic currents. Potency ratios determined in this fashion, however, are not based on a comparison of equieffective amino acid *concentrations*, since the concentration of a particular amino acid near the membrane receptors depends on both its rate of ejection from the micropipette and the rate of removal from the vicinity of receptors by uptake and other factors.

Thus, the high potencies of N-methyl-D-aspartate, D-homocysteate (23), β-N-oxalyl-L-α,β-diaminopropionate (86), ibotenate (48), kainate (47,81), and quisqualate and domoate (8) may to a considerable extent reflect the absence of appropriate uptake mechanisms, as has been demonstrated *in vitro* for D-homocysteate (8) and kainate (49). With amino acids such as these, cotransport with sodium is unlikely to contribute significantly to the depolarization. A similar consideration may apply to the relative toxicity of amino acids; the parallelism demonstrated between excitant and toxic potencies (70,71,77) suggests that excitation *per se* may be more important for the destruction of cells than intracellular metabolic disturbances produced by an amino acid after uptake.

A recent investigation indicates that the depolarization of spinal motoneurones by D-homocysteate, unlike that induced by L-glutamate, is associated with a *decrease* in membrane conductance (54), and other differences between excitatory amino acids have become apparent in both *in vivo* and *in vitro* studies of the effects of magnesium ions and of changes in external sodium and potassium ion concentrations on amino acid excitation. Using the hemisected frog or immature rat spinal cord *in vitro*, in the presence of either procaine or tetrodotoxin, amino acids that depolarize motoneurones have been classified into three types on the basis of the effects of altering the potassium and sodium ion concentrations in the bathing medium (28). Depolarization by group I amino acids (D- and L-aspartate, L-glutamate, L-cysteate and L-cysteine sulfinate) was enhanced in K^+-free media and reduced in low-Na^+ media; that by group II amino acids (N-methyl-D-aspartate, D-homocysteate, and quisqualate and kainate) was unaltered in K^+-free media and reduced in low-Na^+ media; that by group III amino acids (D-glutamate, L-homocysteate, and L-homocysteine sulfinate) was either unaffected or increased in K^+-free media and *enhanced* in low-Na^+ media. Such results suggest the importance of both an induced permeability increase to sodium ions (group I and II) and ion-dependent uptake (unlikely with group II) to the depolarizing effect of amino acids, but the significance of these *in vitro* studies to the *in vivo* situation requires further investigation.

Under similar *in vitro* conditions, low concentrations of magnesium (0.5–1.00

mM) depressed the depolarizing effects of N-methyl-D-aspartate, D-glutamate, D-aspartate, and L-homocysteate, but had little or no effect on the actions of L-glutamate, kainate, or quisqualate (29). The action of L-aspartate was reduced less than that of N-methyl-D-aspartate, and to a significantly greater degree than that of L-glutamate. Since this effect of magnesium does not correlate with known characteristics of amino acid uptake, and, being independent of calcium concentration, is unlikely to result from a presynaptic transmitter-releasing effect of excitant amino acids, Evans et al. (29) suggested the presence of magnesium-sensitive and magnesium-insensitive postsynaptic receptors for excitant amino acids. A subsequent study using cat spinal neurones *in vivo* established similar differences in sensitivity to the elevation of extracellular magnesium ion concentrations between the excitant effects of N-methyl-D-aspartate (reduced by magnesium) and kainate (unaffected), and the action of L-aspartate was reduced to a greater extent than that of L-glutamate (25).

Although these observations are generally consistent with the presence of at least two major types of excitant amino acid receptors (aspartate-preferring, magnesium-sensitive; glutamate-preferring, magnesium-insensitive), the finding that high magnesium concentrations also reduced the sensitivity of Renshaw cells to acetylcholine (25) suggests that the interpretation of the effects of magnesium ions in terms of an influence at excitant amino acid receptors may be an oversimplification.

DEPOLARIZATION OF AFFERENT TERMINALS BY AMINO ACIDS

The recent demonstration that aspartate, glutamate, and other acidic amino acids depolarize primary afferent terminals in the ventral and dorsal horns of the cat spinal cord (19) is also relevant to the possible transmitter functions of these amino acids. Afferent terminals are depolarized by GABA, and this effect, blocked by bicuculline, is consistent with the participation of GABA as a depolarizing transmitter at axo-axonic synapses concerned with "presynaptic" inhibition (see ref. 16). GABA also depolarizes dorsal root ganglion cells at bicuculline-sensitive nonsynaptic sites. In contrast, although dorsal root ganglion cells are insensitive to acidic amino acids both *in vivo* (P. Feltz, *personal communication*) and in culture (56,75), primary afferent terminals *are* depolarized. The effect is not blocked by bicuculline, and, contrary to the preliminary report (19), the relative depolarizing potencies of different acidic amino acids appear to be similar to that of their actions as neuronal excitants. In particular, kainic acid has a very powerful and reversible depolarizing action (Curtis and Lodge, *unpublished*), yet apparently is not neurotoxic to synaptic terminals (31,72). Depolarization of terminals by these excitants appears not to involve an increase in membrane conductance, as is the case with depolarization of GABA.

Further investigation is required to determine the significance of this depolarization, and particularly whether it releases transmitter, or enhances or decreases transmitter release by presynaptic impulses. It will be necessary to ascertain whether other excitatory axo-axonic synapses occur in addition to those at which GABA is a transmitter. The depolarization may be nonsynaptic, but nevertheless an indication that the depolarized terminals *are* aspartergic or glutamergic. Such an action may be entirely nonphysiological. On the other hand, however, it may be important in the control of transmitter release (42,55). In the case of Renshaw cells, a release of acetylcholine from axon collateral terminals by excitant amino acids seems most unlikely, since the excitation of these cells by amino acids is not reduced significantly by dihydro-β-erythroidine, an acetylcholine antagonist.

EXCITANT AMINO ACID ANTAGONISTS

Specific antagonists effective at aspartate- and glutamate-preferring receptors are required not only for studying the actions of excitant amino acids, but also for distinguishing aspartergic from glutamergic synaptic excitation.

The initial studies in this field were concerned predominantly with distinguishing amino acid-induced excitation from that produced by other substances, especially by acetylcholine, the effects of which could be selectively blocked by either dihydro-β-erythroidine or atropine (15). Several compounds have been proposed as relatively specific excitant amino acid antagonists, but consistent results have rarely been obtained by different investigators, and until comparatively recently, none has been generally accepted as suitable for providing unequivocal evidence that a particular synaptic excitation involved one or the other excitant amino acid as the transmitter.

This type of investigation has been limited by technical problems (15). Many agents having depressant effects on the CNS, possibly arising from a selective action at excitatory synapses, are of low aqueous solubility, and hence are unsuitable for testing microelectrophoretically. More importantly, using microelectrophoretic techniques, it is difficult to investigate membrane receptors located on the dendrites of neurones, and many possible antagonists have been studied under conditions where a relatively high perisomatic concentration may have obscured specific effects at dendritic synapses, particularly those of morphologically complex neurones such as motoneurones and pyramidal and Purkinje cells. There has also been a tendency to extrapolate results from one kind of neurone in a particular species, even of invertebrates, to all neurones in the mammalian CNS.

Compounds proposed as antagonists include (+)-lysergic acid diethylamide, α-methyl-DL-glutamate, L-glutamate diethyl ester, L-methionine-DL-sulfoximine, 9-methyoxyaporphine, apomorphine, 5,6-dimethoxyaporphine (nuciferine), and 1-hydroxy-3-aminopyrrolid-2-one (15,63). More recently, other substances tested include SP-III, a water-soluble derivative of delta-tetrahydrocannibinol (80), 2-amino-4-phosphonobutyrate (87), D-α-aminoadipate (5,35), and α,ε-diaminopimelate (5). Unfortunately, until recently, few investigators have made use of amino acid *agonists* having relatively specific effects at aspartate- and glutamate-preferring receptors when testing these antagonists.

Although initial experiments indicated that 5,6-dimethoxyaporphine was of little use in distinguishing amino acid from cholinergic excitation of Renshaw cells in the rat (27) and cat (D. R. Curtis, *unpublished*), some selectivity towards amino acid excitation has been observed in the thalamus (3,65), cuneate nucleus (40), optic tectum of the pigeon (30), and spinal cord (74).

Considerable use has been made of L-glutamate diethyl ester as an excitant amino acid antagonist (17,33,34,62,63,65,78,82,83). Although the usefulness of this substance has been questioned in that there is relatively little difference between concentrations selectively blocking amino acid effects and those having a nonselective action (see refs. 1,10,17,65), this small difference may be more a reflection of the problems of the administration of antagonists microelectrophoretically (15) than of the small degree of selectivity of this glutamate analog. In a more recent study in the rat thalamus (39), L-glutamate diethyl ester has been shown to be less effective as an antagonist of N-methyl-D-aspartate (and kainate) than of L-glutamate.

In contrast to these results, D-α-aminoadipate appears to be a more effective antagonist of the excitatory action of N-methyl-D-aspartate than of L-glutamate and kainate (5,6,7,35,39,57). The diamino acid, α,ϵ-diaminopimelate has a similar but slightly different spectrum of antagonism (5,6). Although there is some controversy regarding the effect of D-α-aminoadipate on the sensitivity of Renshaw cells to acetylcholine (6,57), this amino acid and analogs of relatively simple structure may prove of considerable value in assessing the relative importance of aspartate and glutamate receptors on particular kinds of neurone.

PHARMACOLOGICAL INVESTIGATION OF AMINO ACID PATHWAYS

Although virtually all neurones so far tested *in vivo* are excited by L-aspartate and L-glutamate, certain differences in the relative sensitivities of some neurones to these excitants suggest a predominance of one type of amino acid-releasing termination on them. It is unfortunate, however, that in few of these studies has use been made of agonists that may have selective effects on either aspartate- or glutamate-preferring receptors. This type of investigation, together with the use of antagonists, has provided some information regarding amino acid excitatory pathways.

In the cat ventrolateral thalamus, neurones excited synaptically from the brachium conjunctivum were more sensitive to L-glutamate (and acetylcholine), relative to DL-homocysteate and N-methyl-DL-aspartate, than cells located dorsally, thus suggesting that the excitatory fibers of the brachium may be glutamergic (35,64). This proposal gained support from the reduction of both the sensitivity of these neurones to L-glutamate and their firing by impulses in the brachium conjunctivum by L-glutamate diethyl ester (35).

Antagonism by L-glutamate diethyl ester has also been used to support claims for the involvement of excitant amino acids as transmitters of a number of other pathways: the commissural pathway to the rat hippocampus (78), a corticostriatal pathway in the rat (82), and the perforant pathway to the rat dentate gyrus (67). Additionally, reduction by l-hydroxy-3-aminopyrrolid-2-one of the excitation of cuneate neurones by pyramidal tract volleys (83), and by SP-III of responses generated in the rat hippocampus by commissural volleys (80), have also been used to support the involvement of amino acid excitatory transmitters.

In the spinal cord of the cat, Renshaw cells, which probably are not excited monosynaptically by impulses in dorsal root primary afferent fibers but only polysynaptically after excitation of excitatory interneurones, were less sensitive to L-glutamate than to L-aspartate, dorsal horn interneurones having the opposite order of sensitivity (26). This observation, and the subsequent finding that dorsal horn interneurones were more sensitive than Renshaw cells to kainate, relative to N-methyl-D-aspartate (59), is consistent with the presence of glutamergic primary afferents and aspartergic excitatory spinal interneurones. This conclusion is also supported by recent findings that D-α-aminoadipate selectively reduced the long-latency synaptic firing of Renshaw cells by dorsal root volleys, without altering the cholinergic excitation produced by ventral root stimulation (6,7,57). In the rat, however, although dorsal horn interneurones have been reported to be more sensitive to L-glutamate than to L-aspartate, and Renshaw cells to be equally sensitive to these amino acids (9), the results of another study indicate that the two kinds of neurone cannot be distinguished on the basis of sensitivity to amino acids (44).

NEUROCHEMICAL EVIDENCE FOR ASPARTERGIC AND GLUTAMERGIC PATHWAYS

Although it is difficult to distinguish transmitter from metabolic pools of aspartate and glutamate within neurones, if indeed there are separate pools of these types for each amino acid, the association of these amino acids with particular neurones may be revealed by the changes in distribution that follow specific nucleus or pathway lesions. Furthermore, after such lesions, there may be alterations in the uptake of transmitter amino acids *in vitro* as a consequence of the loss of excitatory terminals, and a reduction of the *in vitro* depolarization-induced release of amino acids, a process that while not necessarily directly related to synaptic release *in vivo*, may nevertheless provide a measure of the presynaptic store of amino acid transmitters.

In the spinal cord of the cat, the high levels of glutamate in dorsal roots and the dorsal horn relative to those ventrally, the higher levels of aspartate in the ventral horn than dorsally, and the correlation between reduced aspartate levels and the loss of central interneurones produced by temporary cord hypoxia suggested that glutamate may be the transmitter of some primary afferents, and aspartate that of some excitatory spinal interneurones (24,45,46). Although dorsal root section had little effect on dorsal horn glutamate levels, glutamergic primary afferent terminals may contain a very small fraction of the total glutamate within the cord (46), and the proposal regarding the functions of these two amino acids are consistent with the neuropharmacological investigations mentioned above.

After unilateral ablation of the cochlear, and degeneration of primary synapses within the ventral cochlear nucleus of the guinea pig, aspartate and glutamate levels in this region were significantly reduced, those of aspartate paralleling the morphological changes (88,89). Some cochlear afferents within the auditory nerve may thus be aspartergic (see also ref. 32).

When neonatal hamsters were infected intracerebrally with a rat parvovirus, there was a more than 90% loss of cerebellar granule cells, together with a profound fall in cerebellar glutamate levels by as much as 40% and the uptake of aspartate and glutamate by cerebellar homogenates (91). Cerebellar levels of glutamate in the rat were significantly reduced by prior X-irradiation, which selectively reduced the number of granule cells (85). Cerebellar glutamate levels were also low in mutant mice having a reduced number of granule cells (43,76), and all of these observations suggest that the granule cell synapses with Purkinje cell dendrites may be glutamergic. In contrast, cerebellar cortical aspartate levels were low in rats after destruction of the inferior olive with 3-acetylpyridine (68), and in humans with inherited olivopontocerebellar atrophy (73), suggesting that climbing fibers, which also synapse on Purkinje cell dendrites, are aspartergic.

In the rat, lesions of corticostriatal pathways produced a significant reduction in the uptake of glutamate by tissue from the caudate-putamen (61), and frontal cortical ablation resulted in a decrease in striatal glutamate levels (50), suggesting the glutamergic nature of a corticostriatal pathway. There appears to be a requirement for this tract to be intact for intrastriatal kainate to destroy neurons (60), although, in the cerebellum, kainic acid apparently destroyed all cells except granule cells, which may be glutamergic (38). The former observation suggests that glutamergic *terminals* need to be present in order for kainate to be neurotoxic, possibly via the release of glutamate, whereas the latter suggests a requirement for postsynaptic glutamate *receptors*. Granule cells, however, are excited by DL-homocysteate (12) and L-glutamate (58), and thus presumably bear glutamate receptors.

In the hippocampal formation, the distribution of glutamate; of zinc, which may have a role in relation to moss fiber terminals; and of labeled glutamate in both normal tissue and after lesions of afferent pathways suggested that the perforant pathway, the granule cells, and the pyramidal cells of areas CA3 and 4 may be aspartergic or glutamergic (84). Studies *in vitro* of the uptake and efflux of aspartate and glutamate by hippocampal slices, and of the effects of chronic lesions of commissural fibers and the perforant path, indicate the possibility that the former (from pyramidal cells) is aspartergic; the latter, glutamergic (36,69). Although the excitation of granule cells in hippocampal slices by stimulation of the perforant path was reduced by DL-2-amino-4-phosphonobutyrate (90), this phosphonic analog of glutamate could not be demonstrated to be a glutamate antagonist in either the cat spinal cord (87) or the rat hippocampus *in vivo* (79).

The substantial reduction in the levels of aspartate and glutamate in the guinea pig olfactory cortex following removal of the olfactory bulb has suggested the presence of aspartergic and glutamergic fibers in the lateral olfactory tract (37). Similarly, changes in the glutamate content of the cerebral cortex after undercutting may be associated with the destruction of afferent glutamergic pathways (51).

CONCLUSION

Although the evidence for transmitter roles of aspartate and glutamate is far from convincing, there is sufficient information to make such functions seem probable. Aspartergic and glutamergic fibers may constitute the main central afferent and efferent pathways within the mammalian CNS, pathways that appear not to involve any other accepted transmitter, such as acetylcholine or the biogenic "amines."

In view of the primary function of the brain as an information processing system, it would not be unexpected that substances hitherto considered mainly of metabolic significance could be important in the actual transfer of "data" at synapses. The use of commonly available and relatively simple molecules would provide reliability in a system in which effectiveness and subtlety depended on the morphology and location of synapses, rather than on the nature of the transmitters. Furthermore, uptake may provide an efficient method for transmitter inactivation, and recycling, without the necessary participation of catabolic and anabolic enzymes.

Blood-brain barriers isolate central synapses from circulating synaptically important amino acids, and extraneuronal concentrations are maintained at relatively low levels by cellular uptake mechanisms that are specific for particular classes of amino acid. Defects in either these barriers or uptake processes, together with abnormally high systemic or local concentrations of natural or unnatural excitant amino acids, could result in excessive discharge of neurones, and the consequent disturbances of the normally well-regulated intracellular ion and metabolite levels may result in cell death.

REFERENCES

1. Altmann, H., Bruggencate, G. ten, Pickelmann, P., and Steinberg, R. (1976): Effects of glutamate, aspartate, and two presumed antagonists on feline rubrospinal neurones. *Pfluegers Arch.*, 364:249–255.
2. Balcar, V. J., and Johnston, G. A. R. (1972): The structural specificity of the high affinity uptake of L-glutamate and L-aspartate by rat brain slices. *J. Neurochem.*, 19:2657–2666.
3. Ben-Ari, Y., and Kelly, J. S. (1975): Specificity of nuciferine as an antagonist of amino acid and synaptically evoked activity in cells of the feline thalamus. *J. Physiol. (Lond.)*, 251:25–27P.
4. Bernardi, G., Floris, V., Marciani, M. G., Morocutti, C., and Stanzione, P. (1976): The action of acetylcholine and L-glutamic acid on rat caudate neurons. *Brain Res.*, 114:134–138.
5. Biscoe, T. J., Davies, J., Dray, A., Evans, R. G., Francis, A. A., Martin, M. R., and Watkins, J. C. (1977): Depression of synaptic excitation and of amino acid induced excitatory responses of spinal neurones by D-α-aminoadipate, α,ϵ-diaminopimelic acid and HA-966. *Eur. J. Pharmacol.*, 45:315–316.
6. Biscoe, T. J., Davies, J., Dray, A., Evans, R. H., Martin, M. R., and Watkins, J. C. (1978): D-α-Aminoadipate, α,ϵ-diaminopimelic acid and HA-966 as antagonists of amino acid-induced and synaptic excitation of mammalian spinal neurones *in vivo*. *Brain Res.*, 148:543–548.
7. Biscoe, T. J., Evans, R. H., Francis, A. A., Martin, M. R., and Watkins, J. C. (1977): D-α-Aminoadipate as a selective antagonist of amino acid-induced and synaptic excitation of mammalian spinal neurones. *Nature*, 270:743–745.
8. Biscoe, T. J., Evans, R. H., Headley, P. M., Martin, M. R., and Watkins, J. C. (1976): Structure-activity relations of excitatory amino acids on frog and rat spinal neurones. *Br. J. Pharmacol.*, 58:373–382.
9. Biscoe, T. J., Headley, P. M., Lodge, D., Martin, M. R., and Watkins, J. C. (1976): The sensitivity of rat spinal interneurones and Renshaw cells to L-glutamate and L-aspartate. *Exp. Brain Res.*, 26:547–551.

10. Clarke, G., and Straughan, D. W. (1977): Evaluation of the selectivity of antagonists of glutamate and acetylcholine applied microiontophoretically onto cortical neurones. *Neuropharmacology*, 16:391–398.
11. Cox, D. W. G., Headley, P. M., and Watkins, J. C. (1977): Actions of L- and D-homocysteate in rat CNS: A correlation between low affinity uptake and the time courses of excitation by microelectrophoretically applied L-glutamate analogues. *J. Neurochem.*, 29:579–588.
12. Crawford, J. M., Curtis, D. R., Voorhoeve, P. E., and Wilson, V. J. (1966): Acetylcholine sensitivity of cerebellar neurones in the cat. *J. Physiol. (Lond.)*, 186:139–165.
13. Curtis, D. R. (1964): Microelectrophoresis. In: *Physical Techniques in Biological Research*, Vol. 5, edited by W. L. Nastuk, pp. 144–190. Academic Press, New York.
14. Curtis, D. R. (1965): The actions of amino acids upon mammalian neurones. In: *Studies in Physiology*, edited by D. R. Curtis and A. K. McIntyre, pp. 34–42. Springer-Verlag, Heidelberg.
15. Curtis, D. R. (1976): The use of transmitter antagonists in microelectrophoretic investigations of central synaptic transmission. In: *Drugs and Central Synaptic Transmission*, edited by P. B. Bradley and B. N. Dhawan, pp. 7–35. Macmillan, London.
16. Curtis, D. R. (1978): Pre- and non-synaptic activities of GABA and related amino acids in the mammalian nervous system. In: *Amino Acids as Chemical Transmitters*, edited by F. Fonnum, pp. 55–86. Plenum Press, New York.
17. Curtis, D. R., Duggan, A. W., Felix, D., Johnston, G. A. R., Tebēcis, A. K., and Watkins, J. C. (1972): Excitation of mammalian central neurones by acidic amino acids. *Brain Res.*, 41:283–301.
18. Curtis, D. R., Duggan, A. W., and Johnston, G. A. R. (1970): The inactivation of extracellularly administered amino acids in the feline spinal cord. *Exp. Brain Res.*, 10:447–462.
19. Curtis, D. R., Lodge, D., and Brand, S. J. (1977): GABA and spinal afferent terminal excitability in the cat. *Brain Res.*, 130:360–363.
20. Curtis, D. R., Perrin, D. D., and Watkins, J. C. (1960): The excitation of spinal neurones by the ionophoretic application of agents which chelate calcium. *J. Neurochem.*, 6:1–20.
21. Curtis, D. R., Phillis, J. W., and Watkins, J. C. (1960): The chemical excitation of spinal neurones by certain acidic amino acids. *J. Physiol. (Lond.)*, 150:656–682.
22. Curtis, D. R., and Watkins, J. C. (1960): The excitation and depression of spinal neurones by structurally related amino acids. *J. Neurochem.*, 6:117–141.
23. Curtis, D. R., and Watkins, J. C. (1963): Acidic amino acids with strong excitatory actions on mammalian neurones. *J. Physiol. (Lond.)*, 166:1–14.
24. Davidoff, R. A., Graham, L. T., Shank, R. P., Werman, R., and Aprison, M. H. (1967): Changes in amino acid concentrations associated with loss of spinal interneurones. *J. Neurochem.*, 14:1025–1031.
25. Davies, J., and Watkins, J. C. (1977): Effect of magnesium ions on the responses of spinal neurones to excitatory amino acids and acetylcholine. *Brain Res.*, 130:364–368.
26. Duggan, A. W. (1974): The differential sensitivity to L-glutamate and L-aspartate of spinal interneurones and Renshaw cells. *Exp. Brain Res.*, 19:522–528.
27. Duggan, A. W., Lodge, D., Biscoe, T. J., and Headley, P. M. (1973): Effect of nuciferine on the chemical excitation of Renshaw cells in the rat. *Arch. Int. Pharmacodyn. Ther.*, 204:147–149.
28. Evans, R. H., Francis, A. A., and Watkins, J. C. (1977): Effects of monovalent cations on the responses of motoneurones to different groups of amino acid excitants in frog and rat spinal cord. *Experientia*, 33:246–248.
29. Evans, R. H., Francis, A. A., and Watkins, J. C. (1977): Selective antagonism by Mg^{2+} of amino acid-induced depolarization of spinal neurones. *Experientia*, 33:489–491.
30. Felix, D., and Frangi, U. (1977): Dimethoxyaporphine as an antagonist of chemical excitation in the pigeon optic tectum. *Neurosci. Lett.*, 4:347–350.
31. Frankhuyzen, A. L., and Mulder, A. H. (1977): release of radiolabeled dopamine, serotonin, acetylcholine and GABA from slices of rat striatum after intrastriatal kainic acid injections. *Brain Res.*, 135:368–373.
32. Godfrey, D. A., Carter, J. A., Berger, S. H., Lowry, O. H., and Matschinsky, F. M. (1977): Quantitative histochemical mapping of candidate transmitter amino acids in cat cochlear nucleus. *J. Histochem. Cytochem.*, 25:417–431.
33. Haldeman, S., Huffman, R. D., Marshall, K. C., and McLennan, H. (1972): The antagonism of the glutamate-induced and synaptic excitation of thalamic neurones. *Brain Res.*, 39:419–425.
34. Haldeman, S., and McLennan, H. (1972): The antagonistic action of glutamic acid diethylester towards amino acid-induced and synaptic excitations of central neurones. *Brain Res.*, 45:393–400.

35. Hall, J. G., McLennan, H., and Wheal, H. V. (1977): The actions of certain amino acids as neuronal excitants. *J. Physiol. (Lond.)*, 272:52–53P.
36. Hamberger, A., Chiang, G., Nylén, E. S., Scheff, S. W., and Cotman, C. W. (1978): Stimulus evoked increase in the biosynthesis of the putative neurotransmitter glutamate in the hippocampus. *Brain Res.*, 143:549–555.
37. Harvey, J. A., Scholfield, C. N., Graham, L. T., and Aprison, M. H. (1975): Putative transmitters in denervated olfactory cortex. *J. Neurochem.*, 24:445–449.
38. Herndon, R. M., and Coyle, J. T. (1977): Selective destruction of neurons by a transmitter agonist. *Science (Wash.)*, 198:71–72.
39. Hicks, T. P., Hall, J. G., and McLennan, H. (1978): Ranking of excitatory amino acids by the antagonists glutamic acid diethylester and D-α-aminoadipic acid. *Can. J. Physiol. Pharmacol.* (in press).
40. Hind, J. M., and Kelly, J. S. (1975): The blockade of synaptic and amino acid evoked firing in the cuneate nucleus by the "glutamic acid antagonist" nuciferine (L,5,6-dimethoxyaporphine). *J. Physiol. (Lond.)*, 246:97–98P.
41. Hösli, L. Andrés, P. F., and Hösli, E. (1976): Ionic mechanisms associated with the depolarization by glutamate and aspartate on human and rat spinal neurones in tissue culture. *Pfluegers Arch.*, 363:43–48.
42. Hubbard, J. I. (1970): Mechanism of transmitter release. In: *Progress in Biophysics and Molecular Biology, Vol. 21.*, edited by J. A. V. Butler and D. Noble, pp. 33–124. Pergamon Press, Oxford.
43. Hudson, D. B., Valcana, T., Bean, G., and Timiras, P. S. (1976): Glutamic acid: A strong candidate as the neurotransmitter of the cerebellar granule cell. *Neurochem. Res.*, 1:73–81.
44. Hutchinson, G. B., McLennan, H., and Wheal, H. V. (1978): The responses of Renshaw cells and spinal interneurones of the rat to L-glutamate and L-aspartate. *Brain Res.*, 141:129–136.
45. Johnson, J. L. (1972): Glutamic acid as a synaptic transmitter in the nervous system. A review. *Brain Res.*, 37:1–19.
46. Johnson, J. L. (1977): Glutamic acid as a synaptic transmitter in the dorsal sensory neuron: Reconsiderations. *Life Sci.*, 20:1637–1644.
47. Johnston, G. A. R., Curtis, D. R., Davies, J., and McCulloch, R. M. (1974): Spinal interneurone excitation by conformationally restricted analogues of L-glutamic acid. *Nature*, 248:804.
48. Johnston, G. A. R., Curtis, D. R., de Groat, W. C., and Duggan, A. W. (1968): Central actions of ibotenic acid and muscimol. *Biochem. Pharmacol.*, 17:2488–2489.
49. Johnson, G. A. R., Kennedy, S. M. E., and Twitchin, B. (1979): Action of the neurotoxin kainic acid on the high affinity uptake of L-glutamic acid in rat brain slices. *J. Neurochem.* (in press).
50. Kim, J.-S., Hassler, R., Haug, P., and Paik, K.-S. (1977): Effect of frontal cortex ablation on striatal glutamic acid level in rat. *Brain Res.*, 132:370–374.
51. Koyama, I., and Jasper, H. (1977): Amino acid content of chronic undercut cortex of the cat in relation to electrical after-discharge: Comparison with cobalt epileptogenic lesions. *Can. J. Physiol. Pharmacol.*, 55:523–536.
52. Krnjević, K. (1974): Chemical nature of synaptic transmission in vertebrates. *Physiol. Rev.*, 54:418–540.
53. Krnjević, K., and Schwartz, S. (1967): Some properties of unresponsive cells in the cerebral cortex. *Exp. Brain Res.*, 3:306–319.
54. Lambert, J. D. C., Flatman, J. A., and Engberg, I. (1977): Aspects of the actions of excitatory amino acids on cat spinal motoneurones and on interaction of barbiturate anaesthetics. In: *Iontophoresis and Transmitter Mechanisms in the Mammalian Central Nervous System,* edited by R. W. Ryall and J. S. Kelly, pp. 375–377. Elsevier, New York.
55. Langer, S. Z. (1977): Presynaptic receptors and their role in the regulation of transmitter release. *Br. J. Pharmacol.*, 60:481–497.
56. Lawson, S. N., Biscoe, T. J., and Headley, P. M. (1976): The effect of electrophoretically applied GABA on cultured dissociated spinal cord and sensory ganglion neurones of the rat. *Brain Res.*, 117:493–497.
57. Lodge, D., Headley, P. M., and Curtis, D. R. (1978): Selective antagonism by D-α-aminoadipate of amino acid and synaptic excitation of cat spinal neurones. *Brain Res.*, 152:603–608.
58. McCance, I., and Phillis, J. W. (1968): Cholinergic mechanisms in the cerebellar cortex. *Int. J. Neuropharmacol.*, 7:447–462.
59. McCulloch, R. M., Johnston, G. A. R., Game, C. J. A., and Curtis, D. R. (1974): The differential

sensitivity of spinal interneurones and Renshaw cells to kainate and N-methyl-D-aspartate. *Exp. Brain Res.*, 21:515–518.
60. McGeer, E. G., McGeer, P. L., and Singh, K. (1978): Kainate-induced degeneration of neostriatal neurons: Dependency upon corticostriatal tract. *Brain Res.*, 139:381–383.
61. McGeer, P. L., McGeer, E. G., Scherer, U., and Singh, K. (1977): A glutamatergic corticostriatal path. *Brain Res.*, 128:369–373.
62. McLennan, H. (1974): Actions of excitatory amino acids and their antagonism. *Neuropharmacology*, 13:449–454.
63. McLennan, H. (1975): Excitatory amino acid receptors in the central nervous system. In: *Handbook of Psychopharmacology, Vol. 4, Amino Acid Neurotransmitters*, edited by L. L. Iversen, S. D. Iversen, and S. H. Snyder, pp. 211–228. Plenum Press, New York.
64. McLennan, H., Huffman, R. D., and Marshall, K. C. (1968): Patterns of excitation of thalamic neurones by amino-acids and by acetylcholine. *Nature*, 219:387–388.
65. McLennan, H., and Wheal, H. V. (1976): The specificity of action of three possible antagonists of amino acid-induced neuronal excitations. *Neuropharmacology*, 15:709–712.
66. Martin, D. L. (1976): Carrier-mediated transport and removal of GABA from synaptic regions. In: *GABA in Nervous System Function*, edited by E. Roberts, T. N. Chase, and D. B. Tower, pp. 347–386. Raven Press, New York.
67. Miller, J. J., and Wheal, H. V. (1977): The identification of two discrete excitatory systems in the dentate gyrus of the rat. *J. Physiol. (Lond.)*, 270:58–59P.
68. Nadi, N. S., Kanter, D., McBride, W. J., and Aprison, M. H. (1977): Effects of 3-acetylpyridine on several putative neurotransmitter amino acids in the cerebellum and medulla of the rat. *J. Neurochem.*, 28:661–662.
69. Nadler, J. V., Vaca, K. W., White, W. F., Lynch, G. S., and Cotman, C. W. (1976): Aspartate and glutamate as possible transmitters of excitatory hippocampal afferents. *Nature*, 260:538–540.
70. Olney, J. W. (1976): Brain damage and oral intake of certain amino acids. In: *Transport Phenomena in the Nervous System. Physiological and Pathological Aspects.* edited by G. Levi, L. Battistin, and A. Lajtha, pp. 497–506. Plenum Press, New York.
71. Olney, J. W., and de Gubareff, T. (1978): Glutamate neurotoxicity and Huntington's chorea. *Nature*, 271:557–559.
72. Olney, J. W., and de Gubareff, T. (1978): The fate of synaptic receptors in the kainate-lesioned striatum. *Brain Res.*, 140:340–343.
73. Perry, T. L., Currier, R. D., Hansen, S., and MacLean, J. (1977): Aspartate-taurine imbalance in dominantly inherited olivopontocerebellar atrophy. *Neurology*, 27:257–261.
74. Polc, P., and Haefely, W. (1977): Effects of intravenous kainic acid, N-methyl-D-aspartate, and (-)-nuciferine on the cat spinal cord. *Naunyn-Schmiedeberg's Arch. Pharmacol.*, 300:199–203.
75. Ransom, B. R., Bullock, P. N., and Nelson, P. G. (1977): Mouse spinal cord in cell culture. III. Neuronal chemosensitivity and its relationship to synaptic activity. *J. Neurophysiol.*, 40:1163–1177.
76. Roffler-Tarlov, S., and Sidman, R. L. (1978): Concentrations of glutamic acid in cerebellar cortex and deep nuclei of normal mice and weaver, staggerer and nervous mutants. *Brain Res.*, 142:269–283.
77. Schwarcz, R., Scholz, D., and Coyle, J. T. (1978): Structure-activity relations for the neurotoxicity of kainic acid derivatives and glutamate analogues. *Neuropharmacology*, 17:145–152.
78. Segal, M. (1976): Glutamate antagonists in rat hippocampus. *Br. J. Pharmacol.*, 58:341–345.
79. Segal, M. (1977): An acidic amino acid neurotransmitter in the hippocampal commissural pathway. In: *Iontophoresis and Transmitter Mechanisms in the Mammalian Central Nervous System*, edited by R. W. Ryall and J. S. Kelly, pp. 384–387. Elsevier, New York.
80. Segal, M. (1978): The effects of SP-111, a water-soluble THC derivative, on neuronal activity in the rat brain. *Brain Res.*, 139:263–275.
81. Shinozaki, H., and Konishi, S. (1970): Actions of several anthelmintics and insecticides on rat cortical neurons. *Brain Res.*, 24, 368–371.
82. Spencer, H. J. (1976): Antagonism of cortical excitation of striatal neuron by glutamic acid diethyl ester: Evidence for glutamic acid as an excitatory transmitter in the rat striatum. *Brain Res.*, 102:91–101.
83. Stone, T. W. (1976): Blockade by amino acid antagonists of neuronal excitation mediated by the pyramidal tract. *J. Physiol. (Lond.)*, 257:187–198.
84. Storm-Mathison, J. (1977): Localization of transmitter candidates in the brain: The hippocampal formation as a model. *Prog. Neurobiol.*, 8:119–181.

85. Valcana, T., Hudson, D., and Timiras, P. S. (1972): Effects of X-irradiation on the content of amino acids in the developing rat cerebellum. *J. Neurochem.*, 19:2229–2232.
86. Watkins, J. C., Curtis, D. R., and Biscoe, T. J. (1966): Central effects of β-N-oxalyl-α,β-diaminopropionic acid and other Lathyrus factors. *Nature*, 211:637.
87. Watkins, J. C., Curtis, D. R., and Brand, S. J. (1977): Phosphonic analogues as antagonists of amino acid excitants. *J. Pharm. Pharmacol.*, 29:324.
88. Wenthold, R. J. (1978): Glutamic acid and aspartic acid in subdivisions of the cochlear nucleus after auditory nerve lesion. *Brain Res.*, 143:544–548.
89. Wenthold, R. J., and Gulley, R. L. (1977): Aspartic acid and glutamic acid levels in the cochlear nucleus after auditory nerve lesion. *Brain Res.*, 138:111–123.
90. White, W. F., Nadler, J. V., Hamberger, A., Cotman, C. W., and Cummins, J. T. (1977): Glutamate as transmitter of hippocampal perforant path. *Nature*, 270:356–357.
91. Young, A. B., Oster-Granite, M. L., Herndon, R. M., and Snyder, S. H. (1974): Glutamic acid: Selective depletion by viral induced granule cell loss in hamster cerebellum. *Brain Res.*, 73:1–13.
92. Zieglgänsberger, W., and Puil, E. A. (1972): Tetrodotoxin interference of CNS excitation by glutamic acid. *Nature [New Biol.]*, 239:204–205.
93. Zieglgänsberger, W., and Puil, E. A. (1973): Actions of glutamic acid on spinal neurones. *Exp. Brain Res.*, 17:35–49.

Glutamic Acid: Advances in Biochemistry and Physiology, edited by L. J. Filer, Jr., et al.
Raven Press, New York © 1979.

Central Nervous System Receptors for Glutamic Acid

Graham A. R. Johnston

Department of Pharmacology, John Curtin School of Medical Research, Australian National University, Canberra City, ACT 2601 Australia

L-Glutamic acid, and structurally related acidic amino acids such as L-aspartic acid, excite most neurons in the mammalian CNS (16). A variety of membrane receptors seem to be involved in such excitation, and it is the purpose of this review to discuss evidence for the multiplicity of CNS receptors for L-glutamic acid from *in vivo* studies on excitant amino acid agonists and antagonists, and from *in vitro* studies of ligand binding to isolated receptors.

There is increasing neurochemical evidence that both L-glutamic acid and L-aspartic acid may function as excitatory synaptic transmitters in the mammalian CNS. For example, in the rat cerebellum, L-glutamic acid is associated with granule cells (29) and L-aspartic acid with climbing fibers (28); and in the cat spinal cord, L-glutamic acid appears to be associated with some primary afferent fibers (32) and L-aspartic acid with interneurones destroyed by temporary aortic occlusion (6). This review presents evidence that L-glutamic acid and L-aspartic acid activate different but overlapping populations of excitatory receptors.

EXCITANT AMINO ACID AGONISTS

Many substances appear to act as L-glutamic acid agonists when administered microelectrophoretically near CNS neurons. Most of these excitants are structurally related to L-glutamic acid (16) and show a remarkable variety of moieties that can mimic the ω-carboxylic acid group of L-glutamic acid (4). The naturally occurring L-isomers of glutamic and aspartic acids are only marginally more potent excitants of CNS neurones than the corresponding D-isomers (5,16). Some agonists are more potent excitants than L-glutamic acid, e.g. kainic, quisqualic, and D-homocysteic acids. The potency of microelectrophoretically administered excitants, however, can be modified by sodium-dependent active transport systems that remove certain substances from the extracellular synaptic environment, both "high-" and "low-affinity" systems being involved (5), and this must be considered when attempting to assess the relative potencies of excitants. Kainic acid (20) and D-homocysteic acid (5) are not actively transported, and this could account for their apparent high potency when compared to L-glutamic acid, which is actively transported. The high excitant potency of substances that are not actively transported supports other evidence that excitation and acidic amino acid transport are separate processes (1).

L-Glutamic acid can exist in a variety of low-energy conformations in aqueous solution as judged by proton magnetic resonance spectroscopy (13). In order to gain insight into the likely active conformation(s) of L-glutamic acid during interaction with excitatory receptors, it is necessary to study conformationally restricted analogs of L-glutamic acid, e.g., kainic acid, where the flexibility of that part of the molecule equivalent to L-glutamic acid is restricted by incorporation into a pyrrolidine ring structure. Such studies have led us to propose that there are at least two types of receptors for excitant amino acids in the cat spinal cord: "L-glutamic acid-preferring" and "L-aspartic acid-preferring" receptors (18). L-Glutamic acid was proposed to interact with L-glutamic acid-preferring receptors in partially extended conformations, and also with L-aspartic acid-preferring receptors in partially folded conformations. On the other hand, L-aspartic acid was proposed to interact only poorly with L-glutamic acid-preferring receptors, because the carbon chain of L-aspartic acid is one atom shorter than that of L-glutamic acid and thus could not stretch out sufficiently to interact efficiently with receptors requiring the extended conformations of L-glutamic acid. Kainic acid was proposed as a selective agonist for L-glutamic acid-preferring receptors since it is an analog of L-glutamic acid in an extended conformation, and N-methyl-D-aspartic acid was proposed as a selective agonist with high affinity for L-aspartic acid-preferring receptors. Experiments on the differential sensitivity of cat spinal neurones to kainic and N-methyl-D-aspartic acids provide support for these proposals (25), as do the results of ligand-binding studies with [^{3}H] kainic acid that will be discussed later (20,34).

Kainic acid is of particular interest as a probable selective agonist of L-glutamic acid. This anthelmintic, found in marine algae, is a potent excitant of CNS neurones (4,18,33) and, like L-glutamic acid, can act as a neurotoxin (30). On direct injection into certain brain regions, kainic acid appears to be specifically neurotoxic to neurones having cell bodies in the vicinity of the injection site while sparing axons of passage and nerve terminals arising from neurones distant from the injection site. The structurally related domoic acid is also a potent excitant (4), but dihydrokainic acid and α-*allo*-kainic acid are much less potent than kainic acid (4,18).

ANTAGONISTS OF AMINO ACID-INDUCED EXCITATION

If different populations of excitatory receptors exist for L-glutamic and L-aspartic acids, it seems reasonable to expect that certain antagonists may be able to antagonize selectively the excitation mediated by one or other of these receptor populations. Most of the known antagonists of amino acid-induced excitation, e.g., 1-hydroxy-3-aminopyrrolidone-2 and nuciferine, show little, if any, selectivity for particular amino acids (16,23). Recent studies using magnesium ions (7), D-α-aminoadipic acid (12), and L-glutamic acid diethylester (GDEE) (14) have revealed, however, relatively selective antagonistic actions, which should be of great value in distinguishing synaptic excitation mediated by L-glutamic or L-aspartic acid when used in appropriate conjunction with other evidence.

Magnesium ions antagonize the L-aspartic acid-induced excitation of cat spinal interneurones more effectively than L-glutamic acid excitation (7). Excitations induced by N-methyl-D-aspartic acid and by acetylcholine were antagonized by magnesium ions. Kainic acid-induced excitation was not influenced by magnesium ions on any of the spinal neurones tested. It seems likely that magnesium ions interfere with some ionic events common to the activation of L-aspartic acid-preferring receptors and certain acetylcholine receptors. Differences in the ionic events associated with the actions of various amino acid excitants have been noted in other types of experiments (9,21).

The prediction that D-α-aminoadipic acid "might be a useful antagonist of the action of the excitatory amino acids" (12) has been amply justified by three groups of workers, who have shown that D-α-aminoadipic acid is a significantly more effective antagonist of excitation induced by L-aspartic acid and N-methyl-D-aspartic acid than of excitation induced by L-glutamic acid and kainic acid (3,22,24). D-α-Aminoadipic acid is particularly useful when used in conjunction with GDEE, since the order of susceptibility of excitant amino acids differs for these two antagonists (14). α,ϵ-Diaminopimelic acid also shows selectivity as an excitant amino acid antagonist (2). Experiments indicate that D-α-aminoadipic acid and α,ϵ-diaminopimelic acid act in a like manner, whereas magnesium ions act at a different site as antagonists of excitation induced by N-methyl-D-aspartic acid (10).

LIGAND-BINDING STUDIES

Until relatively recently, the investigation of CNS receptors for neurotransmitters has had to rely almost exclusively on electrophysiological procedures. Following the development of methods for studying the binding of insulin to isolated receptors (15), and the dramatic advances resulting from the application of similar methods to the binding of opiates (35), ligand-binding studies are becoming an ever-increasing part of amino acid neuropharmacology (17), complementing and extending the results available from electrophysiological studies. Ligand-binding studies can be used to aid in the purification and characterization of receptor macromolecules, in the development of new drugs that act specifically on receptors, in the investigation of abnormal brain function, in the study of receptor regulation and turnover, in the discovery of endogenous ligands that might modulate receptor activity, and to gain a better understanding of binding dynamics with respect to cooperativity and multiple-site interactions.

Appropriate ligands for binding studies are substances that act as potent and specific agonists or antagonists with respect to the receptor(s) under investigation. The ligands need to be labeled to high specific activity, and numerous criteria need to be met before the observed binding can be considered as physiologically relevant. The specificity, affinity, and number of binding sites must be evaluated by careful and detailed comparison with the *in vivo* activity of the ligand (15,35). Absolute correspondence between *in vivo* and *in vitro* data can be achieved rarely, and it is probably the differences between the two sets of data that will provide the most interesting advances in our understanding of receptors.

Ligand-binding studies of excitant amino acid agonists are only at a very early stage of development and studies on the binding of amino acid antagonists have yet to be reported. The probable overlapping populations of receptors binding both L-glutamic acid and L-aspartic acid, together with receptors specific for either L-glutamic acid or L-aspartic acid, complicate binding studies with either of these ligands to relatively crude membrane preparations from CNS tissue. Selective agonists or antagonists for each type of CNS receptor for excitant amino acids need to be developed. To date, only kainic acid seems to exhibit appropriate selectivity, binding to some 8 to 10% of the total sites that bind L-glutamic acid in a crude preparation of rat brain membranes, such binding being relatively insensitive to inhibition by L-aspartic acid (20,34).

Since uptake processes for excitant amino acids are sodium dependent, ligand-binding studies are usually carried out in the absence of sodium ions in order to avoid confusing sodium-independent binding to receptors and sodium-dependent binding to transport carriers (1,35).

L-GLUTAMIC ACID BINDING

Michaelis et al. (27) have described sodium-independent binding of L-glutamic acid to a synaptic membrane subfraction from rat brain. Binding was biphasic to a high-affinity site having a dissociation constant (K_d) of 0.2 μM and a maximal binding capacity (B) of 0.002 nmoles/mg protein, and a low affinity site, $K_d = 4$ μM, $B = 0.009$ nmoles/mg. Binding was stereospecific in that 0.2 mM D-glutamic acid did not influence high-affinity binding. L-Aspartic acid was a competitive inhibitor ($K_i = 1$ μM) of high-affinity L-glutamic acid binding, consistent with L-aspartic acid binding to the high-affinity sites with about 20% of the affinity of L-glutamic acid binding to these sites. The excitant amino acid antagonist GDEE was a relatively weak inhibitor (24% at 1 μM), whereas GABA was a relatively potent inhibitor (86% at 1 μM), acting in a noncompetitive manner ($K_i = 1$ μM). The L-glutamic acid binding system could be solubilized by treatment with 0.5 to 1% Triton X-100 detergent at a briefly elevated pH (9.5 for 1 min) at 0 to 4° C and purified some 200-fold to afford a small (MW 14,000) acidic glycoprotein fraction (26). Binding of L-glutamic acid to this soluble receptor preparation was not influenced by D-glutamic acid or GABA (1 μM) and appeared to be associated with a single class of binding sites, $K_d = 0.8$ μM, $B = 66$ nmoles/mg. This binding was inhibited by L-aspartic acid (36%) and GDEE (30%) at 1 μM.

Roberts (31), using a synaptic membrane fraction from rat brain, found that

L-glutamic acid bound to a single population of binding sites, $Kd = 8$ μM, B approximately 0.03 nmoles/mg, in a sodium-independent manner. This binding was inhibited only weakly by D-glutamic acid (11%) but strongly by L-aspartic acid (63%) and GDEE (60%) at 1 mM.

DeRobertis and Fiszer de Plazas (8) have purified a proteolipid fraction from rat cerebral cortex, by extraction with chloroform-methanol and Sephadex column chromatography, that bound L-glutamic acid in a sodium-independent manner. Three binding sites appear to be involved: high affinity, $Kd = 0.3$ μM, $B = 0.5$ nmoles/mg; medium affinity, $Kd = 5$ μM, B = 32 nmoles/mg; and low affinity, Kd 55 μM, $B = 166$ nmoles/mg. D-Glutamic acid (0.1 mM) did not inhibit the high-affinity site, but greatly inhibited the medium-affinity site. GDEE (65%), L-aspartic acid (55%), and nuciferine (54%) inhibited the high-affinity site at 20 μM.

The above studies show that L-glutamic acid can bind to receptors isolated from rat brain in a manner that may represent the interaction of L-glutamic acid with excitatory receptors on neurones. The observed binding is stereoselective in that D-glutamic acid has little influence on L-glutamic acid binding. This is particularly interesting in view of the fact that L- and D-glutamic acid have very similar excitant actions on CNS neurones *in vivo* (5,16). The binding studies indicate that L- and D-glutamic acid act on different excitatory receptors, or perhaps the medium-affinity sites described by DeRobertis and Fiszer de Plazas (8). In most of the above studies, L-aspartic acid and GDEE inhibit the binding of L-glutamic acid to a similar extent, suggesting that these three substances can act on the same binding sites, L-glutamic acid binding with a 5- to 10-fold higher affinity than do L-aspartic acid and GDEE. Since L-glutamic acid and L-aspartic acid have very similar excitant actions *in vivo* (5,16), L-aspartic acid appears likely to be able to bind to sites other than those preferentially binding L-glutamic acid.

L-ASPARTIC ACID BINDING

Fiszer de Plazas and DeRobertis (11) have studied the sodium-independent binding of L-aspartic acid to the proteolipid fraction that bound L-glutamic acid as described above (8). As with L-glutamic acid binding, the binding of L-aspartic acid appeared to involve three kinetically distinct sites: high affinity, $Kd = 0.2$ μM, $B = 3$ nmoles/mg; medium affinity, $Kd = 10$ μM, $B = 132$ nmoles/mg; and low affinity, Kd 50 μM, $B = 617$ nmoles/mg. Extrapolating back to fresh tissue, it was estimated that there were some eight times more high-affinity L-aspartic acid binding sites (8 nmoles/g fresh tissue) than high-affinity L-glutamic acid binding sites. D-Aspartic acid did not inhibit the high-affinity binding of L-aspartic acid, but strongly inhibited medium-affinity binding. L-Glutamic acid was only a relatively weak inhibitor of high-affinity L-aspartic acid binding (28% at 40 μM), as were GDEE (27%) and nuciferine (18%), whereas N-methyl-D-aspartic acid (45%) was about 50% as potent as was L-aspartic acid itself (92%). Kainic acid (40 μM) did not influence L-aspartic acid binding. The following can be suggested on the basis of these results: (a) D- and L-aspartic acids bind to different sites, though they excite CNS neurones in a similar fashion (5); (b) there are sites preferring L-aspartic acid to L-glutamic acid, and vice versa; and (c) kainic acid and L-aspartic acid bind to different sites.

KAINIC ACID BINDING

Simon et al. (34) described the sodium-independent binding of kainic acid to synaptic membranes from rat brain, with a Kd of 0.06 μM. From the apparent maximal number of kainic acid binding sites ($B = 0.001$ nmoles/mg) and cross-inhibition studies with L-glutamic acid, they estimated that there was a definite difference between the kainic acid binding sites and the bulk of the L-glutamic acid binding sites in their preparations, with the density of the L-glutamic acid binding sites being 8 to 10 times greater than the density of the kainic acid binding sites. This indicates that kainic acid binds to only a relatively small population of L-glutamic acid binding sites. We (20) are in general agreement with most of the observations made by Simon et al. (34). L-Glutamic acid was an order of magnitude weaker than kainic acid in displacing bound radioactive kainic acid, and L-aspartic acid was at least 500 times weaker than L-glutamic acid. D-Glutamic acid and D-aspartic acid appeared to be weaker than L-aspartic acid in displacing kainic acid. N-Methyl-D-aspartic acid did not inhibit kainic acid binding. Kainic acid binding was not inhibited by GDEE, magnesium ions, or a variety of other known antagonists of excitant amino acid action (1-hydroxy-3-aminopyrrolidone-2, LSD, L-methionine-DL-sulfoximine, 9-methoxyaporphine, nuciferine), but DL-α-aminoadipic acid was a weak inhibitor (20,34). Specific kainic acid binding appeared to be localized to CNS tissues and showed a fivefold regional variation in the brain, with the highest binding being associated with the striatum and the lowest with the midbrain and pons medulla (34). Subcellular fractionation indicated that most of the binding activity was associated with synaptosomal membrane fractions.

The affinity of kainic acid for the above binding sites appeared to be some four orders of magnitude higher than its affinity for the sodium-dependent, high-affinity L-glutamic acid transport carrier, kainic acid being a weak competitive inhibitor ($Ki = 250$ μM) of this carrier (20). Dihydrokainic acid was about 500 times less potent than kainic acid as an inhibitor of kainic acid binding, but was approximately equipotent with kainic acid as an inhibitor of L-glutamic acid transport.

D-ASPARTIC ACID BINDING

We have done some preliminary experiments on the sodium-independent binding of D-aspartic acid to a Triton-extracted membrane preparation from rat brain (19). Bound radioactive D-aspartic acid could be displaced most potently by N-methyl-D-aspartic acid followed by D-aspartic acid itself, L-aspartic acid, L-glutamic acid, and DL-α-aminoadipic acid, and less potently by kainic acid. D-Aspartic acid appears to bind to a relatively nonspecific population of sites, which show little selectivity for L-glutamic acid or D- or L-aspartic acid, but interact strongly with N-methyl-D-aspartic acid.

MULTIPLICITY OF BINDING SITES FOR EXCITANT AMINO ACIDS

The above binding studies on various membrane preparations from rat brain with L-glutamic acid, L-aspartic acid, kainic acid, and D-aspartic acid as ligands suggested that several binding sites are involved in the binding of each of these ligands. Certainly, they support the concept of L-glutamic acid-preferring and L-aspartic acid-preferring populations of receptors developed on the basis of *in vivo* studies of amino acid-induced excitation of CNS neurones. Binding studies provide information on the relative affinity of the interaction of ligands with binding sites, but it needs to be emphasized that binding studies do not provide information on the efficacy of such interactions with respect to changes induced in membrane permeability. Indeed, not all binding need be functional. Furthermore, on the basis of binding studies per se it is difficult to distinguish between agonist and antagonist binding phenomena. With these reservations in mind, it is possible to propose that at least four populations of excitant amino acid agonist *binding sites* are associated with rat brain membranes, the classification being based on the naturally occurring excitants L-glutamic and L-aspartic acids:

1. L-Glutamic acid-preferring (extended) binding sites. These sites show a preference for kainic acid over L-glutamic acid and interact poorly with L-aspartic acid, D-aspartic acid, and N-methyl-D-aspartic acid. They may represent sites that preferentially bind extended conformations of L-glutamic acid. These are the sites studied using kainic acid as the binding ligand.

2. L-Glutamic acid-preferring (partially folded) binding sites. These sites prefer L-glutamic acid to either kainic acid or L-aspartic acid and may represent sites which preferentially bind partially folded conformations of L-glutamic acid. These are the bulk of the sites studied using L-glutamic acid as the binding ligand.

3. L-Aspartic acid-preferring binding sites. These sites show a strong preference for L-aspartic acid over L-glutamic acid and show little interaction with kainic acid or D-aspartic acid. These are the sites studied using L-aspartic acid as the binding ligand.

4. L-Glutamic and L-aspartic acid binding sites. These sites show little selectivity between L-glutamic, or D- or L-aspartic acids and may be studied using D-aspartic as the binding ligand. N-Methyl-D-aspartic acid interacts very strongly, and kainic acid interacts only poorly with these sites.

It is probably not coincidental that four populations of binding sites are proposed on the basis of studies with four ligands! Examination of more ligands would no doubt produce further classifications of binding sites, and it will be difficult to relate the various populations of binding sites to functional receptors mediating neuronal excitation until more selective agonists and antagonists are developed. The numbers of binding sites on rat brain membranes associated with these four proposed populations appears to decrease in the following order 3 > 2 > 1 > 4. Of the known excitant amino acid antagonists, there appears to be some selectivity with respect to their interaction with the various proposed populations of binding sites. GDEE appears to interact most strongly with population 2, to a lesser extent with population 3, and not at all with populations 1 and 4. DL-α-Aminoadipic acid appears to interact strongly with population 4 and very weakly with population 1, but further tests are necessary using the pure D-α-aminoadipic acid on all the binding site populations. With respect to ligands that merit future study, D-α-aminoadipic acid, GDEE, D-glutamic acid, and N-methyl-D-aspartic acid are obvious choices.

REFERENCES

1. Balcar, V. J., and Johnston, G. A. R. (1972): Glutamate uptake by brain slices and its relation to the depolarization of neurones by acidic amino acids. *J. Neurobiol.*, 3:295–301.
2. Biscoe, T. J., Davies, J., Dray, A., Evans, R. H., Francis, A. A., Martin, M. R., and Watkins, J. C. (1977): Depression of synaptic excitation and of amino acid induced excitatory responses of spinal neurones by D-α-aminoadipate, α,ε-diaminopimelic acid and HA-966. *Eur. J. Pharmacol.*, 45:315–316.
3. Biscoe, T. J., Evans, R. H., Francis, A. A., Martin, M. R., Watkins, J. C., Davies, J., and Dray, A. (1977): D-α-Aminoadipate as a selective antagonist of amino acid-induced and synaptic excitation of mammalian spinal neurones. *Nature*, 270:743–745.
4. Biscoe, T. J., Evans, R. H., Headley, P. M., Martin, M. R., and Watkins, J. C. (1976): Structure-activity relations of excitatory amino acids on frog and rat spinal neurones. *Br. J. Pharmacol.*, 58:373–382.
5. Cox, D. W. G., Headley, P. M., and Watkins, J. C. (1977): Actions of L- and D-homocysteate in rat CNS: A correlation between low-affinity uptake and the time courses of excitation by microelectrophoretically applied L-glutamate analogues. *J. Neurochem.*, 29:570–588.
6. Davidoff, R. A., Graham, L. T., Shank, R. P., Weman, R., and Aprison, M. H. (1967): Changes in amino acid concentrations associated with loss of spinal interneurones. *J. Neurochem.*, 17:1205–1208.
7. Davies, J., and Watkins, J. C. (1977): Effects of magnesium ions on the responses of spinal neurones to excitatory amino acids and acetylcholine. *Brain Res.*, 130:364–368.
8. DeRobertis, E., and Fiszer de Plazas, S. (1976): Isolation of hydrophobic proteins binding amino acids. Stereoselectivity of the binding of L-[^{14}C]glutamic acid in cerebral cortex. *J. Neurochem.*, 26:1237–1243.
9. Evans, R. H., Francis, A. A., and Watkins, J. C. (1977): Effects of monovalent cations on the responses of motoneurones to different groups of amino acid excitants in frog and rat spinal cord. *Experientia*, 33:246–248.
10. Evans, R. H., and Watkins, J. C. (1978): Dual sites for antagonism of excitatory amino acid actions on central neurones. *J. Physiol. (Lond.)*, 277:57P.
11. Fiszer de Plazas, S., and DeRobertis, E. (1976): Isolation of hydrophobic proteins binding amino acids: L-Aspartic acid-binding protein from the rat cerebral cortex. *J. Neurochem.*, 27:889–894.
12. Hall, J. G., McLennan, H., and Wheal, H. V. (1977): The actions of certain amino acids as neuronal excitants. *J. Physiol. (Lond.)*, 272:52–53P.
13. Ham, N. S. (1974): NMR studies of solution conformations of physiologically active amino-acids. In: *Molecular and Quantum Pharmacology*, edited by E. D. Bergmann and B. Pullman, pp. 261–268. Reidel, Dordrecht.
14. Hicks, T. P., Hall, J. G., and McLennan, H. (1978): Ranking of excitatory amino acids by the antagonists glutamic acid diethylester and D-α-aminoadipic acid. *Can. J. Physiol. Pharmacol. (in press)*.
15. Hollenberg, M. D., and Cuatrecasas, P. (1975): Biochemical identification of membrane receptors: Principles and techniques. In: *Handbook of Psychopharmacology*, Vol. 2, edited by L. L. Iversen, S. D. Iversen, and S. H. Snyder, pp. 129–177. Plenum Press, New York.
16. Johnston, G. A. R. (1978): Amino acid receptors. In: *Receptors in Pharmacology*, edited by J. R. Smythies and R. J. Bradley, pp. 295–333. Dekker, New York.

17. Johnston, G. A. R. (1978): Neuropharmacology of amino acid inhibitory transmitters. *Ann. Rev. Pharmacol.*, 18:269–289.
18. Johnston, G. A. R., Curtis, D. R., Davies, J., and McCulloch, R. M. (1974): Spinal interneurone excitation by conformationally restricted analogues of L-glutamic acid. *Nature*, 248:804.
19. Johnston, G. A. R., and Kennedy, S. M. E.: *unpublished observations*.
20. Johnston, G. A. R., Kennedy, S. M. E., and Twitchin, B. (1979): Action of the neurotoxin kainic acid on high affinity uptake of L-glutamic acid in rat brain slices. *J. Neurochem. (in press)*.
21. Lambert, J. D. C., Flatman, J. A., and Engberg, I. (1977): Aspects of the actions of excitatory amino acids on cat spinal motoneurones and on interaction of barbiturate anaesthetics. In: *Iontophoresis and Transmitter Mechanisms in the Mammalian Central Nervous System*, edited by R. W. Ryall and J. S. Kelly, pp. 375–377. Elsevier, New York.
22. Lodge, D., Headley, P. M., and Curtis, D. R. (1978): Selective antagonism by D-α-aminoadipate of amino acid and synaptic excitation of cat spinal neurones. *Brain Res.*, 152:603–608.
23. McLennan, H., and Wheal, H. V. (1976): The specificity of action of three possible antagonists of amino acid-induced neuronal excitations. *Neuropharmacology*, 15:709–712.
24. McLennan, H., and Hall, J. G. (1978): The action of D-α-aminoadipate on excitatory amino acid receptors of rat thalamic neurones. *Brain Res.*, 149:541–545.
25. McCulloch, R. M., Johnston, G. A. R., Game, C. J. A., and Curtis, D. R. (1974): The differential sensitivity of spinal interneurones and Renshaw cells to kainate and N-methyl-D-aspartate. *Exp. Brain Res.*, 21:515–518.
26. Michaelis, E. K. (1975): Partial purification and characterization of a glutamate-binding membrane glycoprotein from rat brain. *Biochem. Biophys. Res. Commun.*, 65:1004–1012.
27. Michaelis, E. K., Michaelis, M. L., and Boyarsky, L. L. (1974): High-affinity glutamic acid binding to brain synaptic membranes. *Biochim. Biophys. Acta.*, 367:338–348.
28. Nadi, N. S., Kanter, D., McBride, W. J., and Aprison, M. H. (1977): Effects of 3-acetylpyridine on several putative neurotransmitter amino acids in the cerebellum and medulla of the rat. *J. Neurochem.*, 28:661–662.
29. Nadi, N. S., McBride, W. J., and Aprison, M. H. (1977): Distribution of several amino acids in regions of the cerebellum of the rat. *J. Neurochem.*, 28:453–455.
30. Olney, J. W., Rhee, V., and Ho, O. L. (1974): Kainic acid: A powerful neurotoxic analogue of glutamate. *Brain Res.*, 77:507–512.
31. Roberts, P. J. (1974): Glutamate receptors in the rat central nervous system. *Nature*, 252:399–401.
32. Shank, R. P., Graham, L. T., Werman, R., and Aprison, M. H. (1967): Distribution of some synaptic transmitter suspects in cat spinal cord. Glutamic acid, aspartic acid, γ-aminobutyric acid, glycine and glutamine. *J. Neurochem.*, 14:465–472.
33. Shinozaki, H., and Konishi, S. (1970): Actions of several anthelmintics and insecticides on rat cortical neurones. *Brain Res.*, 24:368–371.
34. Simon, J. R., Contrera, J. F., and Kuhar, M. J. (1976): Binding of [^{3}H] kainic acid, an analogue of L-glutamate, to brain membranes. *J. Neurochem.*, 26:141–147.
35. Snyder, S. H., and Bennett, J. P. (1976): Neurotransmitter receptors in the brain: Biochemical identification. *Ann. Rev. Physiol.*, 38:153–175.

Glutamic Acid: Advances in Biochemistry and Physiology, edited by L. J. Filer, Jr., et al.
Raven Press, New York © 1979.

Glutamate in the Striatum

E. G. McGeer, P. L. McGeer, and T. Hattori

Kinsmen Laboratory of Neurological Research, Department of Psychiatry, University of British Columbia, Vancouver, Canada V6T 1W5

Glutamate, aspartate, and the related amino acids have long been favorite neurotransmitter candidates of the neurophysiologists because of their clear-cut excitatory properties. They have, however, been the despair of neurochemists because their multiple roles in the central nervous system have made it difficult to identify chemically a neurotransmitter pool, if such exists. Glutamic acid, in particular, is incorporated into proteins and peptides, is involved in fatty acid synthesis, contributes (along with glutamine) to the regulation of ammonia levels and the control of osmotic or anionic balance, serves as a precursor for GABA and for various Krebs cycle intermediates, and is incorporated as a constituent of at least two important cofactors (glutathione and folic acid). In view of these many roles, it is not surprising that L-glutamic acid should be the most plentiful amino acid in the adult CNS and show a fairly even distribution.

The excitatory effects of glutamate and aspartate on cerebral cortical cells were first demonstrated more than a quarter of a century ago, but it is only very recently that convincing evidence has been accumulated defining a few glutamergic and/or aspartergic tracts. The high concentrations of glutamate and aspartate in most areas of brain, however, suggest, in conjunction with their physiological properties, that they may be the common excitatory neurotransmitters of the brain just as GABA seems to be the common inhibitory neurotransmitter.

In this chapter, we will describe the apparent neuronal roles of glutamate in the neostriatum and mention, in particular, its possible importance in the etiology of Huntington's chorea.

NEURONAL LOCALIZATIONS OF GLUTAMATE IN THE NEOSTRIATUM

In Glutamergic Neurons

Chemical identification of glutamergic neurons has been difficult because of the lack of a specific chemical index of the neurotransmitter pool. The demonstration of a specific, high affinity, sodium-dependent uptake system for glutamate and aspartate in nerve endings (28) represented a major advance. Although separate uptake

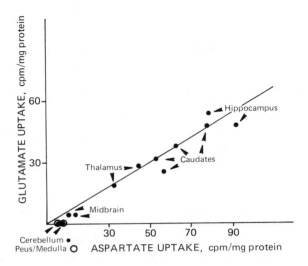

FIG. 1. Correlation between aspartate and glutamate uptakes in various samples of rat brain tissue.

systems for the two amino acids have not yet been distinguished (Fig. 1) and there appears to be high-affinity uptake of glutamate into glial cells as well as into synaptosomes (1,27), it does provide a tool, when combined with selective lesions, for the tentative identification of neurons using these amino acids. These techniques have been used, for example, to suggest that the corticostriatal, entorhinal cortex-hippocampal, retinotectal, primary sensory afferent, and cerebellar granule cell pathways are glutamergic.

Spencer (26) initially suggested that the massive corticostriatal tract might be glutamergic in nature; this suggestion was based on antagonism by diethyl glutamate of the excitation of striatal cells following cortical stimulation. Subsequently, McGeer et al. (22) and Divac et al. (6) found a drop of 40 to 50% in high affinity of glutamate uptake into the synaptosomal fraction of the striatum after ablation or undercutting of the cortex; GABA and dopamine uptake were not affected (Table 1). Both laboratories suggested that the input was glutamergic rather than aspartergic on the basis of the much higher concentration of glutamate as compared with aspartate in the neostriatum, the reported high level of kainic acid (a glutamate analog) binding in the striatum, and Spencer's physiological studies. Subsequently, Kim et al. (15) demonstrated a fall in glutamate levels in the striatum after lesions of the cortex.

It was originally supposed that the thalamostriatal input might also be glutamergic, but large lesions of the thalamus caused no deficit in high-affinity glutamate uptake (Table 1).

Hattori et al. (12) have used axonal transport and autoradiographic methods to define the fine morphology of the probable glutamergic nerve endings in the

TABLE 1. *The accumulation of some putative neurotransmitters and CAT and GAD levels in homogenates of P_2 pellets of caudate-putamen[a]*

	Glutamate (%)	GABA (%)	Dopamine (%)	CAT (%)	GAD (%)
Lobotomized rats	59 ± 4[b]	99 ± 6	110 ± 20	108 ± 8	99 ± 8
Thalectomized rats	105 ± 8	98 ± 4	83 ± 17	100 ± 4	107 ± 3
Lobotomized and thalectomized	58 ± 3[b]	—	—	95 ± 8	95 ± 8

[a] Lesioned side data given as percentage of intact side. CAT, choline acetyltransferase.
[b] $p < 0.001$ for comparison of lesioned and intact sides.

striatum. Figure 2 shows autoradiographs of rat neostriatal terminals labeled by axoplasmic transport following the administration of tritiated proline to the neocortex. The terminals are presumably glutamergic, and all show the same morphology with the common round vesicles and asymmetric contacts typical of Type I excitatory synapses.

The probable importance of this corticostriatal glutamergic tract in the etiology of Huntington's chorea will be discussed later in the section on neurotoxicity of glutamate analogs in the neostriatum.

In GABA-ergic Neurons

The specific localization of glutamic acid decarboxylase (GAD) in GABA-ergic neurons in brain indicates that glutamate is the immediate precursor of this inhibitory neurotransmitter. The brain content of GABA is 200- to 1,000-fold greater than that of neurotransmitters such as dopamine, noradrenaline, acetylcholine, and serotonin. Some of the highest levels of GABA and GAD in brain are found in the basal ganglia, particularly in the globus pallidus and substantia nigra, suggesting a prominent role for this transmitter in the extrapyramidal system. Early uptake studies on radioactive GABA into slices of the caudate-putamen, globus pallidus, and substantia nigra suggested that the uptake was primarily into nerve endings in the substantia nigra, whereas it was into cell bodies as well as nerve endings in the other areas (Table 2) (13).

At least three pathways of the basal ganglia utilize GABA and therefore contain glutamate as a GABA precursor.

Neostriatal interneurons. The probability that much of the GABA in the neostriatum is contained in interneurons is suggested by the small and usually insignificant losses of GABA or GAD following lesions of all known afferents to the neostriatum (20).

Striatonigral pathway. It has been known for many years that a prominent striatonigral path exists with highly preferential innervation of the pars reticulata. More recently, it has been shown that lesions of the neostriatum or hemitransections anterior to the globus pallidus will cause some reduction in nigral GAD (9,14).

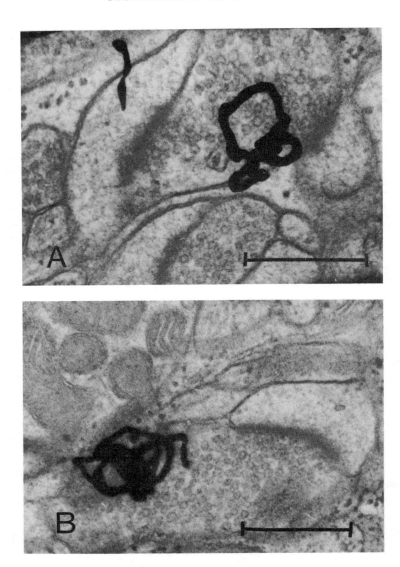

FIG. 2. Two examples of labeled boutons in control rat striatum 7 days after [³H]proline injections into the frontal cortex. Synaptic vesicles are uniformly distributed throughout the bouton. Bars indicate 0.5 μm.

Selective lesions (8,23) suggest that the cell bodies of these reticulata afferents originate in the caudal and lateral aspects of the neostriatum. Since the pathway traverses the globus pallidus, it has been suggested that there may be some terminations within the pallidum itself (8).

Pallidonigral pathway. Compelling evidence in favor of a descending pallidoni-

TABLE 2. Uptake of [³H]GABA into slices of various regions of the rat brain

	Caudate-putamen		Globus pallidus		Substantia nigra	
	% Area	% Grains	% Area	% Grains	% Area	% Grains
Cell soma	47	28	48	42	43	6
Nerve terminal	17	52	15.5	27	18	70
Dendrite	22	15	21	19	24	17
Axon	6.5	2	7.5	6	10.5	4
Glial space and others	7.5	3	8.0	6	4.5	3

From Hattori et al., ref. 13.

gral GABA-containing pathway came originally from studies indicating that electrolytic lesions of the globus pallidus or hemitransections of the brain at the level of the subthalamus led to losses of up to 80% in the GAD activity in the substantia nigra, whereas transections anterior to the globus pallidus led to much smaller losses in GAD (19). The existence of this pathway was established by anterograde axoplasmic flow studies following the injection of ³H-leucine into the globus pallidus. Comparative studies were done following similar injections into the caudate-putamen (NCP). Protein labeling of the nigra resulted from both these injections, there being preferential transport to the zona compacta following pallidal injections and to the zona reticulata following NCP injections (11).

The existence of the pallidonigral tract has been confirmed by retrograde transport of HRP (3,10) and the findings relative to GAD duplicated by Mroz et al. (23). Axonal transport of radioactive GABA in this tract has also been demonstrated (19).

More recently, additional evidence supporting the existence of both GABA-ergic interneurons of the neostriatum and striatonigral GABA-ergic tracts has come from studies of GABA-ergic indices in extrapyramidal nuclei following striatal injections of kainic acid. This analog of glutamate appears to cause destruction only of neurons with perikarya in the injected area and to spare axons of passage and afferent nerve terminals. The biochemical changes in extrapyramidal nuclei following striatal injections of kainic acid (Table 3) are consistent with this hypothesis of its selectivity of action, when considered in conjunction with the presumed biochemical neuroanatomy of the extrapyramidal system (Fig. 3). In particular, the severe losses of GAD, GABA, and GABA uptake in the neostriatum and the substantia nigra are consistent with destruction of GABA-ergic striatal interneurons and striatonigral neurons. Similarly, the lack of a deleterious effect of kainic acid injections on glutamate uptake in the striatum is consistent with the supposition that such uptake reflects the integrity of the corticostriatal glutamergic path.

NEUROTOXICITY OF GLUTAMATE ANALOGS IN THE NEOSTRIATUM

Initial reports on the biochemical deficits resulting from local injections of kainic acid into the neostriata of rats were of particular interest because the morphological

TABLE 3. *Biochemical changes reported in rats given intrastriatal injections of kainic acid and in Huntington's chorea*

Area and biochemical index	Striatal kainic acid	Huntington's chorea
Neostriatum		
GABA-ergic indices[a]	Decreased markedly	Decreased markedly
Cholinergic indices[a]	Decreased markedly	Decreased markedly
Dopaminergic indices[a]	Normal or elevated	Normal or elevated
Serotonergic indices[a]	Normal	Normal
Noradrenergic indices[a]	Normal	Normal
Angiotension converting enzyme	Decreased markedly	Decreased markedly
Receptors[b] for:		
AcCh (muscarinic)	Decreased markedly	Decreased markedly
Serotonin	Decreased markedly	Decreased markedly
GABA	Increased or decreased	Normal or decreased
Substantia nigra		
GABA-ergic indices[a]	Decreased markedly	Decreased markedly
Substance P levels	Decreased markedly	Decreased markedly
Tyrosine hydroxylase levels	Normal	Normal

[a] Neuronal indices used include activity of the synthetic enzyme and levels, uptake, and release of transmitter.
[b] "Receptor" activity refers to sodium-independent binding in synaptic membrane fractions and does not imply pre- or postsynaptic localization or physiological activity.
From Coyle et al., ref. 4.

and biochemical changes were very similar to those reported for Huntington's chorea. Further studies on the biochemical pathology of persons dying with Huntington's chorea and of rats injected intrastriatally with kainic acid have produced additional support for the use of this preparation as an animal "model" of Huntington's chorea (Table 3). Psychopharmacological data so far reported also support the analogy (7).

The fact that kainic acid is recognized as an analog of glutamate, and that glutamate and aspartate, as well as other excitatory amino acids, also have neurotoxic properties, has led to the supposition that Huntington's chorea and other human diseases involving general neuronal loss in specific brain loci may involve glutamate and/or aspartate neuronal systems in their etiology. In particular, it has been suggested that the neostriatal cell death in chorea may be due to chronic functional overactivity of the corticostriatal glutamergic tract. Several possibilities that could have a genetic basis might be considered. For example, there may be enhanced release of glutamate or impaired reuptake of intrasynaptic glutamate causing excessive stimulation of excitatory receptors and ultimately neuronal death. Alternatively, supersensitivity of excitatory receptors or some more general membrane or metabolic disturbance that becomes manifest in cells undergoing glutamate-induced depolarization could be operative. Although these mechanisms are purely speculative, they do suggest new directions for research in the causes of such neuronal degeneration as occurs not only in Huntington's disease, but in senile

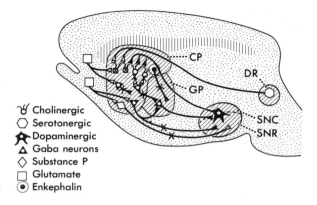

FIG. 3. Hypothesized biochemical neuroanatomy of extrapyramidal system with × indicating neurons affected by intrastriatal injections of kainic acid. Hypothesized encephalin path based on evidence of Cuello and Paxions (5); evidence for other tracts has been recently reviewed (21). CP, caudate-putamen; GP, globus pallidus; DR, dorsal raphe; SNC and SNR, zona compacta and zona reticulata of the substantia nigra.

dementia and some of the ataxias. It is clear that any information shedding more light on the mechanism of kainic acid-induced toxicity is of great importance to a study of this hypothesis, as well as to the application of kainic acid and its analogs as selective lesioning tools in neurobiology.

Olney et al. (24) initially suggested that the toxic actions of glutamate and its analogs, including kainic acid, were through their excitatory action at the glutamate receptor, i.e., in excess, these compounds excite the cells to death (Fig. 6a). According to the theory, no particular toxic action of the administered agent would be necessary beyond persistently activating excitatory receptors: damage would be consequent to ionic shifts exceeding the capacity of membrane pumps to restore and maintain the normal resting potential gradient between the inside and the outside of the cells. Thus, the level of sodium ions would become persistently high inside the cell, whereas potassium ions would leak to the outside. Other ionic changes would also occur, shifting the intracellular ionic balance to a state incompatible with the continued existence of the neuron.

Morphological data from our laboratory (12) offer some support for the hypothesis that glutamergic receptor sites in the striatum are particularly sensitive to the toxic effects of kainic acid. In double labeling experiments, ^{3}H-proline was injected into the frontal cortex of rats, and 7 days later, 1.25 nmoles of kainic acid were injected into the ipsilateral striatum. The animals were sacrificed 10 hr after the kainic acid injection and the control and kainic acid-injected striata examined by light and electron microscopy. In both striata, boutons were the only tissue compartment that was specifically labeled by the radioactive proteins axonally transported from the frontal cortex. Eighty percent of the labeled terminal boutons in the kainic acid-injected neostriata made asymmetrical synaptic contacts with degenerating dendritic spines, and the relative grain density in such boutons was 3.25 times

TABLE 4. *Distribution of silver grains in terminal and preterminal boutons of the kainic acid-affected striatum 7 days after [³H]-proline injection into the ipsilateral frontal cortex* [a]

	% Total grains in boutons	% Total area of boutons	Relative grain density (% grain/% area)
Asymmetric terminal boutons			
On normal dendrites	7	19	0.4
On degenerating dendrites	42	33	1.3
Symmetrical boutons (all on normal dendrites)	3	10	0.3
Preterminal boutons	48	38	1.3

[a] A total of 88 grains was counted in boutons.

greater than that in boutons in synaptic contact with normal dendritic elements (Table 4). In the control striata the boutons showed the uniform distribution of synaptic vesicles evident in Figure 2; in the kainic acid-injected striata, on the other hand, the labeled boutons showed clusters of vesicles that were sometimes, but not always, close to the presynaptic membrane. There was invariably considerable empty space in each such labeled bouton (Fig. 4). The preferential labeling of boutons in contact with degenerating dendritic elements supports the view that neuronal elements carrying many glutamate receptors may be particularly sensitive to the toxic effects of kainic acid (12).

Our experiments on the striatum, however, have led us to propose that kainic acid does not itself act directly on the glutamate receptor since injections of kainic acid do not appear to have much neurotoxic action in the striatum after degeneration of the corticostriatal glutamergic tract (Table 5) (18).

By contrast, thalamic lesions fail to alter the kainic acid-induced neurotoxicity. This indicates that the protection observed after cortical lesions is not a nonspecific effect. Denervation-induced subsensitivity of glutamate receptors has been suggested as a possible explanation. This seems unlikely, both because denervation commonly induces super- rather than subsensitivity and because the time course of the effect (Fig. 5) is that to be expected if the inhibition of kainic acid-induced neurotoxicity depends on degeneration of the presynaptic neurons.

In an attempt to explain these results we are examining three hypotheses.

Loss of excitatory input. One possible hypothesis suggests that removal of the corticostriatal excitatory input reduces the basal level of activity of the neostriatal cells to a point where the additional effects of small doses of kainic acid are no longer able to excite the cells to death. But this simple hypothesis does not explain why lesions of the thalamic input, which is also excitatory, should fail to provide protection. Kainic acid, however, may act selectively on glutamate-receptive neurons to potentiate the action of glutamate. Such potentiation of glutamate-induced excitation, with little or no direct excitatory action of its own, has been reported for kainic acid at the crayfish neuromuscular junction (25).

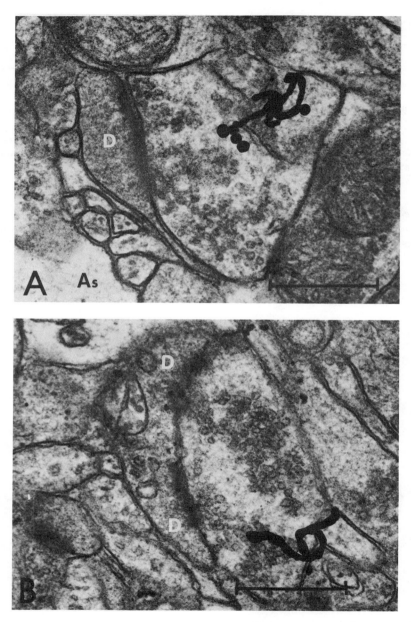

FIG. 4. Two examples of labeled boutons in the kainic acid-affected striatum. Synaptic vesicles form clusters that often appear close to the presynaptic membrane. **As:** astrocyte; **D:** degenerating dendritic spine. Bars indicate 0.5 μm.

TABLE 5. *Protein and enzyme levels in neostriata injected with 2.5 nmoles of kainate as a percentage of those on the contralateral side*

	Protein	GAD	CAT
Unlesioned	92.8 ± 3.6	40.0 ± 0.7	43.8 ± 2.5
Corticostriatal tract lesion	93.4 ± 1.9	95.7 ± 6.3[a]	86.5 ± 5.3[a]
Thalamic lesion	91.4 ± 5.0	40.7 ± 7.8	45.1 ± 8.9

The absolute levels for the contralateral side were comparable in all groups with those found in rats not subjected to any manipulation. Control values were protein, 114.2 ± 3.8 mg/g tissue; GAD, 14.68 ± 0.38 µmoles/hr/100 mg protein; CAT, 34.08 ± 0.81 µmoles/hr/100 mg protein. Six rats per group. All values mean ± SE.
[a] $p < 0.001$ for comparison with data from unlesioned rats.

Release of glutamate or inhibition of its uptake. According to this hypothesis, kainic acid may not act directly, as supposed in Fig. 6a, on the postsynaptic glutamate receptors, but may cause release of glutamate and/or inhibit its reuptake into glutamate nerve endings or glia (Fig. 6b). Either way, the neurotoxicity of

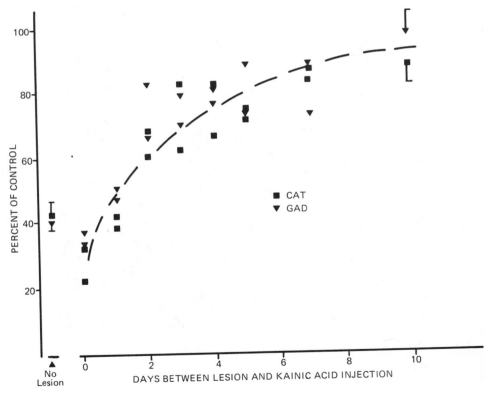

FIG. 5. CAT and GAD activities (as a percentage of control) in the striata of rats injected with 3 nmoles kainic acid at various times after lesions of the corticostriatal tract.

kainic acid would be indirect and due to abnormally high levels of glutamate in the synaptic cleft.

It can be argued that this hypothesis is not tenable because of the very large amounts of glutamate that are required to produce neuronal damage on direct injection into the striatum. However, such exogenous glutamate may well be taken up quickly into glial cells or other compartments so that high concentrations are unable to reach the glutamate receptors.

This hypothesis assumes that kainic acid behaves in glutamate systems much the same way as amphetamine behaves in catecholamine systems. Lakshmanan and Padmanaban (16,17) have previously suggested that the convulsive effects of kainic acid, N-methyl-D-aspartic acid, and β-N-oxalyl-L-α,β-dimethylpropionic acid (ODAP), as well as the neurotoxicity of ODAP (16), might be mediated by glutamate. In support of this hypothesis both they and ourselves have found that kainic acid inhibits the high-affinity, sodium-dependent accumulation of glutamate

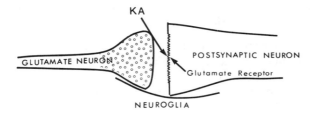

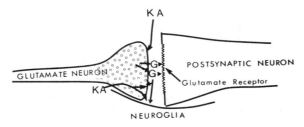

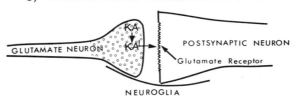

FIG. 6. Various hypothesized mechanisms for the neurotoxic action of kainic acid.

by synaptosomal preparations (Table 6). The effective concentrations are high, but comparable to those which might well occur during intrastriatal injections of this neurotoxin. Similar results are obtained in a calcium-free medium, suggesting kainic acid-induced release may not be involved. Preliminary evidence also suggests that kainic acid does not displace sodium-independent glutamate binding to synaptic membrane preparations of the neostriatum, although glutamate displaces kainic acid binding. Both of these results would be in accord with the hypothesized mechanism shown in Figure 6b.

Metabolite hypothesis. The metabolism of kainic acid in brain tissue has not yet been worked out. It is quite possible that some material formed from kainic acid could be the toxic agent. If this is the case, it seems necessary, in view of the protective effects of the corticostriatal lesion, to postulate that the metabolism of kainic acid producing the toxic material occurs in the glutamergic nerve endings (Fig. 6c). There is little or no evidence to support this hypothesis, but it seems worth further exploration, as do all avenues of research that might lead to an understanding of the mechanism of the neurotoxicity induced by excitatory amino acids.

GLUTAMINASE IN THE STRIATUM

The probable localization of much of the glutaminase activity of brain in the synaptosomal fraction (2) suggests that it might play an important role in the maintenance of the transmitter pool of glutamate, as well as in the formation in GABA-ergic nerve endings of glutamate to be used as a precursor of GABA. The data indicating that the glutamergic and GABA-ergic systems of the neostriatum can be selectively affected by lesions of the corticostriatal tract or by intrastriatal kainic acid injections suggested that this system might be used to explore the relative

TABLE 6. *Effect of kainic acid on sodium-dependent accumulation of [^{14}C]glutamate, [^{14}C]GABA, or [^{3}H]dopamine by synaptosomal fraction of rat neostriatal homogenates*

Concentration of kainate (M)	% Uptake[a]		
	Glutamate accumulation	GABA accumulation	Dopamine accumulation
10^{-3}	24 ± 2	92 ± 9	109 ± 11
3.16×10^{-4}	61 ± 2	112 ± 8	104 ± 6
10^{-4}	74 ± 4	119 ± 10	114 ± 10
3.16×10^{-5}	79 ± 9	102 ± 7	107 ± 8
10^{-5}	94 ± 5		
10^{-6}	93 ± 7		

All accumulation studies were done as previously described (22) on the P$_2$ fraction of rat neostriatal homogenates using 10^{-6} M of the radioactive material; the kainate solution was made up immediately before use and was present during the 5-min preincubation and the 5-min exposure to radioactive material. Control accumulations (in μmoles/5 min/g protein) were 1.67 ± 0.12, glutamate; 1.34 ± 0.09, GABA; and 0.11 ± 0.04, dopamine.

[a] As a percentage of that observed with the same homogenates in the absence of kainate.

amounts of glutaminase associated with these two neuronal types. Glutaminase activity in the neostriatum was therefore measured in rats that had received intrastriatal injections of kainic acid or lesions of the corticostriatal tract at least 10 days before sacrifice.

Figure 7 illustrates typical results on the animals injected with varying amounts of kainic acid. The decrement in glutaminase correlated significantly with the decrement in GAD, and the intercept on the x-axis was such as to suggest that some 60% of the glutaminase activity in the neostriatum is located in GABA-ergic structures (or at least some structures destroyed by the kainic acid injection), whereas 40% is in some unaffected compartment(s). Of the systems so far known to be affected by kainic acid (Table 3), only the GABA-ergic system would appear likely to contain glutaminase. The glutamate uptake in these kainic acid-injected animals was not depressed and indeed, in some, was raised somewhat above the contralateral control striatum. An attempt to take the individual glutamate uptakes into account by calculating for the data a line of correlation of the form

glutaminase activity $= a + b$ (GAD activity) $+ c$ (glutamate uptake)

was unsuccessful; the coefficient of correlation was much lower than for the line of regression shown in Fig. 7.

On the other hand, in animals with lesions of the corticostriatal tract, we could

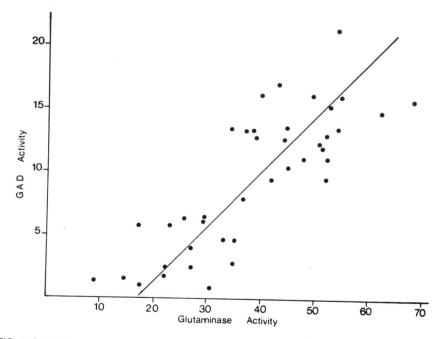

FIG. 7. Correlation between glutaminase and GAD activities (in μmoles/hr/100 mg protein) in the neostriata of rats sacrificed 10 days after injection of 0 to 10 μmoles kainic acid.

not demonstrate a significant correlation between glutaminase activity in the neostriatum and glutamate uptake in that structure. In these animals, GAD activity was not significantly affected by the lesion. These results were disappointing since we had hoped that glutaminase activity might be a convenient index, used alone or together with measurements of GAD, for the integrity of glutamergic systems. These data, however, offer no support for the supposition that the neurotransmitter pool of glutamate may be derived directly from glutamine in the glutamergic nerve endings.

Experiments on the subcellular localization of glutaminase in control rat neostriata indicated that approximately 60% was in the P_2 synaptosomal fraction. This figure is higher than previously reported from other brain areas (2), but is consistent with the results from kainic acid-injected rats. Taken together, these data suggest that about 40% of the glutaminase in the neostriatum is associated with glia and about 60% with GABA-ergic systems.

CONCLUSION

Glutamate certainly plays important multiple roles in the neostriatum, as it probably does in most regions of brain. It remains a challenge to neuroscientists to develop and apply techniques capable of sorting out its complex functions. The very real possibility that glutamergic systems may play a key role in the mechanisms underlying neuronal losses such as occur in Huntington's chorea, senile dementia, and ataxias emphasizes the importance of concentrated research in this field.

ACKNOWLEDGMENTS

This work was supported by the Huntington's Chorea Foundation, Inc., the W. Garfield Weston Foundation, and the Medical Research Council of Canada.

REFERENCES

1. Balcar, V. J., and Johnston, G. A. R. (1977): *J. Neurochem.*, 20:529–539.
2. Bradford, H. F., and Ward, H. K. (1976): *Brain Res.*, 110:115–125.
3. Bunney, B. S., and Aghajanian, G. K. (1976): *Brain Res.*, 117:423–435.
4. Coyle, J. T., McGeer, E. G., McGeer, P. L., and Schwarcz, R. (1978): Neostriatal injections: A model for Huntington's chorea. In: *Kainic Acid,* edited by E. G. McGeer, J. W. Olney, and P. L. McGeer, pp. 123–138. Raven Press, New York.
5. Cuello, A. C., and Paxinos, G. (1978): *Nature,* 271:178–180.
6. Divac, I., Fonnum, F., and Storm-Mathisen, J. (1977): *Nature,* 266:377–378.
7. Fibiger, H. C. (1978): Kainic acid lesions of the striatum: A pharmacological and behavioral model of Huntington's disease. In: *Kainic Acid,* edited by E. G. McGeer, J. W. Olney, and P. L. McGeer pp. 161–236. Raven Press, New York.
8. Fonnum, F., Gottesfeld, Z., and Grofova, I. (1978): *Brain Res.,* 143:125–138.
9. Fonnum, F., Grofova, I., Rinvik, E., Storm-Mathison, J., and Waldberg, F. (1974): *Brain Res.,* 71:77–92.
10. Grofova, I. (1975): *Brain Res.,* 91:286–291.
11. Hattori, T., Fibiger, H. C., and McGeer, P. L. (1975): *J. Comp. Neurol.,* 162:487–504.
12. Hattori, T., McGeer, E. G., and McGeer, P. L. (1977). *Neurosci. Abstr.,* 7:38.

13. Hattori, T., McGeer, P. L., Fibiger, H. C., and McGeer, E. G. (1973): *Brain Res.*, 54:103–114.
14. Kim, J. S., Bak, I. J., Hassler, R., and Okada, Y. (1971): *Exp. Brain Res.*, 14:95–104.
15. Kim, J. S., Hassler, R., Haug, P., and Paik, K. S. (1977): *Brain Res.*, 132:370–374.
16. Lakshmanan, J., and Padmanaban, G. (1974): *Nature*, 249:469–471.
17. Lakshmanan, J., and Padmanaban, G. (1974): *Biochem. Biophys. Res. Commun.*, 58:690–698.
18. McGeer, E. G., McGeer, P. L., and Singh, K. (1978): *Brain Res.*, 139:381–383.
19. McGeer, P. L., Fibiger, H. C., Maler, L., Hattori, T., and McGeer, E. G. (1974): Evidence for descending pallido-nigral GABA-containing neurons. In: *Advances in Neurology*, Vol. 5, edited by F. H. McDowell and A. Barbeau, pp. 153–160. Raven Press, New York.
20. McGeer, P. L., and McGeer, E. G. (1975): *Brain Res.*, 91:331–335.
21. McGeer, P. L., and McGeer, E. G., and Hattori, T. (1979): Transmitters in the basal ganglia. In: *Amino Acids as Chemical Transmitters*, edited by F. Fonnum, pp. 123–142. Raven Press, New York.
22. McGeer, P. L., McGeer, E. G., Scherer, U., and Singh, K. (1977): *Brain Res.*, 128:369–373.
23. Mroz, E. A., Brownstein, M. J., and Leeman, S. E. (1977): *Brain Res.*, 125:305–311.
24. Olney, J. W., Sharpe, L. G., and de Gubareff, T. (1975): *Neurosci. Abstr.*, 5:371.
25. Shinozaki, H. (1978): The discovery of novel actions of kainic acid and related compounds. In: *Kainic Acid*, edited by E. G. McGeer, J. W. Olney, and P. L. McGeer, pp. 17–36. Raven Press, New York.
26. Spencer, H. J. (1976): *Brain Res.*, 102:91–101.
27. Stewart, R. M., Martuza, R. L., Baldessarini, R. J., and Komblith, P. L. (1976): *Brain Res.*, 118:441–452.
28. Wolfsey, A. R., Kuhar, M. J., and Snyder, S. H. (1971): *Proc. Natl. Acad. Sci. USA*, 68:1102–1106.

Glutamate Toxicity in Laboratory Animals

R. Heywood and A. N. Worden

Huntingdon Research Centre, Huntingdon, Cambridgeshire, PE18 6ES England

Since the report by Lucas and Newhouse (16) of retinal degeneration and that by Olney (20) of necrosis of the hypothalamic neurons following the administration of monosodium glutamate (MSG) to the neonate mouse, there have been several attempts to assess the toxicity of this substance in a variety of species. With the use of different strains as well as species, and with varying experimental conditions, some interlaboratory discrepancies might be expected. The published papers on the toxicity of MSG indicate good concordance, and it appears to be the interpretation of the data, rather than the findings themselves, that is open to debate.

The classical toxicological approach to food additives is to administer the test material at three dosage levels to a rodent and a nonrodent species, using control groups for comparison. The highest dose should produce a minimal toxic effect, such as 10% weight loss, or minimal target organ toxicity, and may require adjustment to become an "effect level." The low-dosage level is set at a simple multiple of the intended daily intake. The middle dose should be the arithmetical mean of the high- and low-dose levels. For food additives, the high-dosage level is limited by the amount of compound that can be incorporated in the diet without causing nutritional imbalance or dietary disturbance. The animals used for such studies should be given the new diet as soon as possible after weaning.

CHRONIC ANIMAL STUDIES

Rat

Ebert (6) fed MSG at dietary levels of 0.1 and 0.4% w/w to Sprague-Dawley rats for 2 years; body weight gain, food consumption, hematological values, gross and histopathological observations, tumor incidence, and survival rate in the dosed animals did not differ from those of the controls. Owen et al. (27) fed MSG for 2 years to Charles River CD rats at levels of 1, 2, and 4% w/w of the diet and compared them with two control groups (one undosed and one receiving 2.05% w/w sodium propionate). The additive did not cause any adverse effects on body weight gain, food consumption, hematology, biochemistry, organ weights, or survival. Water consumption, urine volume, and sodium excretion were increased in animals receiving 4% MSG and 2.05% sodium propionate. These physiological changes were accompanied by an increased and earlier occurrence of spontaneous subepithelial basophilic deposits in the renal pelvis. This change is interpreted as reflecting an exacerbation of a spontaneous condition and may be associated with the increased urinary output and sodium excretion. The fact that the reference standard diet (2.05% sodium propionate) induced the same degree of change as in the rats receiving 4% MSG rules out the possibility of this being a specific adverse effect of MSG. Focal mineralization at the renal corticomedullary junction occurred with equal frequency in all groups, including those receiving the basal diet.

Mouse

Ebert (6) fed MSG at dietary levels of 1 and 4% w/w to C-57 black male mice for 2 years without significant adverse effects. In a two-generation study in mice in which MSG was administered in the diet at concentrations of 1 and 2% w/w, Semprini et al. (34) recorded no effect on the development of the animals. The average weight of the treated animals at weaning was slightly higher than that of the controls. In further studies (33), no histopathological abnormalities of the CNS were detected.

Rabbit

Ebert (6) found no adverse effects on reproduction, nor was teratogenicity induced, in New Zealand white rabbits fed 0.1, 0.825 or 8.25% w/w MSG in the diet for 2–3 weeks during the gestation period.

Dog

A long-term feeding study was carried out in purebred beagle dogs with MSG incorporated in the diet at levels of 2.5, 5, and 10% w/w for 2 years by Owen et al. (28). Two control groups were used, the first receiving a standard dog diet (Purina Chow) and the second receiving the basal diet plus 5.13% w/w sodium propionate. MSG did not appear to cause any adverse effects on body weight gain, general behavior, ECG, ophthalmological findings, hematology, blood chemistry, organ weights, or mortality. Urinary volume and sodium excretion were slightly raised in dogs receiving sodium propionate and MSG at all dosage levels, but the ability to concentrate the urine was unimpaired. No morphological changes attributable to the administration of sodium propionate or MSG were detected in any of the tissues examined histologically. Foci of mineralization were seen in the lumen of the medullary tubules of the kidneys of most of the dogs examined, including the untreated controls. This finding is noted commonly in the kidneys of laboratory-maintained beagles and is a spontaneous change.

Monkey

The effects of large dietary supplements of MSG have been studied in infant monkeys by Wen et al. (40). Ten infant squirrel monkeys were fed either 4.8, 9.1, or

16.7% MSG formula diet for 9 weeks and evaluated clinically and histopathologically. None of the monkeys developed hypothalamic or retinal lesions. Monkeys receiving the largest supplement showed retarded growth for no apparent clinical reason. Feeding 9.1% MSG formula for 1 year to an infant cynamolgus and an infant bush-baby monkey had no effect when compared with two control infant monkeys. This appears to be the only long-term feeding study in monkeys.

Dietary administration of MSG in these conventional studies was found to be without significant toxic effect over the varying periods of administration.

ACUTE TOXICITY STUDIES

The majority of studies on glutamate toxicity have concerned acute toxicological manifestations.

Mouse

It is well established that lesions can be induced in the arcuate nucleus of the hypothalamus of the young mouse by either oral or subcutaneous administration of MSG (1,15,20,21,23). Mice treated with MSG during infancy were obese, with abnormalities as adults (20). Pizzi et al. (31) gave MSG at dose levels of 2.2 to 4.2 g/kg on days 2 to 11 after birth. When these animals became adults, reproductive dysfunction was seen in both sexes. The treated females had few pregnancies and smaller litters than normal, and treated males showed reduced fertility. The MSG-treated mice showed increased body weight and decreased pituitary, thyroid, ovary, and testis weights.

A fourfold increase in the level of glutamate in the arcuate nucleus of the hypothalamus followed the elevation of plasma glutamate after a single subcutaneous injection of MSG (30). Peak plasma levels occurred after 15 min, and peak levels in the arcuate nucleus were attained after 3 hr. The results indicate that plasma concentrations above a certain level were necessary to induce brain lesions. Stegink et al. (37) state that arcuate nucleus damage does not occur in the mouse at plasma levels below 50 μm/dl. Bizzi et al. (4) recorded that brain plasma glutamic acid levels are not affected by oral administration of MSG until they exceed the basal plasma concentration by a factor of about 20. Qualitative differences in glutamic acid plasma levels following the oral administration of MSG to neonate mice and rats were related to age, total dosage on a body weight basis, and concentration of the solution administered.

Plasma glutamic acid levels were further investigated by O'hara et al. (19) when MSG was administered as single doses of 1 g/kg body weight by the intraperitoneal, subcutaneous, and oral routes in 10- and 23-day-old and 4-month-old mice. Plasma glutamate levels rose rapidly, reaching maximal values 10 to 30 min after dosing and returning to normal by 90 min. Peak values following oral intubation were significantly lower than those recorded with intraperitoneal or subcutaneous administration. With dietary administration, plasma glutamate values never exceeded five times the base-line value. In an attempt to correlate plasma glutamic acid levels with change in the hypothalamus of adult mice following oral administration of MSG at 1.5 g/kg at varying concentrations, James et al. (14) failed to establish a correlation between plasma glutamate levels and the concentration of MSG administered. There was a correlation between plasma sodium and the concentration of MSG administered. Serum glutamate levels were markedly raised, with a peak concentration at 15 min. Despite this increase, no histological damage was induced in the hypothalamus of the mice killed 4 hr after dosing. Further experiments are necessary to correlate plasma levels with brain tissue levels. On this evidence, it must be assumed that the blood-brain barrier (BBB) protected the hypothalamus of adult mice. However, the plasma glutamic acid levels did not exceed the basal values by the suggested 20-fold factor.

In an attempt to obtain further data relevant to safety-in-use, Heywood et al. (9) administered MSG *ad libitum* in the diet or drinking water at levels of 45.5 g/kg/day or 20.9 g/kg/day, respectively, to weanling mice. No hypothalamic lesions were induced. Plasma glutamic acid levels were doubled by giving MSG at 10% w/w in the diet. The threshold level for the neurotoxicity of MSG when administered in the diet has yet to be established.

Rat

The acute neurotoxicity of MSG in the rat has been described by Burde et al. (5) and Everly (7). Palmer (29) found that in the neonatal rat, hypothalamic lesions were induced at a dosage level of 0.5 g/kg body weight MSG given subcutaneously, but four times that dosage was required to produce the same effect when MSG was given by the oral route. Hypothalamic lesions could be induced within 5 hr by the oral administration of MSG at dosage levels of 2 g/kg and above in rats of up to 25 days of age. The incidence and severity of the lesions varied according to dose level and age. The results of the studies are shown in Tables 1 and 2. In rats dosed at the

TABLE 1. *Incidence of hypothalamic lesions induced by MSG in 3-day-old rats*

Dose (g/kg)	Concentration of solution (%)	Route[a]	Hypothalamic lesion	Incidence (%)
0.5	5	p.o.	−	−
1.0	10	p.o.	−	−
2.0	20	p.o.	+	60
4.0	40	p.o.	+	100
0.5	5	s.c.	+	100
1.0	10	s.c.	+	100
2.0	20	s.c.	+	100
4.0	40	s.c.	+	100

[a] p.o., per oral; s.c., subcutaneous.
[b] Animals died during observation period.

TABLE 2. *Incidence of hypothalamic lesions in 3- to 25-day-old rats 5 hr after treatment with MSG*

Age (days)	Dose (g/kg)	Hypothalamic lesion	Incidence (%)
3	2	+	100
	4	+	100
10	2	+	60
	4	+	100
17	2	+	50
	4	+	100
20	2	+	20
	4	+	40
25	2	+	30
	4	+	60

level of 2 and 4 g/kg MSG at 3, 10, and 17 days, and killed 1 or 4 weeks later, lesions were found only in those dosed at 3 days of age.

Guinea Pig

The guinea pig is also a species susceptible to the effect of glutamate at the dosage level of 1 g/kg (22). In a series of experiments in which 2 or 4 g/kg MSG was administered to 3-day-old guinea pigs as a 29% solution by the oral or subcutaneous route, Heywood et al. (11) confirmed these findings. The results of the studies are given in Table 3. Pyknotic nuclei with associated vacuolation and edema were seen in the hypothalamus of some animals dosed with MSG. Some animals from all groups showed a low incidence of pyknotic nuclei in the thalamic region. Although a dose-related effect of MSG on the hypothalamic region could be seen in the guinea pig, the severity of the lesions was not as marked as in other rodent species given similar dosage levels. The guinea pig has a high rate of brain development before birth, whereas the rat brain grows more quickly after birth. It is

TABLE 3. *Incidence of hypothalamic lesions induced by MSG in 2- to 3-day-old guinea pigs*

Dose (g/kg)	Concentration of solution (%)	Route[a]	Hypothalamic lesion	Incidence (%)
2.0	20	p.o.	+	20
4.0	20	p.o.	+	80[b]
2.0	20	s.c.	+	40
4.0	20	s.c.	+	100

[a] p.o., per oral; s.c., subcutaneous.
[b] Animals died during observation period.

not possible, therefore, to establish a relationship between brain development and the incidence and type of lesions induced with MSG in guinea pigs and rats.

Hamster

The hamster appears to be another rodent species susceptible to the effects of MSG at the neonatal stage (38).

Dog

Oser et al. (26) gave 1 g/kg MSG to 3-day-old dogs, then killed animals at 3, 6, and 24 hr, and, in the case of 3 dogs, 52 weeks after dosing. An additional group of 15 dogs, dosed for 10 days starting when they were 35 days old, was given subcutaneous or intragastric MSG at levels between 2.2 and 4.4 g/kg. These animals were then killed at 48 weeks. No adverse effects were observed on growth, appearance, or behavior, and extensive histopathological examination did not show any changes.

Aspects of the relationship between histopathological and biochemical changes were investigated in a larger species. The dog was chosen because of its convenient size, the ease with which adequate serum and CSF samples can be obtained, and its known propensity to glutamate-induced vomiting (17,39). Groups of three adult dogs were given MSG by gastric intubation at 0, 1, 2, and 4 g/kg at a concentration of 10%. Serum samples were collected at 0, 15, 30, 60, 120, 180, 240, and 300 min. These samples were measured for glutamate according to the assay method of Bernt and Bergmeyer (3). Five hours after dosing, the animals were killed and the brain above the third ventricle was taken for biochemical investigation. Histopathological examination was restricted to the cortex and the thalamus. In a further experiment, groups of three dogs were given MSG at 0 and 2 g/kg at a 10% concentration. Serum samples were taken for glutamate estimation at 0, 15, 30, 60, 120, 180, 240, and 300 min. CSF samples were collected for glutamate estimation at 0, 60, 120, and 300 min using the technique described by Heywood et al. (10). At the end of the 5-hr observation period, the brain above the third ventricle, including the arcuate nucleus, was taken for biochemical investigation. In the third experiment, serum samples were taken at 0, 15, 30, 60, 90, and 120 min for glutamate and sodium ion investigation. CSF fluid was obtained at 0, 30, 60, and 120 min. The brain above the third ventricle was taken for histopathological investigation. Samples of liver, kidney, gut, and intestinal contents were taken for glutamate investigation.

Vomiting was a consistent clinical sign; all dogs dosed with MSG vomited within 22 to 140 min. Dogs receiving the highest dose level (4 g/kg) vomited 22 minutes after dosing, but with most dogs vomiting occurred 30 to 40 min after dosing. Dogs in the third experiment, in which anesthesia was repeated at short intervals, vomited rather later (mean, 48 min).

The plasma and final brain glutamic acid concentrations are shown in Table 4.

TABLE 4. Plasma and terminal brain glutamate concentrations in beagle dogs given MSG by oral gavage

Dose (g/kg)	Plasma glutamate (µg/ml)							Brain glutamate (µg/g)
	Minutes after dosing							
	0	15	30	60	120	180	240	300	
0	5.0 (2.7)	4.8 (1.2)	6.2 (2.4)	5.1 (3.1)	3.1 (0.3)	3.0 (2.2)	4.9 (2.1)	5.2 (3.2)	0.992 (0.061)
1	7.2 (2.4)	147.9 (154.1)	261.7 (78.8)	240.9 (18.1)	38.4 (14.2)	14.6 (5.2)	12.1 (3.9)	8.7 (3.7)	0.931 (0.032)
2	8.3 (0.6)	127.3 (32.6)	272.8 (74.2)	247.0 (68.0)	100.7 (66.8)	29.6 (27.6)	35.9 (33.8)	9.8 (2.4)	0.849 (0.031)
4	6.5 (0.9)	212.1 (39.3)	297.0 (93.5)	268.3 (55.2)	89.5 (16.0)	19.2 (2.9)	11.7 (4.5)	9.1 (3.6)	0.932 (0.169)

Values in parentheses = 1 × SD.

Peak levels were reached 30 min after dosing, but no significant dose relationship could be established. The peak serum concentrations appear to correlate with vomiting. No increase in brain glutamate level could be detected. The histopathological examination had to be restricted to the cortex and thalamus, in which no morphological change was detected. The plasma and CSF values obtained from dogs in the second experiment are shown in Table 5. Although plasma levels were significantly raised 60 min after dosing, there was no evidence that the glutamate crossed the BBB.

The results of the third experiment are shown in Table 6, which confirms the previous observations on plasma and CSF levels. The glutamate concentrations in the kidneys, liver, and duodenum were unchanged by the administration of glutamate at the dosage level of 2 g/kg. No morphological changes were detected in the hypothalamus.

Monkey

Olney and Sharpe (24) recorded damage to the hypothalamic neurons in a neonate rhesus monkey, admittedly classified as premature, which had been injected with 2.7 g/kg body weight MSG and observed for 3 hr before it was killed. Electron microscopic examination showed that the tissue components primarily affected were the dendrites and cell bodies of the neurons. Olney et al. (25) subsequently examined the brains of six rhesus monkeys, 1 to 7 days old, which had been dosed with MSG. It was found that those given 1 or 2 g/kg body weight MSG orally showed small focal hypothalamic lesions, and one receiving 4 g/kg subcutaneously exhibited cyanosis and convulsions during the 5 hr before death. Other groups of

TABLE 5. *Comparison of plasma and CSF values with terminal glutamate concentration in beagle dogs given MSG by oral gavage*

Dose (g/kg)	Plasma/CSF	Glutamate concentration (μg/ml)				Brain glutamate (μg/g)
		Minutes after dosing				
		0	60	120	300	
0	Plasma	8.1 (1.8)	7.3 (2.0)	6.8 (1.7)	5.8 (1.1)	0.762 (0.040)
	CSF	0.8 (0.6)	2.9 (1.8)	1.8 (1.2)	1.1 (0.8)	
2	Plasma	10.4 (0.7)	254.5 (85.4)	89.2 (95.7)	10.0 (4.2)	0.722 (0.079)
	CSF	3.4 (1.3)	3.1 (1.0)	2.2 (0.9)	2.9 (1.2)	

Values in parentheses = 1 × SD.

TABLE 6. Comparison of plasma, CSF and tissue glutamate in beagle dogs given MSG by oral gavage

Dose (g/kg)	Plasma/CSF	Glutamate concentration (µg/ml)						Terminal glutamate concentration (mg/g)			
		Minutes after dosing						Gut content	Kidney	Liver	Duodenum
		0	15	30	60	90	120				
0	Plasma	10.5	5.2	7.9	5.8	10.0	4.9	0.714	0.809	0.390	0.374
	CSF	1.2	—	1.3	1.0	—	1.8				
2	Plasma	19.5	94.5	305.7	334.7	265.9	156.8	0.698	0.750	0.377	0.422
	CSF	0.6	—	1.3	1.2	—	1.1				

investigators (1,2,18,32,36) have not detected any change in the arcuate nucleus or the median eminence after MSG was given at various dose levels and by various routes.

Newman and his co-workers (19) gave 4 g/kg/day MSG to six monkeys during the last trimester of pregnancy. The offspring did not show any hypothalamic lesions. Four animals acted as controls. Stegink et al. (35) demonstrated that even with 10- to 20-fold increases in maternal glutamate levels, there was no effect on fetal glutamate levels, suggesting that the primate placenta is virtually impermeable to glutamate.

Heywood (12) delivered three rhesus monkeys by cesarean section on days 146, 150, and 155 of gestation and administered MSG to these premature animals at 2 g/kg body weight as a 20% solution. After a 5-hr observation period, these animals were killed, but no pathological changes were found in the hypothalamus examined by light and electron microscopy.

MSG, together with Aspartame, was studied by Heywood and his colleagues (13). When 2 g/kg Aspartame with 1 g/kg MSG was given orally to a 2-day-old rhesus monkey, no lesions were induced in the hypothalamus.

The only consistent symptom following MSG administration to the rhesus monkey is vomiting. Abraham (1) and Newman et al. (18) did not record vomiting in their acute experiments. Olney et al. (25) found that three monkeys tolerated oral doses of 1 and 2 g/kg without vomiting, whereas three other animals vomited. In studies on glutamate metabolism in neonate monkeys, Stegink et al. (36) found that 5 out of 10 neonate macaques vomited small amounts of frothy bile-colored fluid 7 to 55 min after dosing with 1, 2, or 4 g/kg MSG. In studies to establish serum glutamate levels in adult rhesus monkeys, Heywood et al. (8) found that when pairs of monkeys were given 1, 2, or 4 g/kg MSG, one of each pair vomited 65 to 140 min after dosing. The plasma glutamate levels for these animals are shown in Table 7. Plasma sodium levels were also elevated, particularly in those animals given 4 g/kg MSG.

Stegink and his co-workers (36) showed that infant monkeys had a higher basal plasma glutamate level (12 μmoles/dl) than adults (5 to 10 μmoles/dl). Following the administration of doses of 1 to 4 g/kg MSG, rapid rises in plasma glutamate in the range of 17- to 33-fold were recorded. Peak levels were reached 1 to 2 hr after

TABLE 7. *Plasma glutamate concentrations in rhesus monkeys given MSG by oral gavage*

Dose (g/kg)	Plasma glutamate concentration (μg/ml)							
	Minutes after dosing							
	0	15	30	60	120	180	240	300
0	22.0	17.9	29.3	18.6	15.5	12.4	11.6	19.7
1	21.8	22.5	23.0	85.7	71.8	54.9	21.7	18.1
2	18.4	20.9	38.4	137.4	71.4	35.6	52.1	38.5
4	28.5	25.8	64.5	122.0	202.5	93.2	68.0	44.5

administration of MSG, and the extent of the increase was proportional to the dose administered. Two monkeys had abnormally high base-line values (62 and 72 μmoles/dl).

DISCUSSION

Extrapolation of experimental animal data to man is beset with pitfalls; calculations of risk-to-benefit ratios are usually arbitrarily established when, as so often occurs, species sensitivity varies. Animal experiments cannot alone determine all aspects of safety-in-use.

The toxicologist tries, whenever possible, to establish an effect level and also to determine target organs or systems in which he can produce positive findings. The various studies on MSG reported to date show good agreement. The exception is the failure to confirm the hypothalamic lesions reported by one group of workers in the neonate monkey. Any known substance must be capable of toxic action in the widest sense, although it may be necessary to produce extreme or bizarre circumstances to demonstrate this. With MSG, positive findings could be induced in a variety of rodent species with massive doses or by routes or methods of administration other than those involved in normal use.

It appears that a threshold level for the neurotoxicity of MSG when administered in the diet has yet to be established for any species, either in neonate or in mature individuals. The metabolic data indicate that in mature dogs and monkeys there is little, if any, transfer of glutamate across the BBB and that the primate placenta protects the fetus. In the conventional 2-year feeding studies, diets containing 4% MSG for the rat and up to 10% for the dog have not been associated with any clinical or histopathological evidence of CNS damage. No dietary study reported so far suggests that MSG is unsafe for use as a food additive.

REFERENCES

1. Abraham, R., Dougherty, W., Golberg, L., and Coulston, F. (1971): The response of the hypothalamus to high doses of MSG in mice and monkeys. *Exp. Mol. Pathol.,* 15:43–60.
2. Abraham, R., Swart, J., Golberg, L., and Coulston, F. (1975): Electron microscopic observations of hypothalami in neonatal rhesus monkeys after administration of monosodium-L-glutamate. *Exp. Mol. Pathol.,* 23:203–213.
3. Bernt, E., and Bergmeyer, H. U. (1974): UV assay with glutamate dehydrogenase and NAD. In: *Methods of Enzymatic Analysis,* Vol. 4, edited by H. U. Bergmeyer, pp. 1704–1708. Academic Press, New York.
4. Bizzi, A., Veneroni, E., Salmona, M., and Garattini, S. (1977): Kinetics of monosodium glutamate in relation to its neurotoxicity. *Toxicol. Lett.,* 1:123–130.
5. Burde, R. M., Schainker, B., and Kayes, J. (1971): Acute effect of oral and subcutaneous administration of monosodium glutamate on the arcuate nucleus of the hypothalamus in mice and rats. *Nature,* 233:58–60.
6. Ebert, A. G. (1971): Chronic toxicity and teratology studies of L-monosodium glutamate and related compounds. *Toxicol. Appl. Pharmacol.,* 17:274 (Abstract 6).
7. Everly, J. L. (1971): Light microscopic examination of monosodium glutamate-induced lesions in the brain of fetal and neonatal rats. *Anat. Rec.,* 169:312.
8. Heywood, R., James, R. W., and Salmona, M. (1978): *Unpublished data.*

9. Heywood, R., James, R. W., and Worden, A. N. (1977): The *ad libitum* feeding of monosodium glutamate to weanling mice. *Toxicol. Lett.,* 1:151–155.
10. Heywood, R., Osborne, B. E., and Street, A. E. (1973): The effect of repeated cisternal puncture and withdrawal of cerebro-spinal fluid in the dog. *Lab. Anim.,* 7:85–87.
11. Heywood, R., Palmer, A. K., and Prentice, D. E. (1977): Effects of a single dose of monosodium-L-glutamate on guinea pigs at 2–3 days post partum. Huntingdon Research Centre Report No. 44/75943.
12. Heywood, R., and Prentice, D. E. (1975): The effect of oral administration of monosodium glutamate on premature rhesus monkeys. Huntingdon Research Centre Report No. 36/75994.
13. Heywood, R., Prentice, D. E., and Edwards, P. F. (1975): Effect of oral administration of monosodium glutamate with aspartame to neonate rhesus monkeys. Huntingdon Research Centre Report No. 36/75136.
14. James, R. W., Heywood, R., Worden, A. N., Garattini, S., and Salmona, M. (1978): The oral administration of MSG at varying concentrations to male mice. *Toxicol. Lett.,* 1:195–199.
15. Lemkey-Johnston, N., and Reynolds, W. A. (1974): Nature and extent of brain lesions in mice related to ingestion of MSG. *J. Neuropathol. Exp. Neurol.,* 33:74–97.
16. Lucas, D. R., and Newhouse, J. P. (1957): The toxic effects of sodium-L-glutamate on the inner layers of the retina. *Arch. Ophthalmol.,* 58:193–201.
17. Madden, S. C., Woods, R. R., Shull, F. W., Remington, J. H., and Whipple, G. H. (1945): Tolerance to amino acid mixtures and casein digests given intravenously. *J. Exp. Med.,* 81:439–448.
18. Newman, A. J., Heywood, R., Palmer, A. K., Barry, D. H., Edwards, F. P., and Worden, A. N. (1973): The administration of monosodium-L-glutamate to neonatal and pregnant rhesus monkeys. *Toxicology,* 1:197–204.
19. O'hara, Y., Iwata, S., Ichimure, M., and Sasaoka, M. (1977): Effect of administration routes of monosodium glutamate on plasma glutamate levels in infant, weanling and adult mice. *J. Toxicol. Sci.,* 2:281–290.
20. Olney, J. W. (1969): Brain lesions, obesity and other disturbances in mice treated with monosodium glutamate. *Science,* 164:719–721.
21. Olney, J. W. (1971): Glutamate-induced neuronal necrosis in the infant mouse hypothalamus. *J. Neuropathol. Exp. Neurol.,* 30:75–90.
22. Olney, J. W., Ho, O. L., Rhee, V., and De Gubareff, T. (1973): Neurotoxic effects of glutamate. *N. Engl. J. Med.,* 289:1374–1375.
23. Olney, J. W., Rhee, V., and De Gubareff, T. (1977): Neurotoxic effects of glutamate on mouse area postrema. *Brain Res.,* 120:151–157.
24. Olney, J. W., and Sharpe, L. G. (1969): Brain lesions in an infant rhesus monkey treated with MSG. *Science,* 166:386–388.
25. Olney, J. W., Sharpe, L. G., and Feigin, R. D. (1972): Glutamate-induced brain damage in infant primates. *J. Neuropathol. Exp. Neurol.,* 31:464–488.
26. Oser, B. L., Morgareidge, K., and Carson, S. B. (1975): Monosodium glutamate studies in 4 species of neonatal and infant animals. *Food Cosmet. Toxicol.,* 13:7–14.
27. Owen, G., Cherry, C. P., Prentice, D. E., and Worden, A. N. (1978): The feeding of diets containing up to 4% monosodium glutamate to rats for 2 years. *Toxicol. Lett.,* 1:221–226.
28. Owen, G., Cherry, C. P., Prentice, D. E., and Worden, A. N. (1978): The feeding of diets containing up to 10% monosodium glutamate to Beagle dogs for 2 years. *Toxicol. Lett.,* 1:217–219.
29. Palmer, A. K., Thomas, E. A., and Hague, P. H. (1970): The effect of single doses of MSG on rats. Huntingdon Research Centre Report Nos. 3310/70/132 and 3482/70/294.
30. Perez, V. J., and Olney, J. W. (1972): Accumulation of glutamic acid in the arcuate nucleus of the hypothalamus of the infant mouse following subcutaneous administration of monosodium glutamate. *J. Neurochem.,* 19:1777–1782.
31. Pizzi, W. J., Barnhart, J. E., and Farnslow, D. J. (1977): Monosodium glutamate administration to the newborn reduces reproductive ability in female and male mice. *Science,* 196:452–454.
32. Reynolds, W. A., Lemkey-Johnston, N., Filer, L. J., and Pitkin, R. M. (1971): Monosodium glutamate: Absence of hypothalamic lesions after ingestion by newborn primates. *Science,* 172:1342–1344.
33. Semprini, M. E., Conti, L., Ciofi-Luzzatto, A., and Mariani, A. (1974): Effect of oral administration of monosodium glutamate (MSG) on the hypothalamic arcuate region of rat and mouse: A histological assay. *Biomedicine,* 21:398–403.

34. Semprini, M. E., D'Amicis, A., and Mariani, A. (1974): Effect of monosodium glutamate on fetus and newborn mouse. *Nutr. Metab.*, 16:276–284.
35. Stegink, L. D., Pitkin, R. M., Reynolds, A. W., Filer, L. J., Boaz, D. P., and Brummel, M. C. (1975): Placental transfer of glutamate and its metabolites in the primate. *Am. J. Obstet. Gynecol.*, 122:70–78.
36. Stegink, L. D., Reynolds, W. A., Filer, L. J., Pitkin, R. M., Boaz, D. P., and Brummel, M. C. (1975): Monosodium glutamate metabolism in the neonatal monkey. *Am. J. Physiol.*, 299:246–250.
37. Stegink, L. D., Shepherd, J. A., Brummel, M. C., and Murray, L. M. (1974): Toxicity of protein hydrolysate solutions: Correlation of glutamate dose and neuronal necrosis to plasma amino acid, levels in young mice. *Toxicology*, 2:285–299.
38. Tafelski, T. J. (1976): Effects of monosodium glutamate on the neuroendocrine axis of the hamster. *Anat. Rec.*, 184:543.
39. Unna, K., and Howe, E. E. (1945): Toxic effects of glutamic and aspartic acid. *Fed. Proc.*, 4:138.
40. Wen, C.-P., Hayes, K. C., and Gershoff, S. N. (1973): Effects of dietary supplementation of MSG on infant monkeys, weanling rats and suckling mice. *Am. J. Clin. Nutr.*, 26:803–813.

Glutamic Acid: Advances in Biochemistry and Physiology, edited by L. J. Filer, Jr., et al.
Raven Press, New York © 1979.

Morphology of the Fetal Monkey Hypothalamus After *In Utero* Exposure to Monosodium Glutamate

W. Ann Reynolds,* Naomi Lemkey-Johnston,** and Lewis D. Stegink†

*Departments of Anatomy and Obstetrics and Gynecology, University of Illinois at the Medical Center, Chicago, Illinois 60612; **Illinois Institute for Developmental Disabilities, Chicago, Illinois 60612; and †Departments of Pediatrics and Biochemistry, University of Iowa College of Medicine, Iowa City, Iowa 52242*

Exhaustive studies by light and electron microscopy on the hypothalamus of the infant monkey have yielded no evidence of damage following monosodium glutamate (MSG) administration (1,2,12,20,25,28). Some 59 infant primates received oral loads of MSG ranging from 0.25 to 4 mg/g and developed no indication of neuronal damage in the hypothalamus (Table 1). These studies, by four independent laboratories, contrast with those of Olney and colleagues (15), who reported neuronal death in the subinfundibular region of three infant monkeys who received

TABLE 1. *Infant monkeys receiving MSG*

Investigators	Number	Route	Dose (g/kg)	Neuroanatomical findings
No neuropathologic findings				
Abraham et al. (1975)	3	Oral	1–4	Normal
	2	s.c.[a]	4	Normal
	2	Dietary	0.25–1	Normal
Wen et al. (1973)	8	Dietary	5–20%	Normal
Newman et al. (1973 and unpublished)	26	Oral	2–4	Normal
Reynolds et al. (1971 and unpublished)	18	Oral	1–4	Normal
Total	59			
Neuropathologic findings				
Olney et al. (1972)	3	s.c.[a]	2.7–4	Spreading lesions
	3	Oral	1–4	Small focal lesions

[a] Subcutaneous.

TABLE 2. Incidence of arcuate lesions in neonatal mice after MSG ingestion

Dose[a] (mg/g)	Number examined	Number with lesions	Incidence (%)
4	24	24	100
2	23	20	87
1	40	25	62
0.5	40	9	22
0.25	6	0	0

[a] 20% solution for 4, 2, and 1; 5% for 0.5; and 2.5% for 0.25 dose for nonsuckling animals.

oral dosages of MSG ranging from 1 to 4 mg/g. There have been many attempts to reconcile these differences in findings. The preponderance of infant primates used in the studies involving no observed lesions (Table 1) had been born in primate facilities under optimum environmental conditions. The infants utilized by Olney and colleagues were purchased from a supplier. Three of these infants were used for injection studies involving MSG and NaCl, one was premature, and two were apparently dehydrated due to difficulties in nursing. This contrast in the condition of the experimental primate infants, along with the sheer weight of numbers of infants studied in different laboratories, leads to the conclusion that Olney's claims of the susceptibility of the neonatal primate hypothalamus to high levels of MSG cannot be confirmed by others.

It is now well accepted that the rodent brain does incur damage in response to high loads of MSG (1,3,9,14). The neonatal mouse sustains hypothalamic lesions (Table 2) upon ingesting 0.5 mg/g of MSG about 22% of the time (19); with increasing age, higher intakes of MSG are necessary for damage to occur. This finding has caused the speculation that because of the immaturity of the rodent brain at birth, it is exquisitely sensitive to high levels of MSG. In contrast, the neonatal brain of the monkey and human is relatively mature and therefore less susceptible to damage by MSG and, for that matter, other chemicals in high doses, such as sucrose and NaCl, than is the neonatal rodent (8). Whether the maturational state of the brain or whether a basic species difference is responsible for the lack of susceptibility of primate brain to MSG could be resolved in part by administering the substance *in utero* directly to primate fetuses and searching for alterations in their developing brains.

MATERIALS AND METHODS

We have shown that MSG does not cross the hemochorial placenta (24) except following extreme elevation of maternal plasma glutamate. Thus, it was necessary to inject MSG directly into the umbilical vasculature or into the fetus in order to create elevated MSG levels in the fetal circulation. The experiments are summarized

TABLE 3. *Macaque fetuses receiving MSG in utero*[a]

Animal number	Species	Gestational age (days)	Weight (g)	Length[b] (cm)	Dose (mg/g)	Interval (hr)	Comment
GG	M. irus	38	6	CH = 3.0	16.6	6	IP injection
JJ	M. arctoides	66	11.2	CH = 7.0	3.6	4.25	Heart not beating at birth
KK	M. arctoides	85	47	CR = 7.5 CH = 10.5	6.8	2.25	
II	M. arctoides	89	60	CR = 10.0 CH = 14.5	0.4	6	Fetus alive at birth
I	M. mulatta	100	69	CR = 12.0 CH = 15.2	5.8	2	Ecchymosed
EE	M. irus	142	275	CR = 16.0 CH = 24.0	1.8	6.75	Fetus alive at birth
DD	M. arctoides	142	340	CR = 15.3 CH = 24.3	2.4	5.5	IP injection, fetus alive at birth

[a] All MSG administration was by injection into the umbilical vein unless otherwise indicated.
[b] CR = crown to rump; CH = crown to heel.

in Table 3. Six of the pregnant primates were obtained from the University of Illinois Primate Facility Colony; in these instances, the dates of conception are known and the length of gestation can be pinpointed. For the one pregnant animal (I) obtained from a commercial source, the age of the fetus was estimated from its weight and length, using growth tables. Of necessity, the required dose was estimated, hopefully to achieve a dosage of 4 mg/g. However, as Table 3 indicates, the end result is variable as to the actual dose achieved because of the impossibility of accurately predicting weight *in utero*. All but one of the dosages turned out to be above the minimal 1 mg/g level required to obtain a lesion in the rodent with reliability (19). It should be noted that experiments I and JJ were the last two performed, and the dosage achieved in them was closer to the desired level because we had learned better how to predict weight *in utero* from the preceding experiments.

Under Sernylan (Parke-Davis Co.) and halothane anesthesia, abdominal laparotomy was performed. The gravid uterus was then exposed and transilluminated so as to determine the location of the bipartite macaque placenta. An incision was then made into the uterus avoiding the placenta. The umbilical cord was carefully located and a short segment of it exteriorized, taking care not to interfere with umbilical blood flow. The estimated dose, in a 20% solution (w/v), was then injected toward the fetus into the umbilical vein and the needle hole was sutured. For the smallest fetus, it was necessary to inject intraperitoneally, since the umbilical cord was too small to exteriorize.

We then chose to perform an intraperitoneal injection of MSG for fetus DD so that circulating levels of glutamate could be compared following this route of administration with those after injection into the umbilical vein. The uterus was replaced into the abdomen and the incision closed temporarily with clamps. At the end of the experimental interval, the fetus was delivered by cesarean section.

In two experiments (EE and DD), a tracer dose of 10 μCi ^{14}C-MSG was added to the loading dose of MSG. At the termination of these two experiments, 6.75 and 5.5 hr after dosing with MSG, respectively, fetal plasma, urine, CSF, and amniotic fluid samples were obtained for amino acid analysis. Maternal plasma samples were obtained at intervals throughout the experiment from an indwelling catheter placed in the saphenous vein. These samples were subjected to simultaneous measurement of radioactivity and amino acid composition (23) in order to see if glutamate moved from fetal to maternal circulation. All fluid samples were deproteinized immediately with sulfosalicyclic acid and stored at $-40°$ C. Amino acid analyses were performed on a Technicon NC-1 analyzer using the Efron buffer system (23). Control values were obtained from earlier studies of fetal (24) and neonatal (25) glutamate metabolism in the monkey or from samples obtained from normal macaque fetuses delivered for other research projects.

The fetal head was removed and the cranium carefully dissected away with Ronjour forceps. Each brain was immersed immediately in 10% buffered formalin. After fixation, the brains were dehydrated in an alcohol series, placed in 1% celloidin in methyl benzoate for 2 days, then processed in benzene to paraffin. Serial sections of the entire brain from the smaller animals and of the hypothalamic regions of the larger animals (over 200 g) were cut at 10 μm and then stained with cresyl echt violet solution.

RESULTS

The only fetus that exhibited any abnormality upon delivery was I, which was ecchymosed, although otherwise normal in appearance. It was derived from a pregnant macaque that had been purchased and arrived in Chicago from New York the day before the experiment. *In utero* deaths in pregnant macaques subjected to the stresses of shipping are common, and it is possible that this fetus was traumatized before the experiment. Fetuses II, EE, and DD were vigorous upon delivery and exhibited spontaneous movement, gasping, respiratory activity, and normal heart rates. Fetuses GG, JJ, and KK were too small to exhibit these signs of vigor; GG and KK possessed beating hearts, whereas the heart of JJ was not beating and was refractile to stimulation.

Maternal plasma glutamate levels increased slightly at 30 min and 1 hr (Exp. DD, 6.5 to 11.1 μmoles/dl; Exp. EE, 2.8 to 24.9 μmoles/dl) and in both instances had returned to base-line levels by 2.5 hr. The radioactivity in maternal plasma was primarily incorporated in glucose and lactate (Fig. 1). Small amounts of ^{14}C-labeled glutamate were observed at 15 min in Exp. EE, but had disappeared by 2 hr; this phenomenon was even more transient and less extensive in Exp. DD. Significant

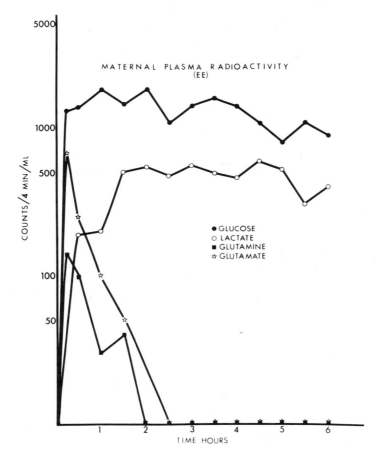

FIG. 1. Most of the counts/min in maternal plasma were found in glucose and lactate. A small, transient increase in labeled glutamate occurred at 15 min and had disappeared by 2 hr. Only when circulating fetal levels of glutamate are exceedingly high does the substance breach the placental barrier.

levels of glutamate radioactivity were still present in cord blood at the end of the experimental period (Exp. DD, 800 counts/4 min; Exp. EE, 900 counts/4 min).

Negligible amounts of radioactivity were associated with glutamate in CSF in both experiments. Similarly, total glutamate levels were not elevated significantly above control values (1.8 ± 1.1 μmoles/dl) in the CSF of fetuses DD and EE).

That both fetuses sustained extensive elevations in circulating glutamate levels is corroborated by the plasma amino acid values encountered at the end of the experiments, 5.5 and 6.5 hr after administration of the dose (Table 4). Glutamate levels were still two to three times those of control animals. Amniotic fluid

TABLE 4. Amino acid levels in fetal plasma and in amniotic fluid

	Aspartate	Glutamine	Glutamate	Proline	Alanine
Plasma					
Control fetus (5)	0.5 ± 0.5	57 ± 18	6 ± 3	18 ± 7	33 ± 10
DD[a]	1.56	63	20.2	24.1	29.1
EE[b]	0.59	148	12.9	11.1	54
Amniotic fluid					
Control pregnancy (5)	0.2 ± 0.2	18.8 ± 4.7	2.4 ± 1.0	9.1 ± 2.0	11.6 ± 2.4
DD[a]	0.9	14.8	22.9	8.8	20.1
EE[b]	0.5	14.8	32.8	8.8	19.0

All values in μmoles/dl.
[a] 5.5 hr after i.p. injection.
[b] 6.8 hr after i.v. injection.

glutamate levels were also markedly elevated (Table 4), probably as a marker of a prior extensive rise in fetal blood levels, clearance to urine, and then micturition into the amniotic fluid. Fetal urine correspondingly contained enormous concentrations of glutamate (Table 5). Most of the radioactivity in fetal urine was associated with glutamate (Table 6).

All brain sections were studied by light microscopy, by which it is possible to

TABLE 5. Amino acid in fetal urine

Amino acid	Control fetus (3)	DD	EE
Glutamate	6.8 ± 4.1	2,650	3,550
Aspartate	2 ± 1	44	35
Glutamine	60 ± 25	62	71
Glycine	75 ± 30	62	70
Histidine	15 ± 9	13	14
Threonine	29 ± 15	35	27

All values in μmoles/dl.

TABLE 6. Percent distribution of total radioactivity in fetal urine

Amino acid	DD	EE
Glutamate	86	91.7
α-Ketoglutarate	4.2	1.3
Aspartate	3.8	2.8
Glucose	2.8	2.0
Lactate	1.0	8.0
Alanine	0.9	0.7
Urea	0.5	0.3
Acetoacetate	0.3	0.05

observe lesions resulting from the ingestion of high levels of MSG in neonatal rodents (9,13).

Ventral hypothalamic morphology was embryonic in fetus GG and quite immature in fetuses JJ, KK, II, and I. The hypothalamic nuclei of GG (Fig. 2) were not yet differentiated. The hypothalamus at this stage is still an evagination of the thalamus, and the optic primordium was still visible in the diencephalon (Fig. 2). The brain of JJ was still very immature, but the median eminence, anterior pituitary, and infundibular nucleus area could be identified. Although fetuses KK, II, and I were more mature (Fig. 3), the median eminence lacked the differentiation it undergoes in later development, probably under the influence of thyroid hormone. As a result of the rapid rate of cell division in the ependymal layer, mantle and marginal layers were packed with cells. However, the cells were still quite undiffer-

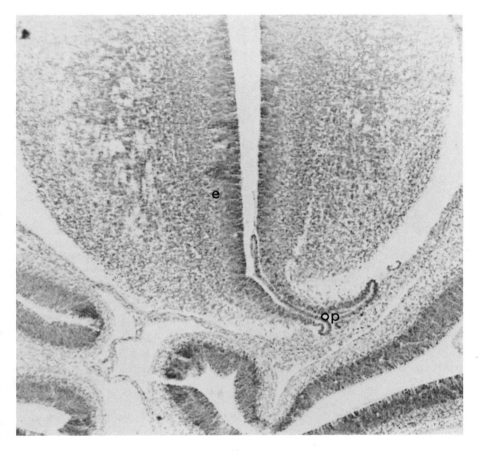

FIG. 2. Section through ventral diencephalic area of fetus GG (38 days gestation). Note the optic primordium *(op)* evaginating from the third ventricle. Cell division in the ependymal *(e)* layers is still occurring at a rapid rate. ×208.

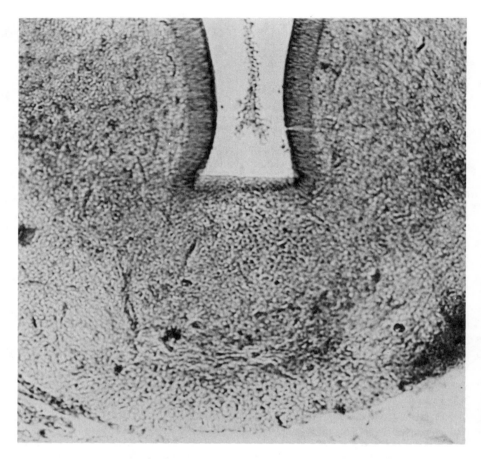

FIG. 3. Section through ventral hypothalamus of fetus KK (85 days). This brain is still immature, but the median eminence is now distinct at the base of the third ventricle. Numerous, undifferentiated cells populate this region. ×524.

entiated, with little cytoplasm, and it was impossible to discern future neurons from glia at this stage (Fig. 4). The cells within the hypothalamic areas of these four fetuses were normal in appearance. There was no evidence of pyknotic nuclei, tissue edema, or cell loss in the subinfundibular region, undifferentiated though it was.

The brains of fetuses DD and EE were more differentiated than those of the younger ones. However, they did not exhibit the degree of myelination, neuronal differentiation, and maturation of the median eminence seen in the neonatal monkey brain (1,2,12,15,19,20). Thus, considerable hypothalamic differentiation was yet to occur in the remaining three to four weeks of gestation. Gestation in the rhesus monkey is 168 ± 7 days (27).

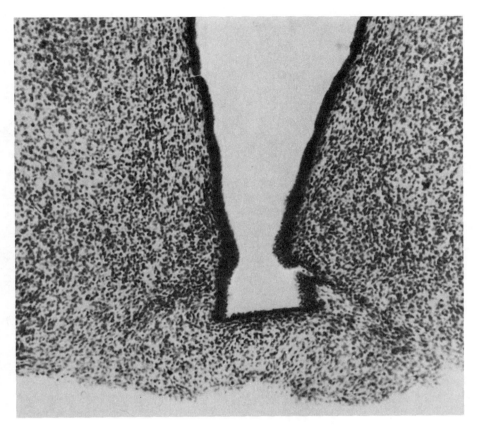

FIG. 4. Section through ventral hypothalamus of fetus II (89 days gestation). The median eminence is not present in its entirety because of mechanical damage. Note the dense population of immature cells, whose differentiative fates cannot yet be determined. ×416.

No neuronal abnormalities were seen in the ventral hypothalamic area of fetus DD or EE (Fig. 5, p. 226). Excessive numbers of cells whose accruing cytoplasm suggests they have a future as neurons are present in this area (Fig. 5). No evidence of cell death or vacuolated cytoplasm, which are characteristics of MSG damage in the neonatal mouse, could be found.

DISCUSSION

A few other studies have involved the fetal effects of MSG. Murakami and Inouye (11) observed hypothalamic damage in the neutromedial and arcuate nuclei of mouse fetuses whose dams received 5 mg/g injections of MSG late in gestation. This is an enormous, acute load, and probably wreaked havoc with the placental barrier. In contrast, when Takasaki (26) fed mice with diets containing up to 15%

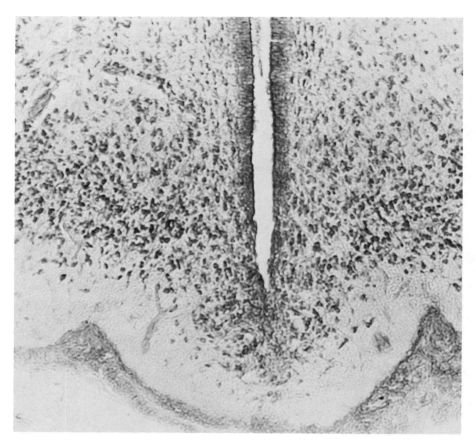

FIG. 5. Section through ventral hypothalamus of fetus DD (112 days gestation). Infundibular area contains cells whose increasing cytoplasm indicates that they are neurons in the process of becoming arranged in the infundibular nucleus. ×524.

MSG or with 5% aqueous solutions of MSG *ad libitum* for up to 4 days during pregnancy, no necrosis of neurons in the fetal brains occurred. Olney (16) double-injected a very large dose of glutamate into a single pregnant monkey and reported a lesion in the fetus.

In contrast Newman et al. (12) administered MSG at a daily level of 4 mg/g in drinking water to six pregnant monkeys during the last third of gestation; the hypothalamic areas of the newborns of these animals exhibited no abnormalities. These latter findings are in line with the fact that glutamate, even injected in large loads, does not cross the neurochorial placenta of the primate (23).

Either intravenous or intraperitoneal administration served to elevate glutamate levels markedly in the fetuses. The early backcrossing of a small amount of glutamate into the maternal circulation (Fig. 1) was similar to that seen previously when 5 mg/g MSG was infused into a fetus, elevating maternal glutamate levels slightly (24). In this instance, fetal levels rose to 400 times base line or 2,000 μmoles/dl. The administration of lesser amounts of glutamate (1.5 or 2.4 mg/g) to the fetus was insufficient to breach the fetomaternal placental barrier. Thus, since glutamate did backcross slightly into the maternal circulation, it would appear that all of these fetuses experienced major elevations in circulating glutamate levels following the injections. These levels would far surpass those achieved by oral administration, where the time required for intestinal absorption attenuates and lowers the peak blood levels of glutamate.

Another phenomenon may have contributed markedly to constantly elevated fetal blood levels of MSG, at least in the more mature fetuses. The monkey fetus forms urine at the rate of about 5 mg/kg/hr (4). In both amniotic fluid (Table 4) and fetal urine (Table 5) the glutamate levels were exceedingly high at the termination of the experiments. The fetuses old enough to drink were probably "recycling" glutamate contained in amniotic fluid at a rate roughly equivalent to 3 ml/kg/hr, extrapolating from the human rate (18).

Even though the fetuses DD and EE were 4 weeks younger than the newborns studied previously (25), their blood-brain barrier impermeability to glutamate was already effective. Neither animal had any increase in CSF glutamate or glutamate labeling with radioisotope. The failure of glutamate to enter CSF in even the fetal monkey may be a critical component of a species difference between the rodent and the primate. We have hypothesized that glutamate may gain access to certain brain regions via CSF, since damage appears to be restricted to areas adjacent to CSF in the mouse (9,19). Many studies have confirmed that there is little if any net transfer of glutamate into CSF in several adult mammals, including mouse (21), rat (10), and dog (7). Thus, lack of a totally effective blood-brain barrier integrity in the young rodent could be a major reason for the susceptibility of its brain to damage by glutamate.

The one embryonic and seven fetal brains studied all exhibited striking cell differentiation and turnover in comparison to that seen in the neonatal primate brain (1,2,12,15,20,28). During the embryonic and early fetal period, the extensive rate of cell division, primarily in the ependymal region, results in a dense cellular population of the neuropil. A minority of this cell population will differentiate into glial and neuronal elements. Most of these cells, however, are destined to die, probably because they fail to make functional connections to the periphery or to adjacent cells (6). Submammalian species have been the most closely studied in this respect; in the frog and chick, less than 50% of the cells persist in given areas after differentiation has occurred (17).

The damage incurred by the neonatal monkey brain following ingestion of MSG is described by Olney (15) as numbering but 2 to 3 neurons per section and totaling 50 to 90 in the rostral subventricular portions of the infundibular nucleus. The lateral portions of the hypothalamus, primarily damaged in rodents (1,14) were not affected in the primate. Paraffin preparations, suitable for discerning lesions in the mouse brain (9,13), would not possess the resolution necessary to discern only 2 to 3 moribund cells per section; plastic sections are required for the study of individual neurons. However, since cell death in the CNS is a normal part of the differentiative process, an entirely different approach would be required to resolve the possibility of "microlesions" involving but a few cells per section. Such analyses require morphometric techniques for painstaking analysis of the actual numbers of viable and nonviable cells present in a given area. It is certainly possible that the CNS experiences microlesions in response to anoxia, drugs, alterations in pH, and other stresses. At present, the search for microlesions awaits the modification of the appropriate statistical and morphometric techniques. It will be a meticulous and difficult task for the primate brain because of the expense and the variability between animals. Further, because of the vasculature and water content of the brain, only fetuses weighing more than 200 g can be perfused adequately so as to prepare epon sections for study.

Even following the extraordinary measure of injecting MSG into the fetal monkey, no damage to the hypothalamus was observed. These seven fetuses, added to 59 neonatal monkeys (Table 1) represent a major search for susceptibility of the primate brain to MSG. Even in the embryonic and fetal period, no hypothalamic damage was found.

For studies of substances involved in human diets, the nonhuman primate is the model of choice. Phocomelia cannot be induced in the fetal period in the rodent by thalidomide; it is readily induced in the fetal primate following maternal administration (5). Insulin is highly teratogenic in the rodent but not in the human (29). The primate brain is not affected by even large dosages of MSG. Differences in glutamate metabolism between the rodent and primate (22), greater integrity of the blood-brain barrier, or specific neuronal susceptibility—all or in concert—could be responsible. Perhaps it is now more appropriate to regard damage to the rodent brain from the ingestion of MSG as an unusual and interesting phenomena, restricted to the order Rodentia.

REFERENCES

1. Abraham, R., Dougherty, W., Goldberg, L., and Coulston, F. (1971): The response of the hypothalamus to high doses of monosodium glutamate in mice and monkeys. Cytochemistry and ultrastructural study of lysosomal changes. *Exp. Mol. Pathol.*, 15:43–60.
2. Abraham, R., Swart, J., Goldberg, L., and Coulston, F. (1975): Electron microscopic observations of hypothalami in neonatal rhesus monkeys (*Macaca mulatta*) after administration of monosodium-L-glutamate. *Exp. Mol. Pathol.*, 23:203–213.
3. Arees, E., and Mayer, J. (1970): Monosodium glutamate-induced brain lesions: Electron microscopic examination. *Science*, 170:549–550.
4. Chez, R. A., Smith, F. G., and Hutchinson, D. L. (1964): Renal function in the intrauterine primate fetus. *Am. J. Obstet. Gynecol.*, 90:128–131.
5. Delahunt, C. S., and Lassen, L. J. (1964): Thalidomide syndrome in monkeys. *Science*, 146:1300–1305.
6. Hamburger, V. (1934): The effects of wing bud extirpation on the development of the central nervous system in chick embryos. *J. Exp. Zool.*, 68:449–494.
7. Kamin, H., and Handler, P. (1951): The metabolism of parenterally administered amino acids. II. Urea synthesis. *J. Biol. Chem.*, 188:193–205.
8. Lemkey-Johnston, N., Butler, V., and Reynolds, W. A. (1975): Brain damage in neonatal mice

following monosodium glutamate administration: Possible involvement of hypernatremia and hyperosmolarity. *Exp. Neurol.,* 48:292–309.
9. Lemkey-Johnston, N., and Reynolds, W. A. (1974): Nature and extent of brain lesions in mice related to ingestion of monosodium glutamate. A light and electron microscope study. *J. Neuropathol. Exp. Neurol.,* 33:74–97.
10. McLaughlan, J. M., Noel, F. J., Botting, H. G., and Knippel, J. E. (1970): Blood and brain levels of glutamic acid in young rats given monosodium glutamate. *Nutr. Rep. Intern.,* 1:131–138.
11. Murakami, U., and Inouye, M. (1971): Brain lesions in the mouse fetus caused by maternal administration of monosodium glutamate (preliminary report). *Congenital Anomalies,* 11:171–77.
12. Newman, A. J., Heywood, R., Palmer, A. K., Barry, D. H., Edwards, F. P., and Worden, A. N. (1973): The administration of monosodium L-glutamate to neonatal and pregnant rhesus monkeys. *Toxicology,* 1:197–204.
13. Olney, J. W. (1969): Brain lesions, obesity and other disturbances in mice treated with monosodium glutamate. *Science,* 164:719–21.
14. Olney, J. W. (1971): Glutamate-induced neuronal necrosis in the infant mouse hypothalamus. *J. Neuropathol. Exp. Neurol.,* 30:75–90.
15. Olney, J. W., Sharpe, L. G., and Feigen, R. D. (1972): Glutamate-induced brain damage in infant primates. *J. Neuropathol. Exp. Neurol.,* 31:464–488.
16. Olney, J. W. (1974): Toxic effects of glutamate and related amino acids on the developing central nervous system. *Heritable Disorders of Amino Acid Metabolism,* edited by William L. Nyhan. John A. May & Sons, Inc., New York.
17. Oppenheim, R. W. (1975): Progress and challenges in neuroembryology. *BioScience,* 25:28–36.
18. Pritchard, J. A. (1965): Deglutition by normal and anencephalic fetuses. *Obstet. Gynecol.,* 25:289–297.
19. Reynolds, W. A., Butler, V., and Lemkey-Johnston, N. (1976): Hypothalamic morphology following ingestion of aspartame or MSG in the neonatal rodent and primate: A preliminary report. *J. Toxicol. Environ. Health,* 2:471–480.
20. Reynolds, W. A., Lemkey-Johnston, N., Filer, L. J., Jr., and Pitkin, R. M. (1971): Monosodium glutamate: Absence of hypothalamic lesions after ingestion by newborn primates. *Science,* 172:1342–1344.
21. Schwerin, P., Bessman, S. P., and Waelsch, H. (1950): The uptake of glutamic acid and glutamine by brain and other tissues of the rat and mouse. *J. Biol. Chem.,* 184:37–44.
22. Stegink, L. D., Reynolds, W. A., Filer, L. J., Jr., Baker G. L., Daabees, T. T., and Pitkin, R.M. (1979): Comparative metabolism of glutamate in the mouse, monkey, and man (*this volume*).
23. Stegink, L. D. (1970): Simultaneous measurement of radioactivity and amino acid composition of physiological fluids during amino acid toxicity studies. *Advances in Automated Analysis, Vol. I, Chemical Analysis,* pp. 591–594. Thurman Associates, Miami.
24. Stegink, L. D., Pitkin, R. M., Reynolds, W. A., Filer, L. J., Jr., Boaz, D. P., and Brummel, M. C. (1975): Placental transfer of glutamate and its metabolites in the primate. *Am. J. Obstet. Gynecol.,* 122:70–78.
25. Stegink, L. D., Reynolds, W. A., Filer, L. J., Jr., Pitkin, R. M., Boas, D. P., and Brummel, M. C. (1975): Monosodium glutamate metabolism in the neonatal monkey. *Am. J. Physiol.,* 229:246–50.
26. Takasaki, Y. (1978): Studies on brain lesions by administration of monosodium L-glutamate to mice. II. Absence of brain damage following administration of monosodium L-glutamate in the diet. *Toxicology* (*in press*).
27. Van Wagenen, G., and Catchpole, H. R. (1965): Growth of the fetus and placenta of the monkey (*Macaca mulatta*). *Am. J. Phys. Anthropol.,* 23:23–33.
28. Wen, C., Hayes, K. C., and Gershoff, S. N. (1973): Effects of dietary supplementation of monosodium glutamate on infant monkeys, weanling rats, and suckling mice. *Am. J. Clin. Nutr.,* 26:803–13.
29. Wilson, J. G. (1973): *Environment and Birth Defects.* Academic Press, New York.

Glutamic Acid: Advances in Biochemistry and Physiology, edited by L. J. Filer, Jr., et al. Raven Press, New York © 1979.

In Utero and Dietary Administration of Monosodium L-Glutamate to Mice: Reproductive Performance and Development in a Multigeneration Study

K. Anantharaman

Experimental Biology Laboratory, Nestlé Products Technical Assistance Co. Ltd., CH-1350 Orbe, Switzerland

Safety evaluation of a food additive or ingredient must necessarily involve a progression from the acquisition of a thorough knowledge of its nonbiological aspects to a basic framework of appropriate biological tests that permits the judicious interpretation of the data. In the context of short- and long-term exposures of the test system, pharmacokinetics, biotransformation, special tests such as those for reproductive function, teratology, and neurotoxicity, and other miscellaneous tests to establish synergism or antagonism, all fall in the category of appropriate biological or toxicological procedures. Certain basic considerations and assumptions underlying these test procedures may be emphasized:

1. All substances can elicit a response in an appropriate biological test system.
2. Such a response is a function of the dose administered and the duration of the exposure.
3. The response may be modified by other factors such as sex, age, nutritional and health status, diet and strain of the animal, and interaction of the test material with other substances to which the animal may be exposed before, during, or after administration.
4. The appropriateness of the test procedures employed is critical. Tests should fit the material. The same experimental design may not be suitable for all, or any other, test substance.

Where a subchronic or prolonged toxicity test is planned, it therefore needs to be recalled that it is truly "a test of (the) measurable harmful effect of the substance on the biological system, occurring as a consequence of administering repeated doses, usually by dosing the system (animal) on a daily basis for 90–120 days, or longer."

Safety-in-Use of Monosodium L-Glutamate (MSG)

The absolute safety of any substance can only be questionably proved to one's own satisfaction. One may therefore appreciate that the literature on the safety-in-use of the ubiquitous glutamates as food-flavoring ingredients, although abundant and clear, continues to cause concern. Why? A brief recapitulation of certain pertinent though contrasting published findings in this context can be useful.

The repeated subcutaneous administration of MSG to newborn mice resulted in severe damage to the inner layers of the retina (5,12). Olney (16), under the same conditions of administration, discovered discrete brain lesions, mainly in the preoptic and arcuate nuclei of the hypothalamus, together with scattered neurons within the median eminence. Olney and Ho (17) further found that the arcuate nuclei were damaged when mice that were 10 to 12 days old were given oral doses of 3 mg MSG/g body weight. Equivalent amounts of NaCl had no effect. The early brain damage in MSG-treated rodents is considered to account, at least in part, for many of the endocrine disturbances observed in later life, such as adult obesity without accompanying hyperphagia (2,3,13,15,16), skeletal stunting (2,16), and reproductive dysfunction and sterility in both sexes (16,22). Djazayery and Miller (6) and Djazayery et al. (7) injected 5 mg MSG/g body weight intraperitoneally to weanling female mice, but with only a moderate success in inducing obesity.

Although administration of MSG to newborn rodents by either s.c. or i.p. injection almost always resulted in brain lesions, by and large, the dietary administration of MSG at even very high doses was not found to result in any of these symptoms, including the endocrine disturbances (3,10,20,26,27,30). Ebert (8) reported that in chronic toxicity trials with rats and mice fed over a 2-year period with two levels of MSG incorporated in diet, there were no abnormalities in body weight gain, food intake, hematology, or histopathology. Owen et al. (20) fed weanling rats diets containing added MSG at 1, 2, or 4% w/w for 2 years and found no adverse effects on body weight gain, economy of food consumption, hematology, blood chemistry, organ weights, or mortality by comparison with control rats receiving the basal diet. Besides focal mineralization at the renal corticomedullary junction occurring with equal frequency in all groups, including the controls, they observed no other histological changes of any significance. It is our experience that the renal change itself is probably due to a mineral (Ca/P) imbalance in the diets.

Different routes of administration of MSG have, therefore, much different effects on even the same test animal species. This is an important point.

Route of Administration

Since MSG is a food-flavoring ingredient, what would be the ideal approach of assessing its safety-in-use, or even potential toxicity in the experimental animal?

The most uncertain aspect of safety evaluation, whatever the primary concept underlying the experimentation—be it nutritional, pharmacological, or physiological—is the relevance of animal data to humans. Indeed, in the case of MSG it would be reasonable to argue that the oral route of administration, especially when admixed with the diet of the animal, represents the only true logical approach to an investigation of its long-term safety, or its cumulative effect on the animal. The present investigations have centered on the assessment of the effects of "*in utero* and preweaning exposures and postweaning dietary administration of MSG to mice on their growth, food and MSG intake, reproductive performance, and brain morphology."

Principal Questions

Why mice? The species differences noticed by a number of workers may be broadly attributed to differences in (a) the effect of the animal on the substance and (b) the effect of the substance on the animal. For the dietary evaluation of the long-term, or subchronic toxicity of MSG, it might be agreed that the mouse appears to be the most sensitive species. The trials involved three generations of mice and attempted to answer such principal questions concerning the safety of MSG as whether it would bring about any of the following:

Reduced reproductive capacity.
Malformation or stunted development of the newborn.
Brain lesions, early or later in life.
Abnormal or excessive growth and food intake during postnatal life.
Adult obesity, with or without accompanying hyperphagia.
Reduced fertility.
Any other pathological change.

DESIGN OF THE MULTIGENERATION STUDY

The organization of the multigeneration trials is set out schematically in Table 1, and the dietary treatments, starting with those for the animals of the F_0 generation, in Table 2.

Among the variety of possible toxic effects that can interfere with the functions of organs and tissues are those occurring during reproduction, i.e., fertility, parturition, and lactation. It is feasible to study these aspects during a period shorter than the animal's lifetime. Information on reproductive performance is essential to an evaluation of a potential hazard, since this complex physiological state is highly susceptible to specific deleterious effects. The fact that previous experimental evidence has shown the increased susceptibility of the newborn, as well as of the embryo and of the fetus to MSG compared to the older animal, made it desirable to investigate the influence of both the *in utero* and dietary administration of this substance on mouse reproduction. The composition of the basal diet is set out in Table 3. MSG was incorporated into this basal diet at two levels: 1 and 4% w/w.

Mice have an acute sense of hearing, and the audiogenic seizures that are provoked by intense noises in certain strains lead to interference with breeding performance (31). To ensure freedom from all stress, including that due to transportation, the F_0 generation of the SPF-derived, CD-1, COBS strain of mice on receipt from Charles-River Farms, France, postweaning, were maintained behind a barrier, as were the future generations. Animals for the feeding trials were housed singly, with

TABLE 1. *Organization of multigeneration protocols for the safety evaluation of MSG in mice*

Generation	Comments
F_0	
(Mating at 12–13 weeks of age)	
$F_{1.1}$	Feeding study until 36 weeks of age. Histopathology of brain and some other organs.
(Mating at 13–14 weeks of age)	
$F_{2.1}$	Feeding study until 27 weeks of age.
(Mating at 16 weeks of age)	
$F_{3.1.1}$	Histopathology of brain at birth, 3, 14, and 21 days.
(Mating at 20–21 weeks of age)	
$F_{2.2}$	Feeding study until 32 weeks of age.
(Mating at 32 weeks of age)	
$F_{3.2.1}$	Until weaning at 21 days.

all receiving food and water supply freely. The mating condition was three females to one male.

Besides the free-feeding trials with the $F_{1.1}$, $F_{2.1}$, and $F_{2.2}$ generations, reproduction studies in the $F_{1.1}$, $F_{2.1}$, $F_{2.2}$, $F_{3.1.1}$, and $F_{3.2.1}$ generations, and food and MSG intake in a voluntary manner, histopathological evaluation of brain tissue was carried out for animals of the $F_{1.1}$ generation and in the newborn 3-, 14-, and 21-day-old mice of both sexes of the $F_{3.1.1}$ generation (Table 4).

TABLE 2. Dietary treatment: schematic representation

Generation	Experimental groups							
	A	B	C	D	E	G	H	
F_0								
On arrival at 3–4 weeks of age until mating at 12 weeks	Basal, reference diet without added MSG							
During gestation, 12–15 weeks of age	Basal	1% MSG	4% MSG	1% MSG	4% MSG	1% MSG	4% MSG	
During lactation, for 3 weeks	Basal	1% MSG	4% MSG	Basal	Basal	1% MSG	4% MSG	
F_1								
Postweaning, to 36 weeks of age	Basal	1% MSG	4% MSG	Basal	Basal	Basal	Basal	
During gestation and lactation	Basal	1% MSG	4% MSG	—	—	—	—	
F_2								
Postweaning, to 27 weeks—$F_{2.1}$ to 32 weeks—$F_{2.2}$	Basal	1% MSG	4% MSG	—	—	—	—	
During gestation and lactation	Basal	1% MSG	4% MSG	—	—	—	—	
F_3								
To dams until pups were weaned	Basal	1% MSG	4% MSG	—	—	—	—	

TABLE 3. Composition of basal diet

Gross energy (on dry matter):	4.63	kcal/g
Protein:[a]	230	g/kg
Total lipids:	50	g/kg
Nonnutritive cellulose:	40	g/kg
Total carbohydrates:	500	g/kg
Composite vitamin-mineral mixture:	60	g/kg
Moisture (maximum):	120	g/kg

Mineral components (mg/kg):		Vitamins:		
Phosphorus (P):	7,800	Vitamin A:	16,800	IU/kg
Calcium (Ca):	8,400	Vitamin D_3:	4,000	IU/kg
Potassium (K):	7,500	Thiamine:	8	mg/kg
Sodium (Na):	3,400	Riboflavin:	13	mg/kg
Magnesium (Mg):	1,700	Pantothenic acid:	27	mg/kg
Manganese (Mn):	67	Pyridoxine:	4	mg/kg
Iron (Fe):	280	Niacin:	88	mg/kg
Copper (Cu):	30	Menadione:	6	mg/kg
Zinc (Zn):	64	Vitamin E:	47	mg/kg
Cobalt (Co):	2	Folic acid:	1	mg/kg
Iodide from		Biotin:	0.1	mg/kg
marine algae		Vitamin B_{12}	0.04	mg/kg
		Choline:	2,100	mg/kg

Note: MSG was added at 1 or 4% w/w to this basal diet, ensuring that all diets were finally isocaloric and isonitrogenous.

[a] Protein (% N × 6.25) as defatted soybean meal, food yeast, fish meal, and milk whey solids.

TABLE 4. Numbers of mice used in trials—all generations

	Control		1% MSG		4% MSG	
Generation	Male	Female	Male	Female	Male	Female
F_0	33	99	17	51	17	51
$F_{1.1}$	370	357	123	133	136	116
$F_{2.1}$	229	219	84	93	91	85
$F_{2.2}$	122	114	59	67	66	63
$F_{3.1.1}$	110	107	58	59	53	57
$F_{3.2.1}$	35	31	27	31	38	27
Total/sex	899	927	368	434	401	399
Total	1,826		802		800	

Note: Numbers of mice in treatment groups D, E, G, and H in the F_0 and $F_{1.1}$ generations are not shown here.

GROWTH AND BODY WEIGHT DISTRIBUTION FREQUENCY

The growth data set out in Figs. 1, 2, and 3 for both sexes of the $F_{1.1}$, $F_{2.1}$, and $F_{2.2}$ generations, respectively, do not need further elaboration. In all the trials, growth curves for the MSG-treatment groups were similar to those for the controls. There were no abnormal developments or abnormal rates of growth in either sex,

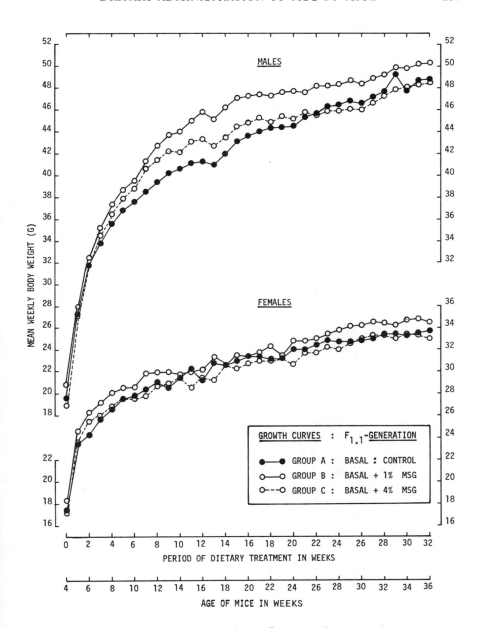

FIG. 1. Growth curves for the $F_{1.1}$ generation.

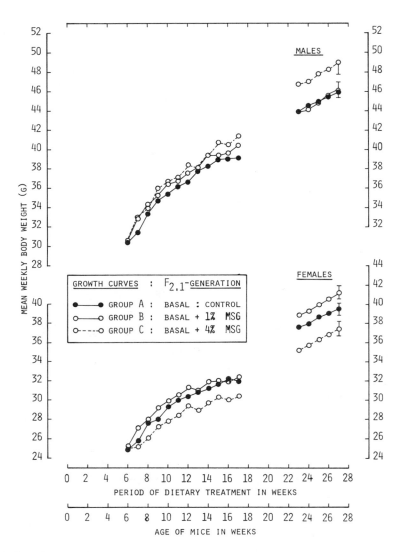

FIG. 2. Growth curves for the $F_{2.1}$ generation. Animals were not weighed between weeks 18 and 23. Terminal values indicated ± SEM.

nor in any of the generations or litters. None of the animals were found to feel fatty or predisposed to obesity.

Females of both the MSG and the control groups in the $F_{2.1}$ and $F_{2.2}$ generations were slightly heavier than those of the parent generation, although the differences in body weight between the treatment groups of the same generation were not statisti-

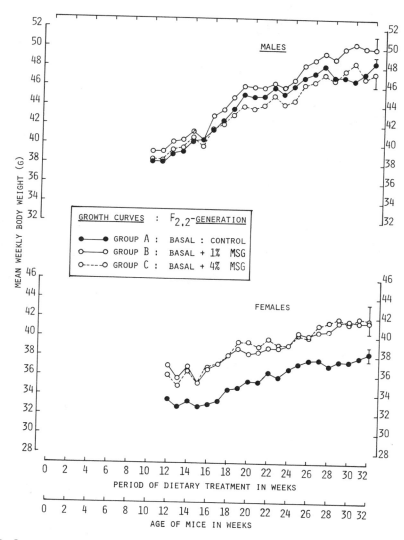

FIG. 3. Growth curves for the $F_{2.2}$ generation. Animals were weighed beginning week 12. Terminal values indicated ± SEM.

cally significant. This is explained as due to the small number of heavier animals in these subsequent generations at the start of trials, chiefly because they were derived from small-sized litters.

As descriptions of the frequency distribution of a series of observations, the most important values are usually the mean and standard deviation. With a normal distribution, only 1 in 20 observations will differ from the mean by more than twice

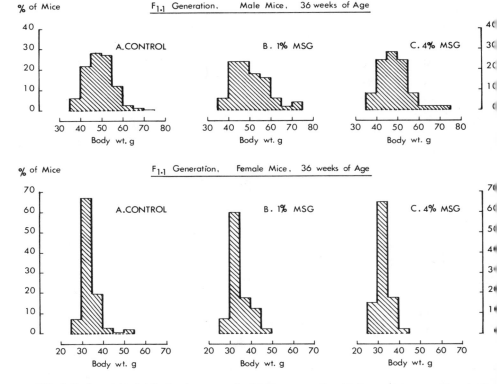

FIG. 4. Body weight distribution frequency for the $F_{1.1}$ generation. Histograms represent terminal values at age 36 weeks.

the standard deviation ($\pm$), and only some 3 in 1,000 will differ from the mean by more than three times the standard deviation ($\pm$).

Body weight distribution frequencies of both sexes of the $F_{1.1}$, $F_{2.1}$, and $F_{2.2}$ generations are represented as histograms in Figs. 4, 5, and 6, respectively. The outliers in each case were again traced to small-sized litters. Otherwise, the histograms permit an eye-fit evaluation of the similarity in the frequency distribution between the controls and the MSG groups.

FOOD INTAKE

Weekly food intake, measured throughout the 32 weeks of trial in the $F_{1.1}$ generation (Fig. 7) showed that, by and large, mice of both sexes of all three groups ingested similarly. Regular fluctuations in food intake occurred in both sexes, although a small but definite increase in food intake in females of all treatment groups was also registered with the progress of the trial, visibly so until the mice

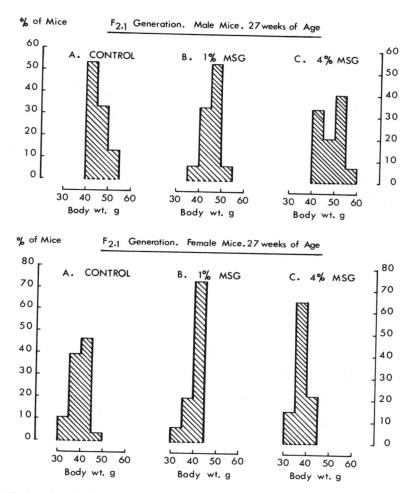

FIG. 5. Body weight distribution frequency for the $F_{2.1}$ generation. Histograms represent terminal values at age 27 weeks.

were 26 weeks of age. A similar pattern of fluctuation was also noted for females of the $F_{2.1}$ generation (Fig. 8), although there was no hyperphagia, nor any significant differences between control and the MSG-treated groups. In this generation, however, food intake was not measured throughout the duration of the trial.

How then explain the progressive, though small, increase with age, as well as the occasional, but definite, decrease in food intake of female mice? A number of investigators have observed that the food intake and energy expenditure of the female cycling rodent varies with the stage of the ovulatory cycle, which is itself distinguished by a waxing and waning of the plasma estradiol concentrations, with the maximum levels reached at proestrus and estrus, and the nadir at diestrus and

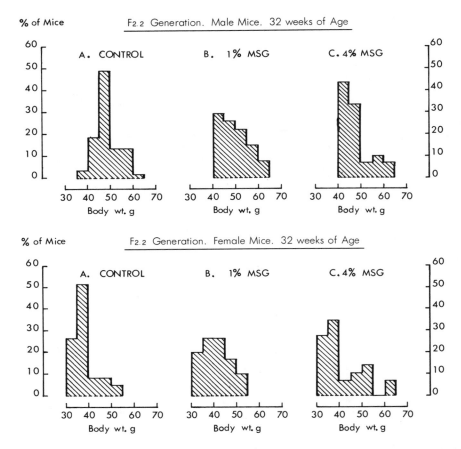

FIG. 6. Body weight distribution frequency for the $F_{2.2}$ generation. Histograms represent terminal values at age 32 weeks.

metestrus (28,29). The food intake of the cycling rodent varies inversely with the plasma estradiol concentration, diminishing at estrus and increasing at diestrus. Any attempt to explain appetite and food intake behavior, and the influence of hormones on these, would be out of place here, though it would seem the logical explanation for the present findings.

MSG INTAKE

Safety evaluation is currently founded on the concept of the "maximum no-effect dose." All approaches are designed to determine the largest daily intake over extended periods that will not produce the injurious effects characteristic of the test substance when given in larger, i.e., toxic amounts. Just as important, these approaches attempt to exclude the possibility that these "subtoxic" amounts will produce some hitherto unsuspected reaction.

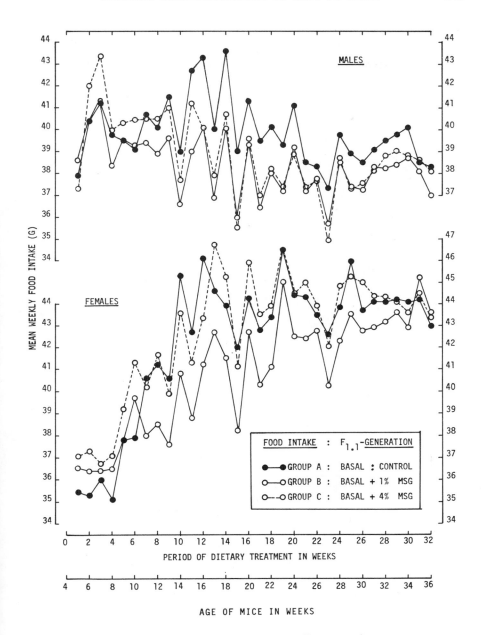

FIG. 7. Food intake curves for the $F_{1.1}$ generation.

The two supplementary levels of MSG were chosen based on a knowledge of the usual food intake of the mouse, i.e., between 5 and 6 g/day, so as to ensure a safe level, and a fairly high level of MSG intake. Now, how much MSG did the animals ingest?

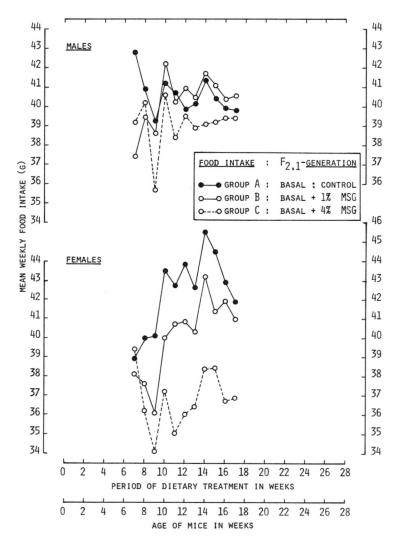

FIG. 8. Food intake curves for the $F_{2.1}$ generation.

MSG Intake During Free Feeding and Lactation

The median intake of MSG as g/kg body weight/day (Table 5) amounted to 1.5 and 6.0 g for males on the 1 and 4% MSG diets, respectively; for the females they were 1.8 and 7.2 g, respectively, due to their lower body weight but similar food intake. The food intake of the adult male, despite the fluctuations noticed, could be considered to be fairly constant over prolonged periods, but that of the dam increases progressively during lactation, averaging 18 to 20 g/day for the 3-week

TABLE 5. *Median MSG intake from diets in growth trials—all generations*

Variable	Males		Females	
	1% MSG	4% MSG	1% MSG	4% MSG
Median body weight (g)	40	40	33	33
Median food intake (g/day)	6.0	6.0	6.0	6.0
Median MSG intake (g/kg/day)	1.5	6.0	1.8	7.2

period. Therefore, the intake of MSG, too, increased during lactation (Table 6). Thus, on the 4% MSG diet, the average intake during the last week of lactation was as high as 25 g/kg body weight/day; yet no adverse effect on the young was observed. Takasaki (26) has reported that ingestion of a 30% w/w MSG diet in a single meal by the lactating dam had no deleterious effect on the dam or on the young. This and other similar observations show that high levels of dietary MSG do not impose any stress or toxic overload on the suckling young, nor cause any abnormal pathology.

When injected subcutaneously or by intragastric means, MSG is absorbed rapidly, leading to elevations of plasma glutamate, whereas dietary ingestion, even in very high amounts, does not lead to such high elevations.

MSG Intake During the Early Postweaning Period

When normally weaned at the age of 21 days, the mouse weighs around 12 g, but its rate of growth during the next 14 days is approximately 1 g/day. The food intake of the mouse during this period is indeed almost near the adult level of intake, which ensures meeting the energy requirements of rapid growth. For a further 4 weeks or so, the rate of weight gain continues at a relatively rapid pace, but with little further increase in food intake. Under these circumstances, the newly weaned mouse would ingest slightly more than 50% of the amount of dietary MSG consumed by the lactating dam during the last week of lactation. The average MSG intake by the newly weaned mouse over a 90-day period is set out graphically in Fig. 9. Thus, on the 4% MSG diet, immediately postweaning, the mouse would ingest around 13 g MSG/kg body weight/day, as against the 25 g/kg body weight/day by the dam.

None of these mice, male or female, developed hyperphagia, hyperactivity, or obesity. Similar observations have been reported by Wen et al. (30) in a study of mice that were injected subcutaneously with varying amounts of MSG from day 6 through 10, and involving a follow-up of the survivors over a 1-year period.

REPRODUCTION PERFORMANCE

A typical reproduction study includes measurement of the following parameters:

Fertility index: the proportion of matings that are successful.
Gestation index: the proportion of pregnancies that result in live litters.

TABLE 6. Mean body weight, and food and MSG intake of dams during lactation weeks 1 to 3: $F_{1.1}$, $F_{2.1}$, and $F_{2.2}$ generations

Generation	Body weight (g)			Food intake (g/day)			MSG intake (g/kg/day)		
	1	2	3	1	2	3	1	2	3
$F_{1.1}$									
A. Control	35	35	35	13.0	18.1	21.9	—	—	—
B. 1% MSG	36	36	36	12.9	18.9	23.9	3.6	5.3	6.6
C. 4% MSG	36	36	36	13.6	18.2	22.4	15.1	20.2	24.9
$F_{2.1}$									
A. Control	36	36	36	13.6	19.2	21.7	—	—	—
B. 1% MSG	37	37	37	13.7	17.3	21.7	3.7	4.7	5.9
C. 4% MSG	35	35	35	14.1	19.2	23.1	16.1	21.9	26.4
$F_{2.2}$									
A. Control	42	42	42	14.4	20.0	23.5	—	—	—
B. 1% MSG	44	44	44	16.1	21.6	25.6	3.7	4.9	5.8
C. 4% MSG	44	44	44	16.0	21.5	26.5	14.5	19.5	24.1

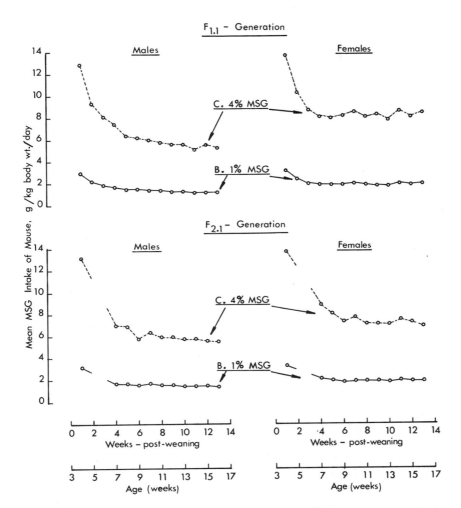

FIG. 9. MSG intake. Values for the $F_{1.1}$ generation are given from postweaning through 90 days of trial. Values for the $F_{2.1}$ generation were measured during 1 week postweaning, and then again from week 7 through 90 days of trial.

Viability index: the proportion of pups born that are alive at 4 days of age.
Lactation index: the proportion of pups alive at 4 days that survive until weaning.

A high viability index usually predisposes a high lactation index, implying a high percentage of weaned young.

The various reproductive parameters for the $F_{1.1}$, $F_{2.1}$, and $F_{2.2}$ generations are set out in Table 7. In all three generations, fertility index, as well as the other parameters, was high (90 to 100%) and identical in all treatment groups. Over 95% of the newborn that were alive on day 4 were weaned by the dam in all groups.

TABLE 7. *Reproductive data in the multigeneration trials: $F_{1.1}$, $F_{2.1}$, and $F_{2.2}$ generations*

	Generation	Fertility index (%)	Gestation index (%)	Viability index (%)	Lactation index (%)
$F_{1.1}$					
	A. Control	77/99 (78)	77/77 (100)	769/814 (95)	727/769 (95)
	B. 1% MSG	28/34 (82)	28/28 (100)	272/283 (96)	256/272 (94)
	C. 4% MSG	28/34 (82)	28/28 (100)	254/281 (90)	252/254 (99)
$F_{2.1}$					
	A. Control	50/60 (83)	50/50 (100)	462/496 (93)	448/462 (97)
	B. 1% MSG	20/20 (100)	20/20 (100)	187/220 (85)	177/187 (95)
	C. 4% MSG	18/20 (90)	18/18 (100)	180/192 (94)	176/180 (98)
$F_{2.2}$					
	A. Control	26/30 (87)	26/26 (100)	239/250 (96)	236/239 (99)
	B. 1% MSG	12/15 (80)	12/12 (100)	133/137 (97)	126/133 (95)
	C. 4% MSG	13/15 (87)	13/13 (100)	140/149 (94)	129/140 (92)

There was no adverse influence on any of the reproductive parameters that was attributable to MSG ingestion. In all these cases, the first mating was initiated (Table 1) when the animals were young, i.e., soon after attaining maturity, in accordance with good breeding practice.

Data for the $F_{3.1.1}$ and $F_{3.2.1}$ generations are set out separately for comparison in Table 8. The sharp differences in the reproductive characteristics exhibited by the $F_{3.2.1}$ generation in contrast to the still high fertility of the $F_{3.1.1}$ generation is striking. Evidently, fertility is at its peak soon after maturation. Current breeding practice goes so far as to recommend breeding mice from around 60 days of age, keeping with the view that the breeder life of the animal could be exploited to maximum. In mice of the $F_{3.2.1}$ generation, for which mating was at 32 weeks of age, fertility dropped below 50%, irrespective of dietary treatment. Again, in this case there was no discernible adverse MSG effect.

Semprini et al. (24) observed no reduction in fertility when consecutive litters were raised on diets containing 1 and 2% w/w MSG. On the other hand, supporting the work of Olney (16), Pizzi et al. (21,22) have reported that MSG administered subcutaneously to newborn mice from day 2 to 11 resulted in a sequence of events that manifested in adulthood as marked reproductive dysfunction in both sexes, with treated females having fewer pregnancies and smaller litters, and treated males showing reduced fertility. In contrast, Adamo and Ratner (1) did not observe any pronounced disturbances in the reproductive function of rats that had been sub-

TABLE 8. *Reproductive data in the multigeneration trials: $F_{3.1.1}$, and $F_{3.2.1}$ generations*

Generation	Fertility index (%)	Gestation index (%)	Viability index (%)	Lactation index (%)
$F_{3.1.1}$				
A. Control	22/30 (73)	22/22 (100)	—	—
B. 1% MSG	12/15 (80)	12/12 (100)	—	—
C. 4% MSG	11/15 (73)	11/11 (100)	—	—
$F_{3.2.1}$				
A. Control	9/20 (45)	9/9 (100)	81/85 (95)	66/81 (81)
B. 1% MSG	8/15 (53)	8/8 (100)	77/79 (97)	58/77 (75)
C. 4% MSG	7/15 (47)	7/7 (100)	65/70 (93)	65/65 (100)

Note: The $F_{3.1.1}$ generation was obtained by mating $F_{2.1}$ at 16 weeks of age. This generation of mice was almost entirely taken up for brain histopathology. The $F_{3.2.1}$ generation was obtained by mating $F_{2.2}$ at 32 weeks of age.

cutaneously treated with MSG when 3 to 4 days old, when later evaluated in adult life. Matsuyama et al. (13), however, observed that newborn mice treated subcutaneously with MSG became obese in adult life, but showed no remarkable changes in their reproductive system or in the sexual cycles of the female, although further generations were not raised for additional evidence. These varying observations lead one to infer that in addition to the importance of the route of administration for a desired effect, one needs to take note of species specificity and age of the animal in making evaluations of MSG. Nevertheless, one point seems clear. Dietary administration of MSG to the gestating or lactating dam, or the newly weaned mouse did not result in reproductive dysfunction, or any other associated disorders (24,26,27).

LITTER SIZE

A variety of factors may adversely affect litter size in a reproduction study. A strain possessing a measure of genetic uniformity and a uniform environment are necessary for producing animals of uniform quality and characteristics. One of the commonest causes of lack of uniformity in the environment, especially in the early environment, is variation in litter size and preweaning influences. In the several hundreds of litters in the present trials, variations in litter size were minimum, although there were some litters as low as 4 and as high as 18 in the population taken as a whole. Nonetheless, these were equally to be seen in the control groups and in the MSG treatment groups.

Individual weights of pups vary inversely with litter size, both at birth and at

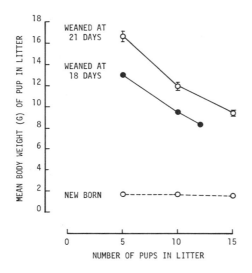

FIG. 10. Graphical representation of litter size as influencing birth and weaning weight of mouse pup. Values for pups weaned at 18 days of age. (From Festing, ref. 9; and Lane-Petter and Lane-Petter, ref. 11.)

weaning. Festing (9) has reported that in inbred mice the average weaning weight goes down by 0.13 ± 0.03 g for every extra pup in the litter. A much greater variation has been recorded for random-bred mice (11). The data in Fig. 10 show the poor performance of large litters compared to optimally sized litters, i.e., around 10 pups to a litter. Especially when early weaning is practiced, there is a compelling reason to ensure uniformly sized litters, for otherwise a pup poorly reared might be expected to continue poorly, postweaning.

With mice treated in early life with MSG subcutaneously, Pizzi et al. (22) observed fewer pregnancies and smaller litters. All through our multigeneration trials, conducted under careful housing and husbandry conditions, the litter size was uniformly about 10 in all groups including the controls, with a mean birth weight of between 1.6 and 1.7 g, and a mean weaning weight of 12 g. Details are presented in Table 9.

HISTOPATHOLOGICAL EXAMINATION OF BRAIN TISSUE

A large number of randomly selected animals of both sexes from the $F_{1.1}$, and the $F_{3.1.1}$ generations were employed for histopathological examinations of brain. The mice of the latter generation were examined at birth (within 90 min after birth), and at 3, 14, and 21 days of age.

The neuronal densities, especially in the arcuate and other nuclei of the hypothalamus, in the basal ganglia, in the hippocampus formation, and the thalamus, as well as in the cortex, of the different treatment groups and the control were compared. Special attention was directed to possible presence of any of the following:

TABLE 9. *Mean litter size and mean body weight at birth and at weaning: $F_{1.1}$, $F_{2.1}$, $F_{2.2}$, and $F_{3.2.1}$ generations*

Generation	No. live pups/litter ($\pm$ SEM)[a]		Body weight (g $\pm$ SEM)[a]	
	At birth	At weaning	At birth	At weaning
$F_{1.1}$				
A. Control	10.32 ± 0.36	9.82 ± 0.35	1.65 ± 0.02	12.02 ± 0.32
B. 1% MSG	9.86 ± 0.70	9.85 ± 0.65	1.63 ± 0.03	12.77 ± 0.55
C. 4% MSG	9.57 ± 0.60	9.69 ± 0.54[b]	1.63 ± 0.04	12.00 ± 0.59
$F_{2.1}$				
A. Control	9.60 ± 0.46	9.14 ± 0.45	1.68 ± 0.03	11.76 ± 0.43
B. 1% MSG	10.25 ± 0.62	9.32 ± 0.68	1.55 ± 0.05	12.67 ± 0.78
C. 4% MSG	10.50 ± 0.38	9.78 ± 0.49	1.62 ± 0.03	11.74 ± 0.58
$F_{2.2}$				
A. Control	9.50 ± 0.60	9.83 ± 0.57[b]	1.72 ± 0.03	12.50 ± 0.39
B. 1% MSG	11.33 ± 0.50	11.45 ± 0.25[b]	1.71 ± 0.05	12.03 ± 0.31
C. 4% MSG	11.38 ± 1.00	11.73 ± 1.03[b]	1.72 ± 0.03	11.80 ± 0.77
$F_{3.2.1}$				
A. Control	9.60 ± 0.41	9.41 ± 0.39	1.71 ± 0.03	13.06 ± 0.32
B. 1% MSG	9.80 ± 0.33	9.58 ± 0.32	1.69 ± 0.03	12.68 ± 0.27
C. 4% MSG	10.00 ± 0.34	9.33 ± 0.45	1.70 ± 0.04	12.88 ± 0.46

[a] Represents mean of litter means.
[b] Higher values compared to at birth would indicate a litter in this group did not survive until weaning.

Ganglial cell degeneration and necrosis
Phagocytosis of decaying ganglial cells
Decreased density in ganglial cells, especially of the hypothalamic nuclei
Disturbed bilateral symmetry of ganglial cell pattern
Glial proliferation, altered glial cells, or evidence of edema and myelin changes.

In addition, the incidence of occurrence of technical artifacts that were observed was compared between the treatment groups, and with the controls. Histopathology, which involved the evaluation by light microscopy of thousands of brain sections from hundreds of mice of different age groups for the possible presence of any of the changes listed above, clearly showed that none of them were present in any of the MSG groups.

The main conclusion from this exercise is, indeed, that the dietary administration of MSG over prolonged periods, including exposure *in utero*, does not cause or provoke the typical brain lesions attributed to the administration of glutamate by different routes (4,14,16,17,25). Other investigators have also underscored the fact that the dietary oral administration of MSG does not cause the brain lesions and other specific changes associated with neuronal damage (18,19,23).

We may conclude that these trials, involving thousands of mice over three generations, have clearly shown the tolerance of the mouse, the most sensitive of the laboratory animal species for this work, to prolonged ingestion of MSG at elevated dietary levels.

SUMMARY

The present study involved the subchronic dietary administration of 1 and 4% w/w MSG admixed with a basal diet; its aim was to investigate the possible adverse cumulative effect(s) of such a diet. The treatment crossover design employed permitted the evaluation of the effects of dietary ingestion of MSG by pregnant dams during gestation only, during gestation through lactation, and subsequently during postweaning.

Median, voluntary food intake was 6 g/day for both sexes, with median body weights of 40 and 33 g for males and females, respectively. MSG intakes under these conditions were 1,500 and 1,800 mg/kg/day and 6,000 and 7,200 mg/kg/day on the 1 and 4% diets for males and females, respectively. The food intake of dams increased considerably in all groups during lactation, with the intakes of dams on the 4% diet rising to 25,100 mg MSG/kg/day. Nevertheless, the pre- and postweaning performance of the young were unaffected.

Reproduction characteristics—fertility, gestation, viability, and lactation indices—were comparable in all groups and in all generations. There was no evidence of male or female sterility attributable to MSG. There was no incidence of hyperphagia or obesity throughout the trial.

No incidence of brain lesions, nor any other pathological change, was encountered in any of the animals of any treatment group. Overall, the dietary administration of MSG was without any untoward incidence, reinforcing the safety-in-use of MSG.

REFERENCES

1. Adamo, N. J., and Ratner, A. (1970): Monosodium glutamate: Lack of effects on brain and reproductive function in rats. *Science*, 169:673–674.
2. Araujo, P. E., and Mayer, J. (1973): Activity increase associated with obesity induced by monosodium glutamate in mice. *Am. J. Physiol.*, 225:764–765.
3. Bunyan, J., Murrell, E. A., and Shah, P. P. (1976): The induction of obesity in rodents by means of monosodium glutamate. *Br. J. Nutr.*, 35:25–39.
4. Burde, R. M., Schainker, B., and Kayes, J. (1971): Acute effect of oral and subcutaneous administration of monosodium glutamate on the arcuate nucleus of the hypothalamus in mice and rats. *Nature*, 223:58–60.
5. Cohen, A. I. (1967): An electron microscopic study of the modification by monosodium glutamate of retinas of normal and "rodless" mice. *Am. J. Anat.*, 120:319–355.
6. Djazayery, A., and Miller, D. S. (1973): The use of gold-thioglucose and monosodium glutamate to induce obesity in mice. *Proc. Nutr. Soc.*, 32:30A–31A.
7. Djazayery, A., Miller, D. S., and Stock, M. J. (1973): Energy balances of mice treated with gold-thioglucose and monosodium glutamate. *Proc. Nutr. Soc.*, 32:31A–32A.
8. Ebert, A. G. (1970): Chronic toxicity and teratology studies of monosodium L-glutamate and related compounds. *Toxicol. Appl. Pharmacol.*, 17:274.
9. Festing, M. (1969): Research (e): Genetic studies. *Lab. Anim. Centre Newsletter*, 37:9.
10. Huang, P.-C., Lee, N.-Y., Wu, T.-J., Yu, S.-L., and Tung, T.-C. (1976): Effect of monosodium glutamate supplementation to low protein diets on rats. *Nutr. Rep. Intern.*, 13:477–486.

11. Lane-Petter, W., and Lane-Petter, M. E. (1971): Toward standardized laboratory rodents: The manipulation of rat and mouse litters. In: *Defining the Laboratory Animal, IV Symposium, International Committee on Laboratory Animals*, pp. 3–12. National Academy of Sciences, Washington, D.C.
12. Lucas, D. R., and Newhouse, J. P. (1957): The toxic effect of sodium L-glutamate on the inner layers of the retina. *A.M.A. Arch. Ophthalmol.*, 58:193–201.
13. Matsuyama, S., Oki, Y., and Yokoki, Y. (1973): Obesity induced by monosodium glutamate in mice. *Natl. Inst. Anim. Health Q., (Tokyo)*, 13:91–101.
14. Murakami, U., and Inouye, M. (1971): Brain lesions in the mouse fetus caused by maternal administration of monosodium glutamate. *Congenital Anomalies*, 11:171–177.
15. Nikoletseas, M. M. (1977): Obesity in exercising, hypophagic rats treated with monosodium glutamate. *Physiol. Behav.*, 19:767–773.
16. Olney, J. W. (1969): Brain lesions, obesity and other disturbances in mice treated with monosodium glutamate. *Science*, 164:719–721.
17. Olney, J. W., and Ho, O.-L. (1970): Brain damage in infant mice following oral intake of glutamate, aspartate or cysteine. *Nature*, 227:609–611.
18. Oser, B. L., Carson, S., Vogin, E. E., and Cox, G. E. (1971): Oral and subcutaneous administration of monosodium glutamate to infant rodents and dogs. *Nature*, 229:411–413.
19. Oser, B. L., Morgareidge, K., and Carson, S. (1975): Monosodium glutamate studies in four species of neonatal and infant animals. *Food Cosmet. Toxicol.*, 13:7–14.
20. Owen, G., Cherry, C. P., Prentice, D. E., and Worden, A. N. (1978): The feeding of diets containing up to 4% monosodium glutamate to rats for 2 years. *Toxicol. Lett.*, 1:221–226.
21. Pizzi, W. J., and Barnhart, J. E. (1976): Effects of monosodium glutamate on somatic development, obesity and activity in the mouse. *Pharmacol. Biochem. Behav.*, 5:551–557.
22. Pizzi, W. J., Barnhart, J. E., and Fanslow, D. J. (1977): Monosodium glutamate administration to the newborn reduces reproductive ability in female and male mice. *Science*, 196:452–454.
23. Semprini, M. E., Conti, L., Ciofi-Luzzatto, A., and Mariani, A. (1974): Effect of oral administration of monosodium glutamate (MSG) on the hypothalamic arcuate region of rat and mouse: A histological assay. *Biomedicine*, 21:398–403.
24. Semprini, M. E., D'Amicis, A., and Mariani, A. (1974): Effect of monosodium glutamate on fetus and newborn mouse. *Nutr. Metab.*, 16:276–284.
25. Takasaki, Y. (1978): Studies on brain lesion by administration of monosodium L-glutamate to mice. I. Brain lesions in infant mice caused by administration of monosodium L-glutamate. *Toxicology*, 9:293–305.
26. Takasaki, Y. (1978): Studies on brain lesions after administration of monosodium L-glutamate to mice. II. Absence of brain damage following administration of monosodium L-glutamate in the diet. *Toxicology*, 9:307–318.
27. Trentini, G. P., Botticelli, A., and Botticelli, C. S. (1974): Effect of monosodium glutamate on the endocrine glands and on the reproductive function of the rat. *Fertil. Steril.*, 25:478–483.
28. Wade, G. N., and Zucker, I. (1970): Development of hormonal control over food intake and body weight in female rats. *J. Comp. Physiol. Psychol.*, 70:213–220.
29. Wade, G. N., and Zucker, I. (1970): Hormonal modulation of responsiveness to an aversive stimulus in rats. *Physiol. Behav.*, 5:269–273.
30. Wen, C. P., Hayes, K. C., and Gershoff, S. N. (1973): Effects of dietary supplementation of monosodium glutamate on infant monkeys, weanling rats, and suckling mice. *Am. J. Clin. Nutr.*, 26:803–813.
31. Zondeck, B., and Tamari, I. (1967): Effects of auditory stimuli on reproduction. In: *Ciba Foundation Study Group No. 26*, edited by G. Wolstenholme and M. O'Conner, pp. 4–19. Little, Brown & Co., Boston.

Glutamic Acid: Advances in Biochemistry and Physiology, edited by L. J. Filer, Jr., et al.
Raven Press, New York © 1979.

Toxicological Studies of Monosodium L-Glutamate in Rodents: Relationship Between Routes of Administration and Neurotoxicity

Yutaka Takasaki, Yoshimasa Matsuzawa, Seinosuke Iwata, Yuichi O'hara, Shinobu Yonetani, and Masamichi Ichimura

Life Science Laboratory, Central Research Laboratories, Ajinomoto Co., Inc., 214 Maldo-cho, Totsuka-Ku, Yokohama-shi, Japan

It was reported that administration of monosodium L-glutamate (MSG) resulted in necrotic changes of neurons in the central nervous system (4,8,20,23) and in the retina (10,26) and induced elevation of some serum hormone levels (22) as acute effects in rodents. Physiological and behavioral abnormalities were demonstrated in rodents following neonatal administration of MSG; namely, stunting (15,20,28), obesity (14,15,17,20,28), precocious puberty (35), female sterility (20), changes in activity level (1,15,25,27), and learning deficits (3,27). However, the abnormalities were observed in animals given a high dose of MSG either parenterally or by forced intubation. Since MSG is used as a food additive, its safety should be evaluated by studies involving the intended condition of use, that is, *ad libitum* dietary feeding. To achieve this end, this study compares the effect of MSG given by parenteral administration or by forced intubation and in *ad libitum* dietary feeding with respect to neuropathological, biochemical, endocrinological, and behavioral effects in mice and rats.

It has been suggested that the repeated ingestion of subneurotoxic doses of MSG in early development might result in functional abnormalities upon reaching maturity (22). The long-range effects of MSG on growth and various physiological parameters were examined in rats given subneurotoxic doses of MSG during the neonatal and infancy stages.

ACUTE EFFECTS

Histological Changes in the Brain

Acute histological changes in the brain following MSG administration were studied in mice, the species most susceptible to damage.

Distribution of Brain Lesions

The regions of histological changes in the brain were examined in infant mice (ICR) injected with a high dose of MSG (4 g/kg body weight) (Fig. 1). The brain was fixed by perfusion with formalin solution and embedded in glycol methacrylate. Serial Nissl-stained sections were examined microscopically. The affected regions were the hypothalamic arcuate nuclei, subfornical organ, preoptic area, area postrema, dentate gyrus, and cerebral cortex (32). Degeneration of neurons and subsequent necrotic changes, such as pyknosis and kariolysis, were observed in these regions. Olney (23) found lesions in the medial habenula, as well as the regions described above. Reynolds et al. (29) also found lesions in the subcommissural organ, fornix, entopeduncular nuclei, amygdala, tectum, and cerebellum after similar treatment. The effects of MSG on the arcuate nuclei were studied, since these nuclei were most sensitive to MSG (29,32).

Relationship Between Hypothalamic Lesions and Plasma Glutamate Level

Parenteral Administration and Forced Intubation

There are few detailed, analytical studies of the relationship between the route of administration and hypothalamic lesions in animals of various ages. Our results on the lowest effective dose (LED) of MSG inducing hypothalamic lesions by various routes of administration in ICR mice are summarized as follows: by intraperitoneal injection (i.p.) and forced intubation (p.o.) in 10-day-old mice, 0.4 (32) and 0.7 g/kg, respectively; by subcutaneous injection (s.c.) and intubation in 23-day-old mice, 0.7 and 2.0 g/kg, respectively; and subcutaneously in adult mice, 1.2 g/kg. These results indicate that hypothalamic lesions are induced by lower doses of MSG by parental administration than by oral administration and that the LED increases with the age of the animals.

We studied in detail the relationship between the LED of MSG for hypothalamic lesions and plasma glutamate levels in infant (10-day-old), weanling (23-day-old), and adult (3- to 4-month-old) ICR mice. Plasma samples were deproteinized with sulfosalicylic acid and then analyzed by an amino acid analyzer (18). As shown in Table 1, the transient peak value of plasma glutamate at LED increased with age

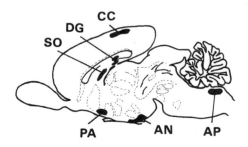

FIG. 1. Distribution of brain damage in 10-day-old mice injected intraperitoneally with 4 g/kg body weight of MSG. AN, arcuate nucleus; AP, area postrema; PA, preoptic area; SO, subfornical organ; DG, dentate gyrus; CC, cerebral cortex.

TABLE 1. *Relationship between dose of MSG for inducing hypothalamic lesions and plasma glutamate level in infant, weanling, and adult mice*

Age	Treatment of MSG			No. of mice affected	Peak level of plasma glutamate (μmoles/100 ml)
		Route	Dose (g/kg)		
Infant (10 days)	Nontreatment		—	0/8	17 ± 1[c]
	p.o.[a]		0.5	0/8	62 ± 6
	p.o.		0.7[d]	2/8	104 ± 18[d]
	p.o.		0.8	4/8	—
Weanling (23 days)	Nontreatment		—	0/8	10 ± 1
	s.c.[b]		0.5	0/8	278 ± 15
	s.c.		0.7[d]	2/8	385 ± 32[d]
	s.c.		1.0	6/8	760 ± 52
Adult (3–4 months)	Nontreatment		—	0/8	10 ± 1
	s.c.[b]		1.0	0/8	539 ± 25
	s.c.		1.2[d]	4/8	631 ± 30[d]
	s.c.		1.5	4/8	841 ± 105

[a] 10% aq. soln.
[b] 4% aq. soln.
[c] Mean ± SE.
[d] Lowest effective dose (LED) of MSG for inducing hypothalamic lesions and peak value of plasma glutamate at LED.

(19). The peak values in infant, weanling, and adult mice were about 6, 40, and 60 times resting levels, respectively. The increase in peak plasma glutamate with age may be related to the development of the blood brain barrier. Olney (21) reported that the peak concentration of plasma glutamate inducing hypothalamic lesions might be 20 times the resting level in infant mice. On the other hand, Stegink et al. (31) estimated that a hypothalamic lesion was induced when plasma glutamate rose to approximately 50 μmoles/dl, based on the plasma glutamate level after administration of hydrolyzed casein and fibrin and on the hypothalamic lesions induced by these hydrolysates (24) in 9- to 11-day-old mice.

The time course of plasma glutamate in infant, weanling, and adult mice was measured after administration of 1 g/kg of MSG by different routes: subcutaneous (aqueous solution), intraperitoneal (aq. soln.), forced intubation (aq. soln. or with diet), and free feeding in a single meal (Figs. 2 and 3) (18). Peak plasma glutamate after subcutaneous injection was highest in the infant mice and lowest in adult mice (Fig. 2). Lengthy retention of glutamate in plasma over base-line values was observed in infant animals. These results suggested that the hepatic capacity to metabolize glutamate developed with age. Peak plasma glutamate after forced intubation was much lower than after parenteral administration (s.c. and i.p.) in mice at any age. This may be due to the intestinal metabolism of glutamate.

MSG dissolved in milk (given to infants) or in soup (given to adults) resulted in lower glutamate levels than MSG given in aqueous solutions (Fig. 3). McLaughlan

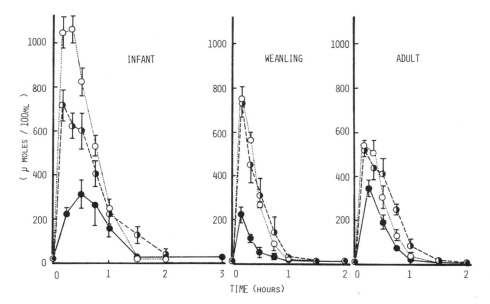

FIG. 2. Time course of plasma glutamate levels after administration of MSG. Infant, weanling, and adult mice were given a single dose of 1 g MSG/kg body weight by s.c. (○), i.p. (◐), or p.o. (●) administration. A 4% (w/v) MSG aq. soln. was used for the s.c. and i.p. routes, and a 10% (w/v) aq. soln. was used for the p.o. administration. Each point represents mean ± SE.

et al. (13) reported that the elevation of plasma glutamate levels in weanling rats following administration of 0.2 g/kg MSG with meat was slower than that in rats given the same dose in aqueous solution, but the peak values were similar in both treatments. In our studies, the peak in mice given 1 g/kg MSG in the diet was much lower than in mice given the same dose in aqueous solution. This discrepancy may be due to a difference in the dose of MSG and/or the kind of food mixed as vehicle.

Dietary Free Feeding

Figure 3 shows the plasma glutamate levels in weanling and adult mice fed a large amount of MSG (1 g/kg) in a commercial laboratory chow *ad libitum* in a single meal (15 to 30 min). In both weanling and adult mice, plasma glutamate levels were lower than after forced intubation, the value being around 40 μmoles/dl. The results indicate that the method of administration has a great influence on the plasma glutamate level and may affect the occurrence of hypothalamic lesions.

There are some reports on the histological effects of MSG following dietary feeding in rodents. Semprini et al. (30) fed a diet containing 1 and 2% MSG *ad libitum* to rats and mice during pregnancy and lactation. Histological examinations of the brains of offsprings at 0, 15, and 30 days of age were made, and no abnormalities were found. Huang et al. (5) reported that histological examination of the hypothalamus of young male rats fed diets containing 2, 4, and 6% MSG *ad*

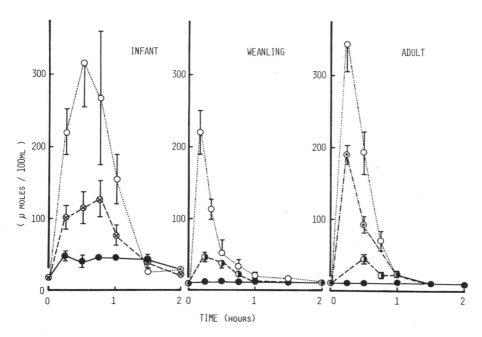

FIG. 3. Effects of MSG load by food accompaniment or dietary administration on time course of plasma glutamate levels in mice. Mice were given a single dose of 1 g MSG/kg body weight with or without food accompaniment by p.o. or dietary administration. Infant mice were given a 10% (w/v) MSG aq. soln. (○) or an infant formula containing 10% (w/v) MSG (⊗) or an infant formula only (●) by p.o. administration. Weanling mice were given a 10% (w/v) MSG aq. soln. (○) by p.o., or were fed with a basal diet containing 10% (w/w) MSG (◐) or a basal diet only (●) by dietary administration. Adult mice were given a 10% (w/v) aq. soln. (○) or a clear soup containing 10% (w/v) MSG (⊙) or a clear soup only (●) by p.o., or were fed with a basal diet containing 10% (w/w) MSG (◐) or a basal diet only (●) by dietary administration. Each point represents mean ± SE.

libitum for 80 days showed no necrosis of neurons. In addition, Wen et al. (36) reported that no abnormality of the hypothalamus was found in weanling rats fed diets containing 20 and 40 g MSG/100 g feed *ad libitum* for 5 weeks.

The effects of MSG administered by dietary feeding were studied. Mice were fed diets containing 5, 10, and 15% MSG or given a 5% aq. soln. of MSG *ad libitum* for 1 to 4 days during pregnancy or lactation or during the weaning stage (33). Since obvious necrosis of the hypothalamic neurons due to MSG disappears within 24 hr of administration of MSG, mice were killed every day within 2 to 3 hr of the end of the feeding period. Histological examination of the hypothalamus showed no necrosis of neurons in pregnant mice or their fetuses, or lactating females or their suckling mice. Weanling mice had no brain lesions following the ingestion of large amounts of MSG (20 to 30 g/kg/day) in diet or water (Table 2).

In order to determine why necrosis of the neurons was not induced in the dietary feeding experiment, feeding and drinking patterns and plasma glutamate levels were

TABLE 2. Daily MSG intake in free-feeding mice

Treatment	MSG intake (g/kg body weight/day)			
	Weanling[a]		Pregnant[b] females	Lactating[c] females
	Male	Female		
5% MSG diet	11	10	6	11
10% MSG diet	23	22	8	23
15% MSG diet	31	35	11	31
5% MSG soln.	20	18	10	21

[a] Mean value of 6 animals in total of 12 days.
[b] Mean value of 2 animals in total of 3 days.
[c] Mean value of 2 animals in total of 6 days.

examined. Figure 4 shows the amount of MSG ingested every 30 min by weanling and adult mice in one day. Mice were maintained in a room with a 12-hr light cycle. In the dark phase, mice ingested 1 to 3 g/kg of MSG per 30 min incessantly (6). Figure 5 shows the individual plasma glutamate levels. In the dark phase, the increased intakes of food and water seem to give higher plasma glutamate levels, but they remained much lower than those required to induce hypothalamic lesions. These low levels of plasma glutamate in mice fed large amounts of MSG in water or in the diet explain why brain lesions were not observed.

Acute Effects of MSG on Some Endocrine Functions

It has been reported that subcutaneous injections of high doses of MSG (1 g/kg) in adult male rats acutely raises serum luteinizing hormone (LH) and testosterone levels (22). It was postulated that these changes were induced by glutamate exerting its effect against the hypothalamic arcuate neurons.

We examined the acute effects of MSG administered parenternally on LH and testosterone levels in adult male Wistar rats. Serum LH was measured by double antibody radioimmunoassay and serum testosterone was measured by radioimmunoassay by the charcoal adsorption method (37). As shown in Fig. 6, the levels of serum LH and testosterone fluctuated in the light or dark phases after subcutaneous injection of MSG (1 g/kg), but did not rise immediately after injection. In the rats injected with NaCl, there was also a significant change 4 hr after injection. These fluctuations appeared to be within the range of circadian variation obtained with nontreated animals and were considered temporary, since serum LH levels the day after the MSG injection were comparable to those of nontreated rats (12).

It was determined whether these variations in hormone levels during the day were affected by free feeding of diets containing large amounts of MSG. Male rats were given diets containing 4 or 8% MSG for 8 days (average MSG intake was 2.8 or 5.2 g/kg/day). On the day 8, serum LH, testosterone, and glutamate were measured every 4 hr (Fig. 7). In control rats fed the basal diet, serum LH levels appeared to fluctuate, but no clear circadian variation was observed. Changes in LH and

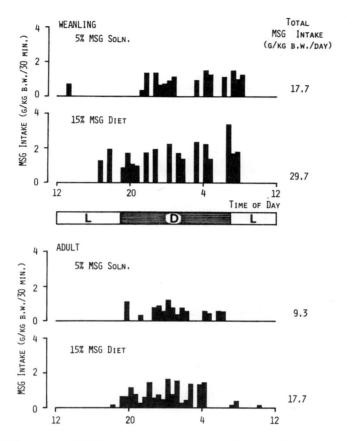

FIG. 4. A 24-hr pattern of MSG intake in weanling and adult mice. Each bar represents the cumulative value taken every 30 min. Mice were fed with diets containing 15% (w/w) MSG or were given 5% (w/v) MSG solution as drinking water *ad libitum*.

testosterone levels in a day were comparable to those reported by Kalra et al. (7). Significant changes of LH levels were partially noted in the rats fed with MSG in the diet, but seemed to be of no biological significance, because no dose-related response was observed and the changes were within the range of daily variation in control rats. Serum glutamate levels in the rats fed the 8% MSG diet increased in the dark phase, but no corresponding changes in serum LH or testosterone levels were observed. These results suggested that ingestion of a large amount of MSG with the diet had no substantial effect on these hormone levels.

LONG-RANGE EFFECTS

The long-range effects of MSG on growth, organ weight, reproductive function, and behavior were studied in rats (34). Wistar rats were given neurotoxic or

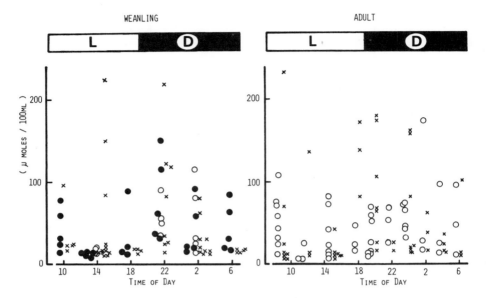

FIG. 5. A 24-hr variation of plasma glutamate concentrations in weanling and adult mice. Mice were fed with diets containing 10% (w/w) MSG (○) and 15% (w/w) MSG (●), or were given 5% (w/v) MSG solution as drinking water (×) *ad libitum*.

subneurotoxic doses of MSG in the neonatal and infant stage and were fed large amounts of MSG in the diet during weaning. Animals were housed in a room maintained at constant temperature (23 ± 2° C) and under controlled lighting (12-hr light and dark periods).

Growth

As shown in Table 3, daily subcutaneous injections of a high dose of MSG (4 g/kg body weight) to neonatal rats resulted in suppression of weight gain in males and suppression of development in body and tail length in both sexes. Autopsy at 3 months in females and 5 months in males revealed that their thigh bones were shorter than those of controls. Lee's index, which has been shown to correlate well with carcass fat (2), increased significantly in these rats, particularly in females. But obesity accompanied with a marked increase of body weight as reported in mice (20) was not observed for 16 weeks. Nikoletseas (17) and Redding et al. (28) also reported obesity without marked weight gain in rats.

In our experiment, the obese rats showed hypophagia in adulthood, as reported previously in mice (20). However, these animals were hyperphagic for 10 days immediately after weaning, which suggested an increase in the number of lipocytes in this stage.

Body weight gain and the development of body and tail lengths in rats injected subcutaneously with 4 g/kg of MSG daily for 10 days in the infant stage were

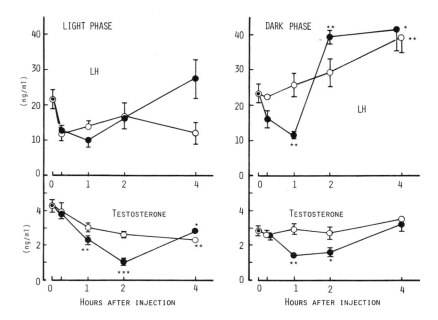

FIG. 6. Effect of subcutaneous injection of MSG or NaCl on serum LH and testosterone concentration in rats in the light and dark phases. Rats were given a single dose of 1 g MSG/kg or 0.35 g NaCl/kg 0.25, 1, 2, and 4 hr before sacrifice. LH and testosterone values were expressed as ng/ml serum using NIAMDD Rat LH-RP-1 and testosterone provided by Sigma Co. as standards. Each point represents mean ± SE. ⊙, control; ●, MSG; ○, NaCl; *$p < 0.05$, **$p < 0.01$, and ***$p < 0.001$ versus control (Student's t-test).

suppressed in comparison with the control. However, Lee's index indicated that these animals were not obese, but tended to be lean, unlike rats injected with MSG as neonates. Similar results were reported in rats treated with a single shot of 4 g/kg (15) or daily injection of 2 g/kg (9) in the neonatal stage. As shown in Fig. 8, neurons of the hypothalamic arcuate nuclei in rats injected with MSG in the neonatal stage disappeared almost completely, whereas considerably numbers of neurons were observed in rats injected in the infant stage. These results suggest that the effect of MSG on growth depends on the severity of hypothalamic damage.

The LED of MSG capable of inducing hypothalamic lesions in 2-day-old rats injected subcutaneously and in 10-day-old rats dosed by forced intubation were 0.4 and 1.4 g/kg, respectively. Daily subcutaneous injections of a subneurotoxic dose of MSG (0.2 g/kg) in the neonatal stage and daily administration by forced intubation of a subneurotoxic dose (0.5 g/kg) in the infant stage did not affect body weight, body length, tail length, or Lee's index. When weanling rats were further fed with diets containing 5% MSG *ad libitum* for 10 days after forced intubation as described above, these rats ingested about 8 g MSG/kg body weight/day. They did not present any sign of obesity or stunted growth, despite the ingestion of large amounts of MSG.

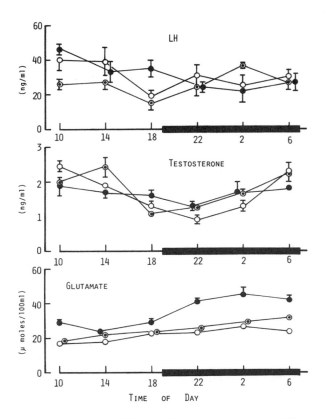

FIG. 7. Effect of dietary feeding of MSG on 24-hr changes of serum LH, testosterone, and glutamate concentration in rats. Each point represents mean ± SE. The solid bars along the abscissa represent the dark period: ○, control; ⊙, 4% (w/w) MSG diet; ●, 8% (w/w) MSG diet.

Organ Weight

Table 4 shows the weights of hormonal and other organs at autopsy of 3-month-old female rats and of 5-month-old male rats. In rats injected repeatedly with high doses of MSG in the neonatal stage, the weights of the anterior pituitary, adrenals, ovaries, uterus, testes, and seminal vesicle were significantly lower than in controls. On the other hand, in rats injected with high doses of MSG in the infant stage, organ weights—excepting the uterus, ovaries, and testes—were not significantly changed. Histological changes were not observed under light microscopy, even in the organs that decreased in weight.

It has been reported that daily injections of high doses of MSG in neonatal rodents resulted in a decrease in the weight of the pituitary, ovaries, uterus, adrenals, and testes (15,35). A significant decrease in weight of the pituitary was reported in mice (14) given a single shot of 4 g/kg in the neonatal stage, but Matsuyama et al. (11)

TABLE 3. Body weight, body length, tail length, and Lee's index of rats at 16 weeks of age

Group	Sex	Body weight (g)	Body length[a] (cm)	Tail length (cm)	Lee's index[b]
Experiment I (s.c., daily 2–11 days of age)					
Saline	M	547 ± 34[e]	25.0 ± 0.7	19.4 ± 0.5	0.325 ± 0.007
	F	317 ± 20	22.0 ± 0.5	17.2 ± 0.3	0.309 ± 0.006
MSG 0.2 g/kg	M	526 ± 44	25.1 ± 0.3	19.5 ± 0.5	0.321 ± 0.008
	F	313 ± 21	22.1 ± 0.4	17.4 ± 0.3	0.308 ± 0.005
MSG 4 g/kg	M	422 ± 37[f]	21.7 ± 0.6[f]	14.6 ± 1.5[f]	0.345 ± 0.007[f]
	F	300 ± 23	19.3 ± 0.5[f]	13.9 ± 0.6[f]	0.348 ± 0.012[f]
MSG 4 g/kg[c]	M	472 ± 39[f]	24.5 ± 0.6	18.4 ± 0.5[f]	0.317 ± 0.008[g]
	F	273 ± 22[f]	20.9 ± 0.4[f]	16.5 ± 0.3[f]	0.310 ± 0.008
Experiment II (p.o., daily 10–19 days of age)					
Saline	M	536 ± 30	25.4 ± 0.4	19.2 ± 0.6	0.320 ± 0.008
	F	300 ± 13	21.7 ± 0.4	17.2 ± 0.2	0.309 ± 0.003
MSG 0.5 g/kg	M	544 ± 30	25.3 ± 0.7	19.4 ± 0.3	0.323 ± 0.005
	F	300 ± 12	21.8 ± 0.4	17.5 ± 0.3	0.307 ± 0.007
MSG 0.5 g/kg[d]	M	513 ± 26	25.2 ± 0.3	19.2 ± 0.3	0.320 ± 0.006
	F	304 ± 25	21.8 ± 0.8	17.5 ± 0.4	0.308 ± 0.006

[a] Nasoanal length.
[b] Lee's index = (body weight)$^{1/3}$/body length.
[c] Rats of this group were treated from 10 to 19 days of age.
[d] Rats of this group were additionally fed with diets containing 5% (w/w) MSG from 20 to 29 days of age after p.o. administration.
[e] Mean ± SD ($N = 12$).
[f] $p < 0.001$ vs saline group.
[g] $p < 0.05$ vs saline group.

failed to obtain these findings in similar experiments. Lengvári reported that there was no decrease in the weight of the pituitary, adrenal or thyroid gland, or gonads in rats injected with a single dose of 2 g/kg in the neonatal stage (9). As with the effects of MSG on growth, the effect of MSG on organ weights may depend on the severity of hypothalamic damage, which varies with the dose and age of the animals at the time of experimentation.

In rats repeatedly treated with subneurotoxic doses as neonates or infants and in rats fed large amounts of MSG in the diet during weaning, the organ weights were not changed compared with those of the controls.

Reproductive Function

The age at vaginal opening, an index of the onset of puberty, was noted, and vaginal smears were taken from 50 to 90 days of age in rats treated with MSG. Serum levels and pituitary contents of LH and follicle stimulating hormone (FSH) were measured in the morning and evening on the day of proestrus in 3-month-old rats. LH and FSH were measured by double antibody radioimmunoassay. As shown in Fig. 9, female rats injected repeatedly with high doses of MSG in the neonatal

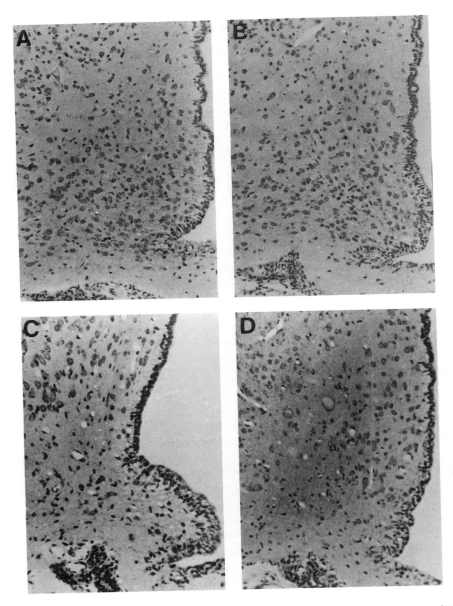

FIG. 8. Hypothalamic arcuate region of 5-month-old rats (× 200). **A:** Control rats were given daily saline from 10 to 19 days of age. **B:** After daily oral administration of 0.5 g MSG/kg from 10 to 19 days of age, rats were fed with large amounts of MSG in the diet from 20 to 29 days of age. No histological changes are seen. Compare with control rats. **C:** Rats were daily injected with 4 g MSG/kg from 2 to 11 days of age. Neurons in the arcuate nucleus disappeared almost completely, and the third ventricle was dilated. **D:** Rats were daily injected with 4 g MSG/kg from 10 to 19 days of age. A considerable number of neurons in the arcuate nucleus are observed.

TABLE 4. Organ weight in adult rats

Group	Sex	Anterior pituitary (mg)	Gonads (g)	Seminal vesicle (g)	Uterus (g)	Adrenals (mg)	Thyroid (mg)
Experiment I (s.c., daily 2–11 days of age)							
Saline	M[c]	8.4 ± 1.7[e]	2.93 ± 0.16	1.94 ± 0.44	—	56.8 ± 7.3	27.3 ± 5.4
	F[d]	11.6 ± 2.1	0.12 ± 0.02	—	0.91 ± 0.07	76.9 ± 10.0	—
MSG 0.2 g/kg	M	8.5 ± 0.5	2.91 ± 0.10	2.05 ± 0.41	—	52.2 ± 5.3	22.0 ± 1.1
	F	12.3 ± 1.8	0.12 ± 0.02	—	0.88 ± 0.11	79.4 ± 6.7	—
MSG 4 g/kg	M	4.4 ± 0.6[f]	1.34 ± 0.10[f]	1.21 ± 0.16[g]	—	42.4 ± 4.2[f]	20.9 ± 4.0
	F	7.5 ± 1.7[f]	0.05 ± 0.02[f]	—	0.63 ± 0.13[f]	49.1 ± 6.5[f]	—
MSG 4 g/kg[a]	M	7.7 ± 0.6	2.69 ± 0.09[f]	1.76 ± 0.30	—	54.1 ± 9.8	21.8 ± 5.1
	F	12.0 ± 2.0	0.10 ± 0.07[g]	—	0.82 ± 0.09[g]	71.0 ± 12.7	—
Experiment II (p.o., daily 10–19 days of age)							
Saline	M	9.2 ± 0.8	3.04 ± 0.08	2.32 ± 0.19	—	56.2 ± 6.9	26.1 ± 3.2
	F	13.6 ± 1.4	0.12 ± 0.01	—	0.88 ± 0.08	81.2 ± 7.9	—
MSG 0.5 g/kg	M	9.5 ± 0.9	3.04 ± 0.13	2.41 ± 0.47	—	54.9 ± 4.8	23.8 ± 4.1
	F	12.9 ± 2.7	0.12 ± 0.02	—	0.89 ± 0.10	85.0 ± 10.0	—
MSG 0.5 g/kg[b]	M	10.1 ± 0.8	2.93 ± 0.10	2.38 ± 0.27	—	54.5 ± 3.0	25.8 ± 3.4
	F	13.9 ± 1.9	0.12 ± 0.01	—	0.86 ± 0.12	79.4 ± 6.8	—

[a] Rats of this group were treated from 10 to 19 days of age.
[b] Rats of this group were additionally fed with diets containing 5% (w/w) MSG from 20 to 29 days of age after p.o. administration.
[c] Five-month-old male rats ($N = 6$).
[d] Three-month-old female rats ($N = 12$).
[e] Mean ± SD.
[f] $p < 0.001$.
[g] $p < 0.01$.

Group	Experiment I (SC, 2-11 days)			Group	Experiment II (PO, 10-19 days)		
	Day of V.O.	Estrous Cycle 50 day 60 79 89			Day of V.O.	Estrous Cycle 50 day 60 79 89	
Saline	31.4[c] ±1.7			Saline	30.2 ±2.3		
MSG 0.2G/Kg	31.8 ±2.0			MSG 0.5G/Kg	31.3 ±1.5		
MSG 4G/Kg	29.0** ±2.7			MSG[b] 0.5G/Kg -Diet	29.8 ±1.6		
MSG[a] 4G/Kg	30.5 ±2.6						

FIG. 9. Day of vaginal opening (V.O.) and estrous cycle in rats. Estrous cycles of 6 rats of each group are shown and illustrated in the following manner:

where 1 = diestrus-1, 2 = diestrus-2, 3 = proestrus, and 4 = estrus. a) Rats of this group were treated from 10 to 19 days of age. b) Rats of this group were additionally fed with diets containing 5% (w/w) MSG from 20 to 29 days of age after p.o. administration. c) Mean ± SD ($N = 12$). **$p < 0.01$.

stage had early vaginal opening and irregular estrous cycles, i.e., continuous estrus or prolongation of estrus or diestrus. In these females, pituitary LH and FSH were significantly lower in the morning than in controls (Table 5). Mean serum LH and FSH in the morning and evening did not significantly differ from those in control animals. Among females of this treated group, rats with prolonged estrus had low serum levels (LH, 29 ng/ml; FSH, 198 ng/ml), indicating no surge of gonadotrophins in the evening.

Rats injected with high doses of MSG in infancy showed no differences in the day of vaginal opening with respect to the regularity of estrous cycles. Although the pituitary FSH content of these rats was significantly diminished, serum LH and FSH levels in the morning and evening did not differ from those in controls. The elevation of serum gonadotrophins in the evening in these rats indicated a normal preovulatory surge of LH. It is noticeable that the rats of this group with hypothalamic damage showed normal secretion of gonadotrophins. It has been reported that injection of high doses of MSG to neonatal rodents induces early vaginal opening (35), irregular estrous cycles (14), and a decrease in the pituitary

TABLE 5. *Serum level and anterior pituitary content of LH and FSH on proestrous day in adult female rats*

Group		Serum level (ng/ml)		Anterior pituitary content (μg/gland)	
		AM[a]	PM[a]	AM[a]	PM[a]
Experiment I (s.c., daily 2–11 days of age)					
Saline	LH	20 ± 7[d]	627 ± 190	146 ± 17	90 ± 28
	FSH	136 ± 26[d]	450 ± 51	4.1 ± 1.6	2.1 ± 1.2
MSG	LH	20 ± 9	750 ± 166	155 ± 12	110 ± 35
0.2 g/kg	FSH	142 ± 16	405 ± 53	5.7 ± 1.6	2.8 ± 1.3
MSG	LH	13 ± 8	368 ± 400	86 ± 46[e]	74 ± 12
4 g/kg	FSH	155 ± 25	308 ± 197	1.5 ± 0.5[f]	1.5 ± 0.9
MSG	LH	23 ± 5	627 ± 169	126 ± 16	85 ± 26
4 g/kg[b]	FSH	145 ± 17	385 ± 56	2.1 ± 0.7[e]	1.9 ± 0.5
Experiment II (p.o., daily 10–19 days of age)					
Saline	LH	19 ± 4	773 ± 157	147 ± 47	101 ± 2.4
	FSH	152 ± 24	430 ± 69	5.3 ± 1.4	2.7 ± 1.7
MSG	LH	14 ± 9	837 ± 146	120 ± 52	102 ± 22
0.5 g/kg	FSH	138 ± 22	449 ± 44	4.8 ± 2.1	2.3 ± 0.8
MSG	LH	22 ± 10	797 ± 129	149 ± 32	112 ± 27
0.5 g/kg[c]	FSH	155 ± 22	396 ± 70	5.3 ± 1.4	2.2 ± 0.7

[a] Rats were sacrificed in the morning (10:30–11:30 a.m.) and evening (5:30–6:00 p.m.).
[b] Rats of this group were treated from 10 to 19 days of age.
[c] Rats of this group were additionally fed with diets containing 5% (w/w) MSG from 20 to 29 days of age after p.o. administration.
[d] Mean ± SD ($N = 6$), expressed as NIH-LH-RP-1 and NIH-FSH-RP-1.
[e] $p < 0.05$.
[f] $p < 0.01$.

LH content (28). However, Matsuyama et al. (11) and Lengvári (9) did not observe abnormalities in the onset of puberty or estrous cycle in rats repeatedly injected with 2 g/kg in the neonatal stage. These facts suggest that the effect of MSG on reproductive function depends on the severity of hypothalamic damage.

Rats given repeated subneurotoxic doses of MSG as neonates or infants and rats given large amounts of MSG in the diet during weaning showed no abnormalities in the onset of vaginal opening or estrous cycle (Fig. 9). Serum levels and pituitary contents of LH and FSH did not differ from those of controls (Table 5). The gonadotrophin secretion system functioned normally in these females.

As shown in Table 6, the pituitary LH content in males injected repeatedly with high doses of MSG in the neonatal stage was significantly lower than in controls, but serum LH and FSH levels did not differ. The later findings appeared to be consistent with the reports by Nemeroff et al. (16). Male rats injected with high doses of MSG in infancy, rats given subneurotoxic doses in the neonatal or infant stage, and those given large amounts of MSG in the diet during weaning did not show any differences in serum levels or pituitary content of LH and FSH compared with controls.

TABLE 6. *Serum level and anterior pituitary content of LH and FSH in adult male rats*

Group	Serum level (ng/ml)		Anterior pituitary content (μg/gland)	
	LH^a	FSH^a	LH^a	FSH^a
Experiment I (s.c., daily 2–11 days of age)				
Saline	26 ± 12	206 ± 13	47 ± 8	35 ± 7
MSG 0.2 g/kg	30 ± 8	213 ± 64	46 ± 10	38 ± 6
MSG 4 g/kg	18 ± 9	185 ± 28	34 ± 10^d	34 ± 5
MSG 4 g/kg^b	16 ± 5	195 ± 30	39 ± 6	34 ± 7
Experiment II (p.o., daily 10–19 days of age)				
Saline	33 ± 23	210 ± 20	51 ± 8	44 ± 12
MSG 0.5 g/kg	24 ± 6	210 ± 29	49 ± 6	42 ± 9
MSG 0.5 g/kg^c	35 ± 11	234 ± 22	53 ± 5	39 ± 6

^{a} Mean ± SD (N = 6), expressed as NIH-LH-RP-1 and NIH-FSH-RP-1.
^{b} Rats of this group were treated from 10 to 19 days of age.
^{c} Rats of this group were additionally fed with diets containing 5% (w/w) MSG from 20 to 29 days of age after p.o. administration.
^{d} $p < 0.05$.

Behavioral Observations

Spontaneous motor activities (SMA) were measured by Animex under the usual housing conditions in male rats treated with MSG. As shown in Table 7, rats injected daily subcutaneously with high doses of MSG in the neonatal stage had significantly less total SMA in one day, decreased SMA in the dark phase, and a greater percentage of light-phase activity. In a 24-hr pattern of SMA (Fig. 10), these rats showed significantly less activity in the dark phase, except for the initial stage, and a significant increase of activity at later stages of the light phase. The decrease in total SMA in a day may be correlated with the decrease in dietary intake during the day in these obese animals. Similar decreases in total SMA in a day have been reported in rats (27) and mice (25). On the other hand, Araujo et al. (1) observed increased SMA in 2 hr in mice. The contradictory results are considered to be due to differences in the experimental conditions, such as the dose of MSG administered, apparatus employed, and time of measurement.

Daily subcutaneous injections of high doses of MSG during infancy, the administration of subneurotoxic doses in the neonatal or infant stages, and large amounts of MSG in the diet during weaning did not elicit any changes in total SMA in a day or in the 24-hr SMA pattern in rats when compared with controls.

Three-month-old male rats were exposed to an open field arena for 3 min, and center latency and ambulation scores were recorded. Observations were carried out

TABLE 7. SMA in a day measured by Animex in 3-month-old male rats

Group	SMA ($\times 10^3$ counts)[a]			% Light-phase activity
	Total	Light phase	Dark phase	
Experiment I (s.c., daily 2–11 days of age)				
Saline ($N = 9$)	31.8 ± 4.7[d]	6.3 ± 1.9	25.5 ± 4.6	19.8 ± 6.3
MSG 0.2 g/kg ($N = 4$)	30.7 ± 2.7	6.1 ± 1.6	24.5 ± 1.8	19.9 ± 4.3
MSG 4 g/kg ($N = 9$)	21.8 ± 7.3[e]	6.9 ± 1.9[e]	14.9 ± 4.2[e]	32.2 ± 8.7[f]
MSG[b] 4 g/kg ($N = 4$)	29.2 ± 2.6	5.4 ± 1.1	23.9 ± 1.6	18.3 ± 2.1
Experiment II (p.o., daily 10–19 days of age)				
Saline ($N = 4$)	27.1 ± 2.1	6.1 ± 1.0	21.0 ± 2.4	22.5 ± 4.4
MSG 0.5 g/kg ($N = 4$)	28.3 ± 3.2	5.7 ± 1.6	22.6 ± 2.5	19.9 ± 4.3
MSG[c] 0.5 g/kg ($N = 4$)	28.0 ± 3.4	5.9 ± 2.0	22.1 ± 2.0	20.8 ± 4.8

[a] Activity was measured in a pair of rats under the illumination schedule of 12 hr light/12 hr dark.
[b] Rats of this group were treated from 10 to 19 days of age.
[c] Rats of this group were additionally fed with diets containing 5% (w/w) MSG from 20 to 29 days of age after p.o. administration.
[d] Mean ± SD.
[e] $p < 0.001$.
[f] $p < 0.01$.

from 1:00 to 5:00 p.m. on two consecutive days. In this open field test, prolongation of center latency and increased ambulation were observed in rats injected daily subcutaneously with high doses of MSG in the neonatal stage (Table 8). These changes suggested that the emotionality of these rats was affected. As for the increase in ambulation, the time when the open field test was carried out coincided with the time of increased SMA (Fig. 10). In other groups treated with MSG, no difference was observed in the open field test compared with the control group.

Nemeroff et al. reported that self-mutilation (tail autoingestion) was observed in 14% of males and 88% of females injected daily with high doses of MSG in the neonatal stage (15). Tail mutilation was observed in a few cases among rats injected subcutaneously with high doses of MSG as neonates.

Berry et al. (3) reported less ability to learn in the water maze test only in rats

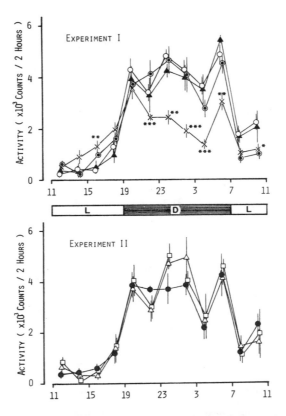

FIG. 10. A 24-hr pattern of SMA in 3-month-old male rats. Activity in a pair of rats was measured by Animex. Each point represents the mean ± SE of cumulative values taken every 2 hr. Rats were daily injected with saline (○), 0.2 g MSG/kg body weight (▲) and 4 g/kg (×) from 2 to 11 days of age, and 4 g/kg from 10 to 19 days of age (⊙), subcutaneously (Experiment I). Rats were daily given saline (●) and 0.5 g MSG/kg (△) from 10 to 19 days of age, orally. After this treatment with MSG, rats were fed with diets containing 5% (w/w) MSG from 20 to 29 days of age (□) (Experiment II). *$p < 0.05$, **$p < 0.01$, and ***$p < 0.001$.

given repeated subcutaneous injections of high doses of MSG in the neonatal stage. Pradhan et al. (27) reported a deficiency in discriminatory learning in a T-maze test in rats given a high dose of MSG repeatedly by forced intubation in the neonatal stage, but their learning ability in a fixed-ratio food reinforcement schedule was not affected. In a Lashley III maze in our experiments, there were no differences in the number of errors or in running time among all groups tested.

To check neuromuscular ability, the rotating-rod test and the inclined-plane test were carried out in 1- and 3-month-old rats. There were no differences among all groups in either test. No abnormalities in corneal reflexes or pinna reflexes were observed in any group.

TABLE 8. *Center latency and ambulation scored in 3-min period by open field test in 3-month-old male rats*

	Center latency (sec)		Ambulation	
Group	1st day	2nd day	1st day	2nd day
Experiment I (s.c., daily 2–11 days of age)				
Saline	16.3 ± 3.3[c]	9.0 ± 2.2	16.4 ± 3.7	28.5 ± 4.4
MSG 0.2 g/kg	20.9 ± 5.3	7.5 ± 1.2	20.0 ± 2.1	30.8 ± 2.6
MSG 4 g/kg	15.3 ± 2.5	24.5 ± 6.1[d]	27.3 ± 3.0[d]	31.4 ± 4.2
MSG 4 g/kg[a]	39.3 ± 20.4	26.3 ± 11.9	17.4 ± 3.8	24.0 ± 3.8
Experiment II (p.o., daily 10–19 days of age)				
Saline	20.5 ± 8.7	10.1 ± 1.3	21.0 ± 3.5	28.8 ± 2.1
MSG 0.5 g/kg	21.8 ± 5.1	18.1 ± 7.2	18.9 ± 3.9	24.9 ± 3.7
MSG 0.5 g/kg[b]	10.1 ± 2.3	11.5 ± 1.5	17.0 ± 2.8	21.3 ± 3.4

[a] Rats of this group were treated from 10 to 19 days of age.
[b] Rats of this group were additionally fed with diets containing 5% (w/w) MSG from 20 to 29 days of age after p.o. administration.
[c] Mean ± SE ($N = 8$).
[d] $p < 0.05$.

SUMMARY

The LED of MSG inducing lesions in the hypothalamic arcuate nuclei and plasma glutamate levels at LED increased with age in mice.

Plasma glutamate was raised less after oral administration of MSG than after parenteral injection. When mice were given MSG orally with food, plasma glutamate rose less than after injections of similar doses.

Weanling, pregnant, and lactating mice fed large amounts of MSG in the diet (10 to 30 g/kg body weight/day) did not develop hypothalamic lesions. Their fetuses and the newborn mice were also unaffected. Plasma glutamate levels in mice fed large amounts of MSG in the diet were much lower than those required to induce brain damage.

Fluctuations of serum LH and testosterone levels in rats fed MSG in the diet were within the normal daily range of control animals, and dietary feeding of large amounts of MSG had no substantial effects on these hormone levels.

Disturbances in growth and reproductive function and some behavioral abnormalities were observed in rats injected repeatedly with high doses of MSG as neonates or infants. These disturbances and abnormalities were more marked when the MSG had been injected during the neonatal period. When neonatal and infant rats were repeatedly given subneurotoxic doses and when weanling rats were fed diets containing large amounts of MSG, no adverse effects were seen in the mature animals.

From these results it can be concluded that MSG, a food additive, does not cause any acute or long-range adverse effects following *ad libitum* feeding in rodents.

ACKNOWLEDGMENTS

We thank Drs. J. Kirimura, Y. Yugari, Y. Sugita, M. Sasaoka, K. Torii, and K. Ishii of Ajinomoto Co., Inc., for helpful advice. Our thanks are also extended to Dr. Reynolds, Dean of the Graduate College, University of Illinois, for kindly reading through our manuscript. The excellent technical assistance of Mr. S. Sekine is also acknowledged.

REFERENCES

1. Araujo, P. E., and Mayer, J. (1973): Activity increase associated with obesity induced by monosodium glutamate in mice. *Am. J. Physiol.*, 225:764–765.
2. Bernardis, L. L. (1972): Hypoactivity as a possible contributing cause of obesity in the weanling rat ventromedial syndrome. *Can. J. Physiol. Pharmacol.*, 50:370–372.
3. Berry, H. K., Butcher, R. E., Elliot, L. A., and Brunner, R. L. (1974): The effect of monosodium glutamate on the early biochemical and behavioral development of the rat. *Dev. Psychobiol.*, 7:165–173.
4. Burde, R. M., Schainker, B., and Kayes, J. (1971): Acute effect of oral and subcutaneous administration of monosodium glutamate on the arcuate nucleus of the hypothalamus in mice and rats. *Nature*, 223:58–60.
5. Huang, P. C., Lee, N. Y., Wu, T. J., Yu, S. L., and Tung, T. C. (1976): Effect of monosodium glutamate supplementation to low protein diets on rats. *Nutr. Rep. Intern.*, 13:477–487.
6. Iwata, S., Torii, K., and O'hara, Y. (1978): in preparation.
7. Kalra, P. S., and Kalra, S. P. (1977): Circadian periodicities of serum androgens, progesterone, gonadotropins and luteinizing hormone-releasing hormone in male rats: The effects of hypothalamic deafferentation, castration and adrenalectomy. *Endocrinology*, 101:1821–1827.
8. Lemkey-Johnston, N., and Reynolds, W. A. (1972): Incidence and extent of brain lesions in mice following ingestion of monosodium glutamate (MSG). *Anat. Rec.*, 172:354.
9. Lengvári, I. (1977): Effect of perinatal monosodium glutamate treatment on endocrine functions of rats in maturity. *Acta Biol. Acad. Sci. Hung.*, 28:133–141.
10. Lucas, D. R., and Newhouse, J. P. (1957): The toxic effect of sodium L-glutamate on the inner layers of the retina. *AMA Arch. Opthalmol.*, 58:193–201.
11. Matsuyama, S., Oki, Y., and Yokoki, Y. (1973): Obesity induced by monosodium glutamate in mice. *Natl. Inst. Anim. Health Q.*, 13:19–101.
12. Matsuzawa, Y., and Yonetani, S. (1978): in preparation.
13. McLaughlan, J. M., Neel, F. J., Botting, H. G., and Knipfel, J. E. (1970): Blood and brain levels of glutamic acid in young rats given monosodium glutamate. *Nutr. Rep. Intern.*, 1:131–138.
14. Nagasawa, J., Yanai, R., and Kikuyama, S. (1974): Irreversible inhibition of pituitary prolactin and growth hormone secretion and of mammary gland development in mice by monosodium glutamate administered neonatally. *Acta Endocrinol.*, 75:249–259.
15. Nemeroff, C. B., Grant, L. D., Bissette, G., Ervin, G. N., Harrell, L. E., and Prange, A. J., Jr. (1977): Growth, endocrinological and behavioral deficits after monosodium L-glutamate in the neonatal rat: Possible involvement of arcuate dopamine neuron damage. *Psychoneuroendocrinology*, 2:179–196.
16. Nemeroff, C. B., Konkol, R. J., Bissette, G., Youngblood, W., Martin, J. B., Brazeau, P., Rone, M. S., Prange, A. J., Jr., Breese, G. R., and Kizer, J. S. (1977): Analysis of the disruption in hypothalamic pituitary regulation in rats treated neonatally with monosodium L-glutamate (MSG): Evidence for the involvement of tuberoinfundibular cholinergic and dopaminergic systems in neuroendocrine regulation. *Endocrinology*, 101:613–622.

17. Nikoletseas, M. M. (1977): Obesity in exercising, hypophagic rats treated with monosodium glutamate. *Physiol. Behav.*, 19:767–773.
18. O'hara, Y., Iwata, S., Ichimura, M., and Sasaoka, M. (1977): Effect of administration routes of monosodium glutamate on plasma glutamate levels in infant, weanling and adult mice. *J. Toxicol. Sci.*, 2:281–290.
19. O'hara, Y. and Takasaki, Y. (1978): *in preparation*.
20. Olney, J. W. (1969): Brain lesions, obesity, and other disturbances in mice treated with monosodium glutamate. *Science*, 164:719–721.
21. Olney, J. W. (1976): Brain damage and oral intake of certain amino acids. *Adv. Exp. Med. Biol.*, 69:497–506.
22. Olney, J. W., Cicero, T. J., Meyer E. R., and de Gurareff, T. (1976): Acute glutamate-induced elevations in serum testosterone and luteinizing hormone. *Brain Res.*, 112:420–424.
23. Olney, J. W., and Ho, O. L. (1970): Brain damage in infant mice following oral intake of glutamate, aspartate, or cysteine. *Nature*, 227:609–610.
24. Olney, J. W., Ho, O. L., and Rhee, V. (1973): Brain-damaging potential of protein hydrolysates. *N. Engl. J. Med.*, 289:391–395.
25. Pizzi, W. J., and Barnhart, J. E. (1976): Effects of monosodium glutamate on somatic development, obesity and activity in the mouse. *Pharmacol. Biochem. Behav.*, 5:551–557.
26. Potts, A. M., Modrell, R. W., and Kingsbury, C. (1960): Permanent fractionation of the electroretinogram by sodium glutamate. *Amer. J. Ophthalmol.*, 50:900–907.
27. Pradhan, S. N., and Lynch, J. F., Jr. (1972): Behavioral changes in adult rats treated with monosodium glutamate in the neonatal stage. *Arch. Int. Pharmacodyn. Ther.*, 197:301–304.
28. Redding, T. W., Schally, A. V., Arimura, A., and Wakabayashi, I. (1971): Effect of monosodium glutamate on some endocrine functions. *Neuroendocrinology*, 8:245–255.
29. Reynolds, W. A., Butler, V., and Lemkey-Johnston, N. (1976): Hypothalamic morphology following ingestion of aspartame or MSG in the neonatal rodent and primate: A preliminary report. *J. Toxicol. Environ. Health*, 2:471–480.
30. Semprini, M. E., Conti, L., Ciofi-Luzzatto, A., and Mariani, A. (1974): Effect of oral administration of monosodium glutamate (MSG) on the hypothalamic arcuate region of rat and mouse: A histological assay. *Biomedicine*, 21:398–403.
31. Stegink, L. D., Shepherd, J. A., Brummel, M. C., and Murray, L. M. (1974): Toxicity of protein hydrolysate solutions: Correlation of glutamate dose and neuronal necrosis to plasma amino acid levels in young mice. *Toxicology*, 2:285–299.
32. Takasaki, Y. (1978): Studies on brain lesion by administration of monosodium L-glutamate to mice. I. Brain lesions in infant mice caused by administration of monosodium L-glutamate. *Toxicology*, 9:293–305.
33. Takasaki, Y. (1978): Studies on brain lesion by administration of monosodium L-glutamate to mice. II. Absence of brain damage following administration of monosodium L-glutamate in the diet. *Toxicology*, 9:307–318.
34. Takasaki, Y., Matsuzawa, Y., Iwata, S., Yonetani, S., and Ichimura, M. (1978): *in preparation*.
35. Trentini, G. P., Botticelli, A., and Botticelli, C. S. (1974): Effect of monosodium glutamate on the endocrine glands and on the reproductive function of the rat. *Fertil. Steril.*, 25:478–483.
36. Wen, C. P., Hayes, K. C., and Gershoff, S. N. (1973): Effects of dietary supplementation of monosodium glutamate on infant monkeys, weanling rats, and suckling mice. *Am. J. Clin. Nutr.*, 26:803–813.
37. Yonetani, S. and Matsuzawa, Y. (1977): Effect of monosodium glutamate on serum luteinizing hormone and testosterone in adult male rats. *Toxicol. Lett.*, 1:207–211.

Glutamic Acid: Advances in Biochemistry and Physiology, edited by L. J. Filer, Jr., et al.
Raven Press, New York © 1979.

Effects of Glutamate Administration on Pituitary Function

Alan F. Sved and John D. Fernstrom

Laboratory of Brain and Metabolism, Program in Neural and Endocrine Regulation, Massachusetts Institute of Technology, Cambridge, Massachusetts 02139

The administration of glutamic acid to neonatal animals in high, repeated doses was originally reported to elicit retinal degeneration (9,16). In 1969, Olney performed experiments to determine whether glutamate injection also caused damage to other portions of the central nervous system (CNS). He anticipated other CNS lesions might be present because of the obesity noted in glutamate animals several months after treatment with the amino acid. Additional lesions were found, particularly in the preoptic area and arcuate nucleus of the hypothalamus (13). Widespread endocrine abnormalities were also suspected, owing to the stunted linear growth of glutamate-treated animals, to their obesity, reduced food intake, and deficient reproductive capacity. It seemed most likely that the hypothalamic lesion modified the normal flow of signals from the brain to the pituitary, thereby producing aberrations in the release of pituitary hormones, and ultimately, in body development and functions. This notion was subsequently explored by Redding et al. (17), who were the first to note abnormalities in pituitary hormone content following glutamate administration. Since 1971, several other investigators have looked for, and found, some abnormalities in pituitary hormone secretion (4,10,11,20). The findings suggest a hypothalamic-pituitary etiology for a few of the peripheral endocrine and metabolic changes associated with glutamate administration. However, the mechanism(s) by which glutamate injection causes the remainder of the abnormalities remains obscure.

This chapter will first review the main alterations in body function found in association with the administration of glutamate to newborn animals. We will then discuss the data currently available on the effects of glutamate administration on the anterior pituitary gland, and on the secretion of its hormones. The likelihood that glutamate administration produces the observed changes in pituitary hormone secretion via the induction of arcuate lesions will also be discussed, along with other possible sites of glutamate action within the CNS that might also elicit such changes in pituitary functions.

In considering these data, it should be borne in mind that, to date, glutamate administration has been shown to produce these gross endocrine abnormalities in adult animals *only* when injected early in postnatal life, in repeated, very large doses. The likelihood that infants of any species might normally be exposed to levels sufficiently high to induce such alterations seems vanishingly small. In particular, infant animals and humans are unlikely to consume amounts sufficient to induce such lesions, either via prepared diets or breast milk (18,19).

EFFECTS OF MSG ON GROWTH AND ENDOCRINE ORGANS

Growth

The short-term administration of high doses of glutamate to neonatal rodents often produces marked alterations in body growth (11,13,14). When newborn animals are injected with the amino acid at a dose level of 1,000 mg/kg/day on postnatal days 1 to 10, they subsequently gain weight at a faster rate than control animals (14). This effect is more readily apparent in females than males. In mice, the difference in body weight is not manifested until after the animals are 50 days of age or older (14), which may explain the failure of some investigators to confirm this phenomenon in younger animals (17,21). In general, the weight abnormality becomes more exaggerated as the animal becomes older (14).

When such glutamate-treated animals are autopsied, the increased body weight has been noted to result mainly from a large increment in total fat stores (11,13), such that they often exceed twice the amount present in normal animals (5). Although observations such as these have led investigators to suggest that the neonatal glutamate treatment somehow induces a hyperphagia syndrome (13), careful study of food intake has revealed the paradoxical result that treated animals actually consume slightly less food each day than control animals (13,17). The only other alternative was that animals treated with glutamate somehow expended less energy than controls. Some data actually support this view, as well. For example, Djazayery et al. (6) have found O_2 consumption and CO_2 production to be reduced in MSG-treated rats, and Pizzi and Barnhart (14) observed gross locomotor activity to be abnormally low. The mechanism by which glutamate administration causes these effects is presently unknown, but might at least in part follow from the reduced levels of thyroid hormones in the blood reported by one laboratory (11).

Obesity is not the only growth abnormality of glutamate-treated animals. Stunting of skeletal growth is also a significant feature (13). In both rats (11) and mice (13), for example, the neonatal administration of glutamate leads to reductions in nasoanal length of about 10%. This effect is also observable when measurements of the long bones are made: in one study, the adult femur length was found to be 14.5 mm in glutamate-treated rats and 16.3 mm in controls (2). Tibia length was also reduced in treated animals. As will be discussed below, at least one possible mechanism has been suggested to account for stunted bone growth in MSG-treated animals, i.e., reduced secretion of growth hormone (12).

Endocrine and Reproductive Organs

Soon after Olney reported that glutamate administration to newborn mice induced hypothalamic lesions (13), a search was begun to identify endocrine abnormalities that might be associated with this treatment. Olney (13) himself observed that glutamate treatment decreased the weights of certain reproductive organs and glands in mice, and also impaired reproductive function. Other investigators (17,21) confirmed these findings: ovaries were noted to be abnormally small in glutamate-treated females and to contain numerous atretic follicles (Table 1) (13). In males (17), the testes appeared histologically normal, but were reduced in size. Some (13,15), but not all (1,21), experimenters found these changes to be accompanied by an impairment of reproductive function. For example, the administration of glutamate early in life to female mice reduced their ability to conceive as adults when mated with normal male mice (15). In male mice treated postnatally with glutamate, some investigators noted a reduced ability to impregnate female animals successfully (15), whereas others found the fertility of the male mouse to be unaltered by this treatment (13). Glutamate-treated animals generally have small anterior pituitaries (10,13,17); the intermediate and posterior portions of the pituitary are of normal size. The weight reduction in the adenohypophysis appears to reflect a general diminution in cell number and size. Thyroid size appears to show inconsistent alterations in adult animals injected with glutamate around the time of birth (10,21); similarly, contradictory effects have been obtained on adrenal size (10,21).

EFFECTS OF NEONATAL GLUTAMATE ADMINISTRATION ON ANTERIOR PITUITARY FUNCTION

Alterations in Hormonal Secretions

Only fragmentary data are available on the effects of neonatal MSG treatment on the secretion of individual hormones by the anterior pituitary. Moreover, for the hormones whose secretions do seem to be modified, essentially no *careful* studies have to date been performed to attempt to identify the underlying mechanisms. Hence, the following summaries should be taken as *very* tentative.

Growth Hormone

Serum growth hormone (GH) levels are markedly reduced in animals treated neonatally with MSG (Table 2) (12,20). This probably reflects a decrease in the hormone's secretion, rather than acceleration in its degradation: the normal, pulsatile secretion of growth hormone has been found to be still present in MSG-treated rats, but the pulse heights are reduced (20). Total pituitary growth hormone content is also diminished (Table 3) (17); however, pituitary size is also decreased, and thus smaller differences are noted when hormone levels are expressed per milligram pituitary (10). The observed decrease in growth hormone release does not appear to reflect changes in the hypothalamic content of somatostatin. Somatostatin levels in the hypothalamus, if at all changed, appear to be reduced (Table 4).

Peripheral signs of inadequate growth hormone secretion appear to characterize in part the symptomology of MSG-treated rodents. For example, abnormally small

TABLE 1. *Changes in body weight, food intake, and gland and reproductive organ weights in rats treated with MSG*

Parameter	Control	MSG
Body weight (g)	358	322
Lee index	0.307	0.322[a]
Food intake (g/day)	24.5	20.0
Reproductive organ and gland weights (mg/100 g body weight)		
Thyroid	3.01	2.73[b]
Adrenals	11.58	8.99[a,b]
Testes	930	800[b]
Anterior pituitary	2.02	1.34[a,b]
Ovaries	32	17[a,b]

Groups of rats received MSG every day from day 2 to day 10 of life, in incremental doses between 2.2 and 4.4 g/kg, s.c. They were killed at 110 days of age. All data are for male rats, except for ovary weights, which are from identically treated female rats. The Lee index is an estimate of obesity.
[a] $p < 0.05$ compared with control.
[b] *Absolute* tissue weight is significantly different from control.
Adapted from Redding et al., ref. 17.

adult bone length is a characteristic of GH deficiency during development. In addition, increased fat deposition may well reflect the absence of normal amounts of GH, since GH normally promotes fat mobilization (lipolysis) from adipocytes (22).

It is widely held that GH does not mediate directly its growth-promoting and metabolic effects on the body, but modulates the availability within the body of the somatomedins, a class of peptide compounds whose biologic actions are essentially identical to those of GH (23). No data are available on the effects of MSG-treatment on this class of compounds. Such information might be of interest in explaining

TABLE 2. *Plasma concentrations of pituitary hormones in adult rats treated with MSG as neonates*

	Males		Females	
Hormone	Control	MSG	Control	MSG
Prolactin	40	156[a]	146	108
Thyrotropin	557	889	616	629
GH	52	29[a]	52	12[a]
LH	27	24	42	38

Data expressed in ng/ml. Group size varied between 6 and 15 animals. Holtzman rat pups received i.p. MSG (4 g/kg) on days 1, 3, 5, 7, and 9 postpartum, and were sacrificed at the age of 18 weeks.
[a] Significantly different from control values, $p < 0.05$.
Reproduced from Nemeroff et al., ref. 12.

TABLE 3. *Anterior pituitary hormone contents in rats treated with MSG*

Hormone	Control	MSG
GH (μg/gland)	289	83[a]
LH (μg/gland)	6.6	1.7[a]
TSH (mU/gland)	608	408

Male rats received MSG each day from days 2 to 10 of life, in incremental doses from 2.2 to 4.4 g/kg (s.c.). They were sacrificed at 40 days of age.
[a] Significantly different from controls, $p < 0.001$.
Adapted from Redding et al., ref. 17.

why, for example, when pituitary GH stores are so markedly reduced, only small overall reductions are observed in body length (the large increase in adiposity may not reflect the absence of GH alone, but also of thyroid hormones and gonadal hormones, although no data are available yet on the effects of MSG treatment on circulating gonadal steroid levels).

Gonadotropins

Serum luteinizing hormone (LH) levels are not abnormally low in animals treated with MSG early in life (Table 2) (12). [One abstract, however, does state that both LH and follicle-stimulating hormone (FSH) levels were low in sera obtained from MSG-treated rats (4), but no data were provided.] The pituitary content of LH, however, does not appear to be significantly decreased (Table 3) (17), but hypothalamic luteinizing hormone release hormone (LHRH) levels are reportedly normal (Table 4) (4,8), and the increase in serum LH levels that follows an injection of LHRH is within the normal range (12). The only indication of a possible impairment of LH secretion is found in ovariectomized rats: in these animals, the increase in serum LH induced by estrogen administration was blunted if MSG was administered early in postuterine life (4). It is difficult to reconcile the apparent *lack* of MSG effects on

TABLE 4. *Effect of neonatal administration of MSG on the concentrations of hypothalamic hormones in adult rats*

Hormone	Males		Females	
	Control	MSG	Control	MSG
TRH	5.5	5.9	5.3	5.6
LHRH	4.2	4.2	3.6	4.6
Somatostatin	—	—	32.0	28.0

Data expressed in ng/mg protein; $N = 6$–12. Holtzman rat pups received i.p. injections of MSG (4 g/kg) on days 1, 3, 5, 7, and 9 of life, and were killed at 18 weeks of age. Data are for *ventrobasal* hypothalamus.
Reproduced from Nemeroff et al., ref. 12.

LH in intact animals with the findings of atrophic gonads, disrupted estrous cycle, and possibly impaired reproductive functioning. [Of course, not all investigators obtained such effects (1,21).] Perhaps FSH levels are low, as suggested by Clemens et al. (4). However, no data have been reported on FSH, and this possibility therefore remains speculative. In addition, it is surprising that no measurements of circulating levels of estrogen, progesterone, or testosterone have been made in the 9 years since MSG-induced reproductive abnormalities were first reported (13). Hence, from the available data, it is not possible to state whether MSG exerts its reputed antireproductive actions via indirect action on the hypothalamus, or by a direct effect on the gonads or reproductive organs.

Thyroid hormones

Thyroid-stimulating hormone (TSH) levels in serum and pituitary fall within the normal range in animals receiving MSG (Tables 2 and 3) (12,17). Moreover, hypothalamic thyrotropin-releasing hormone (TRH) levels (Table 4) and TRH-induced TSH release are both normal in MSG-treated rats (8,12). Consistent with this apparently normal function are the observations that thyroid size is not substantially different from normal (Table 1) (10,17) and that the uptake of ^{131}I by the thyroid gland is normal. However, some evidence suggests that MSG-treated animals are hypothyroid, as indicated by reduced serum T_3 and free thyroxine levels (Table 5) (11). Because of these latter observations, it has been suggested that TSH secretion may actually be subnormal since even though serum TSH concentrations are within the normal range, they are too low given the reductions in circulating thyroid hormone levels (12). Thus, MSG might induce the reductions in T_3 and free T_4 via an action on the hypothalamus-pituitary. However, in the absence of convincing data, if the finding that circulating T_3 and free T_4 levels are diminished holds up to the test of time, it seems equally likely that MSG-treatment might exert these effects by direct action on the thyroid gland; the hypothalamo-pituitary control of thyroid function appears quite normal.

Hypothyroidism would be compatible with some of the gross metabolic and behavioral manifestations of MSG-treated animals. As noted above, these animals show decreased O_2 consumption and CO_2 production (6), and appear lethargic [i.e.,

TABLE 5. *Thyroid status of MSG-treated rats*

Hormone	Control	MSG
Serum T_3 (ng %)	~280	~200[a]
Serum free thyroxine index	~4.5	~3.3[a]

Neonatal rats received i.p. MSG (4 g/kg) every other day for the first 10 days of life. They were killed 30 or 47 weeks later.
[a] $p < 0.05$ compared to control values.
Adapted from Fig. 4, Nemeroff et al., ref. 11.

as indicated by reduced locomotor activity (14)]. Hypothyroidism would also tend to promote the accumulation of adipose tissue (24), and neonatal onset hypothyroidism is known to be associated with retarded skeletal growth (25). Clearly, if hypothyroidism is to be invoked as a causative effect of MSG treatment on linear growth and adiposity, many more data are needed; e.g., a careful study of the hypothalamo-pituitary-thyroid axis *during* the growth and development period of prenatal MSG-treated animals.

Prolactin

Serum prolactin concentrations appear to be normal in MSG-treated animals (4,20), although one report does describe increased concentrations in male rats (Table 2) (12). Pituitary prolactin content, however, has been noted by one group to be significantly reduced (10), and some physiologic signs of inadequate prolactin secretion are indicated by retarded mammary gland development in female rats treated with MSG (10).

Prolactin secretion by the pituitary is believed to be controlled by the hypothalamus, which may secrete dopamine as its prolactin release-inhibiting factor (see ref. 7). A reduction in hypothalamic dopamine release would therefore be expected to elicit a rise in serum prolactin levels. Dopamine levels have been measured in the median eminence and the arcuate nucleus, and appear to be diminished in MSG-treated rats (Table 6) (11,12). In contrast, serotonin and norepinephrine concentrations are unaffected (Table 6). The reduction in hypothalamic dopamine could indicate either diminished synthesis and thus release of the amine, or simply enhanced release with no change in synthesis. In the former case, prolactin secretion would be stimulated, and in the latter, inhibited. Retarded mammary gland development (10) suggests too little prolactin, supporting the idea

TABLE 6. *Median eminence and arcuate nucleus levels of monoamines in adult rats injected with MSG soon after birth*

Brain region	Control	MSG
Median eminence		
Serotonin	31 ± 3	24 ± 3
Dopamine	85 ± 3	52 ± 7[a]
Norepinephrine	14 ± 1	14 ± 1
Arcuate nucleus		
Serotonin	12 ± 1	12 ± 1
Dopamine	17 ± 1	8 ± 1[a]
Norepinephrine	17 ± 1	14 ± 1

Data presented as means ± SE in ng/mg protein; $N = 6$. Newborn male and female rats received i.p. MSG (4 g/kg) on postpartum days 1, 3, 5, 7, and 9, and were killed at 18 weeks of age.
[a] Significantly different from control, $p < 0.05$.
Adapted from Nemeroff et al., ref. 12.

of increased release of dopamine; however, reduced pituitary prolactin content (10) and increased plasma prolactin levels (12) could indicate enhanced prolactin secretion, and therefore inhibition of hypothalamic dopamine synthesis and release. Clemens et al. (4) have reported the unusually large increase in plasma prolactin induced by 5-hydroxytryptophan administration to MSG-treated rats, consistent with the notion that inhibition of prolactin secretion by dopamine might be diminished in these animals. However, again, the data are simply too fragmentary, incomplete, and inconsistent to warrant any firm statements or conclusions.

GENERAL COMMENT AND CONCLUSIONS

Based on the above analysis, it would appear somewhat premature to draw conclusions regarding the effects of MSG administration on pituitary function. Available data are often inconsistent among laboratories attempting to reproduce the same experiments; e.g., although some laboratories note reductions in reproductive capacity in MSG animals (13,15), others do not (1,21). Moreover, results obtained within a single laboratory are not always consistent from study to study, e.g., the effect of MSG on plasma prolactin levels in male rats (12,20). The differences in the times after MSG treatment when samples are collected by the various laboratories is often invoked as a possible explanation for such inconsistencies. However, no one has as yet carefully characterized the time course of any of the hypothesized endocrine abnormalities following neonatal glutamate injections (except, of course, for the growth curves). Not only are there inconsistencies in the literature concerning the ability of MSG to change pituitary hormone secretion, but its ability to produce arcuate lesions also is not generally accepted (1,3).

The "glutamate model" (i.e., the production of arcuate lesions in adult animals by the repeated administration of high doses of the amino acid to newborn pups) has been extant for over 10 years, and the *lack* of published data relating to neuroendocrine abnormalities is surprising. This apparent absence of interest on the part of neuroendocrinologists may derive from the conviction that the model is not scientifically useful, or perhaps from the inconsistencies in the literature concerning the production and effects of the MSG lesions. In either case, the data that are available on pituitary function after MSG administration are too fragmentary to allow the development of a useful hypothesis or model to describe possible mechanisms by which extremely high doses of glutamate can occasionally produce abnormalities in growth and reproduction.

REFERENCES

1. Adamo, N. J., and Ratner, A. (1970): Monosodium glutamate: Lack of effects on brain and reproductive function in rats. *Science,* 169:673–674.
2. Araujo, P. E., and Mayer, J. (1973): Activity increase associated with obesity induced by monosodium glutamate in mice. *Am. J. Physiol.,* 225:764–765.
3. Arees, E. A., and Mayer, J. (1970): Monosodium glutamate-induced brain lesions: Electron microscopic examination. *Science,* 170:549–550.
4. Clemens, J. A., Roush, M. E., and Shaar, C. J. (1977): Effect of glutamate lesions of the arcuate nucleus on the neuroendocrine system in rats. *Soc. Neurosci. Abstr.,* 3:341.

5. Djazayery, A., and Miller, D. S. (1973): The use of gold-thioglucose and monosodium glutamate to induce obesity in mice. *Prod. Nutr. Soc.*, 32:30A–31A.
6. Djazayery, A., Miller, D. S., and Stock, M. J. (1973): Energy balances of mice treated with gold-thioglucose and monosodium glutamate. *Proc. Nutr. Soc.*, 32:31A–32A.
7. Fernstrom, J. D., and Wurtman, R. J. (1977): Brain monoamines and reproductive function. In: *Reproductive Physiology,* Vol. 2, edited by R. O. Greep, pp. 23–55. University Park Press, Baltimore.
8. Lechman, R. M., Alpert, L. C., and Jackson, I. M. D. (1976): Synthesis of luteinising hormone releasing factor and thyrotropin-releasing factor in glutamate-lesioned mice. *Nature,* 264:463–465.
9. Lucas, D. R., and Newhouse, J. P. (1957): The toxic effect of sodium L-glutamate on the inner layers of the retina. *AMA Arch. Ophthalmol.,* 58:193–201.
10. Nagasawa, H., Yanai, R., and Kikuyama, S. (1974): Irreversible inhibition of pituitary prolactin and growth hormone secretion and of mammary gland development in mice by monosodium glutamate administered neonatally. *Acta Endocrinol. (Kbh.),* 75:249–259.
11. Nemeroff, C. B., Grant, L. D., Bissette, G., Ervin, G. N., Harrell, L. E., and Prange, A. J. (1977): Growth, endocrinological and behavioral deficits after monosodium L-glutamate in the neonatal rat: Possible involvement of arcuate dopamine neuron damage. *Psychoneuroendocrinology,* 2:179–196.
12. Nemeroff, C. B., Konkol, R. J., Bissette, G., Youngblood, W., Martin, J. B., Brazeau, P., Rone, M. S., Prange, A. J., Breese, G. R., and Kizer, J. S. (1977): Analysis of the disruption in hypothalamic-pituitary regulation in rats treated neonatally with monosodium L-glutamate (MSG): Evidence for the involvement of tuberoinfundibular cholinergic and dopaminergic systems in neuroendocrine regulation. *Endocrinology,* 101:613–622.
13. Olney, J. W. (1969): Brain lesions, obesity, and other disturbances in mice treated with monosodium glutamate. *Science,* 164:719–721.
14. Pizzi, W. J., and Barnhart, J. E. (1976): Effects of monosodium glutamate on somatic development, obesity, and activity in the mouse. *Pharmacol. Biochem. Behav.,* 5:551–557.
15. Pizzi, W. J., Barnhart, J. E., and Fanslow, D. J. (1977): Monosodium glutamate administration to the newborn reduces reproductive ability in female and male mice. *Science,* 196:452–453.
16. Potts, A. M., Modrell, R. W., and Kingsbury, C. (1960): Permanent fractionation of the electroretinogram by sodium glutamate. *Am. J. Ophthalmol.,* 50:900–905.
17. Redding, T. W., Schally, A. V., Arimura, A., and Wakabayashi, I. (1971): Effect of monosodium glutamate on some endocrine functions. *Neuroendocrinology,* 8:245–255.
18. Stegink, L. D., Filer, L. J., and Baker, G. L. (1972): Monosodium glutamate: Effect on plasma and breast milk amino acid levels in lactating women. *Proc. Soc. Exp. Biol. Med.,* 140:836–841.
19. Stegink, L. D., Pitkin, R. M., Reynolds, W. A., Filer, L. J., Boaz, D. P., and Brummel, M. C. (1975): Placental transfer of glutamate and its metabolites in the primate. *Am. J. Obstet. Gynecol.,* 122:70–78.
20. Terry, L. C., Epelbaum, J., Brazeau, P., and Martin, J. B. (1977): Monosodium glutamate: Acute and chronic effects on growth hormone (GH), prolactin (PRL) and somatostatin (SRIF) in the rat. *Fed. Proc.,* 36:364.
21. Trentini, G. P., Botticelli, A., and Botticelli, C. S. (1974): Effect of monosodium glutamate on the endocrine glands and on the reproductive function of the rat. *Fertil. Steril.,* 25:478–483.
22. Williams, R. H. (1974): *Textbook of Endocrinology,* p. 51. W. B. Saunders, Philadelphia.
23. Williams, R. H. (1974): *ibid.,* p. 51.
24. Williams, R. H. (1974): *ibid.,* p. 153.
25. Williams, R. H. (1974): *ibid.,* p. 1044.

*Glutamic Acid: Advances in Biochemistry
and Physiology,* edited by L. J. Filer, Jr., et al.
Raven Press, New York © 1979.

Excitotoxic Amino Acids: Research Applications and Safety Implications

John W. Olney

Departments of Psychiatry and Neuropathology, Washington University School of Medicine, St. Louis, Missouri 63110

It has been demonstrated repeatedly in recent years that glutamate (Glu), a putative excitatory transmitter and the most abundant amino acid in the mammalian central nervous system, has striking neurotoxic properties. Moreover, it is clear from molecular specificity studies that certain structural analogs of Glu mimic both the neuroexcitatory and neurotoxic effects of Glu and have the same orders of potency for their excitatory and toxic activities. For convenience in referring to those structural analogs of Glu having both neuroexcitatory and neurotoxic activity, we have proposed the term "excitotoxic" amino acids. Because these agents, when injected directly into brain, exert toxic activity against dendrosomal portions of the neuron without damaging axons and because some analogs are much more powerful toxins than Glu itself, neurobiologists are finding excitotoxins quite useful as "axon-sparing" lesioning agents. The ability of excitotoxins to penetrate select regions of the endocrine hypothalamus from blood and to interact with neuroendocrine regulatory units makes these agents also valuable as neuroendocrine investigational tools. Here my major purpose will be to review the development of information pertaining to the neurotoxicity and neuroendocrine interactions of excitatory amino acids. Certain aspects of the food safety issue posed by the use of Glu as a food additive will also be addressed briefly.

SYSTEMIC NEUROTOXICITY OF GLUTAMATE

Background

The first hint that Glu might have neurotoxic potential was provided 20 years ago by Lucas and Newhouse (41) in a report that neurons in the inner layers of the retina rapidly degenerate following subcutaneous (s.c.) administration of Glu to infant mice. Evidence more recently developed includes confirmation of the retinal findings in mice (15,56,90), rats (22,26,33), and rabbits (25); and the demonstration in mice (1,3,4,10,11,29,38,39,47,55,57,70,83,84), rats (3,5,10,11,13,21,51,52,54,

106), guinea pigs (72,106), hamsters (37,101), chicks (99), and rhesus monkeys (59,75,79,81) that nerve cells in the developing brain are also destroyed by systemic Glu administration. Although the subcutaneous route of administration was employed in many of the above studies, Glu-induced brain damage has been demonstrated following oral administration of Glu to mice (39,61,70,106), rats (10, 11,106), guinea pigs (106), and monkeys (81). The lowest effective doses for the two routes of administration are not markedly different (39,61,106).

Other toxic manifestations associated with systemic administration of Glu include convulsions in rats (6,32,100), cats (23), and monkeys (81); vomiting in dogs (42,104), monkeys (81), and man (40); and the "Chinese restaurant syndrome" in human adults (34,93,94,97), which involves intensely disagreeable pain and burning sensations about the face, neck, and torso following ingestion of foods heavily seasoned with Glu. A Glu-intolerance syndrome consisting of "shudder" attacks and headaches has been described in human children (95). Whether or how the latter syndromes in humans relate to the neurotoxicity of Glu demonstrable in experimental animals remains undetermined.

General Features of Glu Neurotoxicity

The brain damage resulting from systemically administered Glu is selective (55,61,76) for certain brain regions, which are said to lie "outside" blood-brain barriers and are known collectively (105) as circumventricular organs (CVOs) (Fig. 1). Both neonatal and adult animals are vulnerable to Glu-induced brain damage, but higher systemic doses are required to damage the brain in adulthood (39,55,61,76). This may be due in part to the greater capacity of the adult liver to metabolize Glu; in any event, the same brain regions are preferentially affected at either age (55,61,76). One CVO region, the arcuate nucleus of the hypothalamus (AH) and contiguous median eminence (ME), has received more attention than others, both because of its particular vulnerability to Glu-induced damage and because of associated neuroendocrine disturbances. Two other CVO regions, the subfornical organ (SFO) and area postrema (AP), are as vulnerable as AH-ME to Glu-induced damage. Future research quite likely will reveal Glu to be a useful tool for probing the functions of these rather obscure brain regions.

Glu lesions, whether in the retina (56) or brain (57,59,71,76,81) and regardless of species, involve rapid swelling of neuronal dendrites and cell bodies followed by acute degenerative changes in intracellular organelles and coarse clumping of nuclear chromatin (Figs. 2 and 3). The reaction is very rapid with onset of dendritic swelling being detectable in 15 to 30 min and phagocytosis of the necrotic neuronal cell body beginning as early as 3 hr after s.c. administration of Glu (56,57). Because axons exhibit no degenerative changes in the acute period, whereas signs of a severe toxic reaction are evident in other neuronal compartments, we have termed this a dendrosomatotoxic but axon-sparing type of cytopathology.

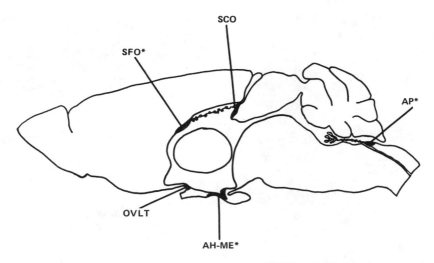

FIG. 1. Diagram of a midsagittal section of rat brain indicates the location of the CVOs, which are special midline periventricular zones that differ from all other brain regions in having fenestrated capillaries that are readily penetrated by various blood-borne substances. Neurons in or near CVOs are subject to damage by systemically administered excitotoxins, whereas other brain regions, even in infancy, are well protected. CVOs include the AP, subcommissural organ (SCO), SFO, organum vasculosum of the lamina terminalis (OVLT), and the AH-ME. Asterisks indicate CVOs most vulnerable to damage by systemically administered excitotoxins.

Molecular Specificity: Clue to Mechanism

In electrophysiological studies, Curtis and colleagues have established that Glu and certain structural analogs of Glu have in common the property of exciting (depolarizing) central mammalian neurons (see chapters by Curtis and Johnston, *this volume*). In a separate series of studies (58,60,65,69,71,73,74,77,80), we administered various Glu analogs s.c. to infant mice and found the excitatory analogs toxic to retinal and hypothalamic neurons, whereas nonexcitatory analogs lacked such toxicity; moreover, it was abundantly clear from these studies that the toxic potency of a given analog parallels its excitatory potency. Coyle et al. (16) in a recent study of the retinotoxic activity of excitatory amino acids, have now corroborated and extended these findings to include additional neuroexcitants not studied by us. Johnston and colleagues (*this volume*) have established that when administered systemically to immature rats, the excitatory amino acids act as convulsants that vary in potency in direct proportion to their known excitatory and toxic activities. Those compounds identified either by us or others as neurotoxic analogs of Glu and that have also been identified as neuroexcitants and convulsants (see chapters by Curtis and Johnston, *this volume*) are depicted in Fig. 4. It is worth emphasizing that these compounds are not merely neurotoxins, but that each appears to exert the same type of neurotoxic action as Glu and that this action, by ultrastructural analysis, is

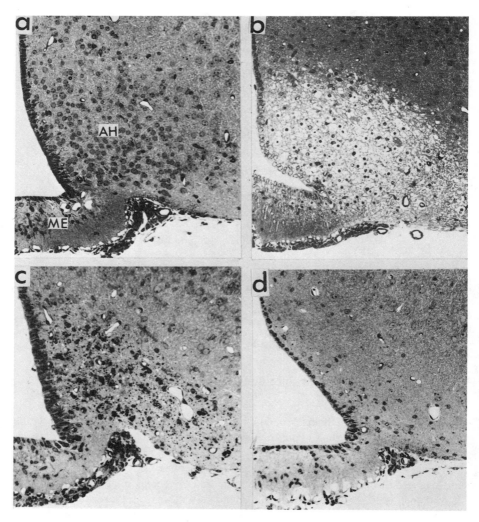

FIG. 2. **a:** The AH and ME regions of the normal 10-day-old mouse hypothalamus. **b:** AH-ME of a 10-day-old mouse 6 hr after a 3 mg/g s.c. dose of Glu. Note the intense edema of cellular components in the AH region, the sharp demarcation of the lesion at the borders of AH with total sparing of other hypothalamic zones and the altered contour of the third ventricle due to AH-ME swelling. **c:** Hypothalamus of an 11-day-old mouse 24 hr after a 3 mg/g s.c. dose of Glu. The AH region is no longer edematous, but is studded with dark-staining dense bodies, which in electron micrographs are recognized as degenerated neurons enclosed within phagocytes. **d:** Hypothalamus of 14-day-old mouse 4 days following a 3 mg/g s.c. dose of Glu. Products of degeneration have disappeared from the AH region, leaving it hypocellular (compare with **a**), and the majority of cells present are nonneuronal. Loss of nerve cell mass is accompanied by compensatory widening of the third ventricle. × 200. (From Olney, ref. 57.)

MICROINJECTION OF EXCITOTOXINS INTO BRAIN

Cysteine-S-Sulfonic Acid

The first Glu analog evaluated by direct microinjection into brain was cysteine-S-sulfonic acid (CSS), an abnormal metabolite associated with the neurodegenerative metabolic disease, sulfite oxidase deficiency. Because CSS resembles the potent excitotoxin, homocysteic acid (HCA) in molecular structure (Fig. 4) and was known to have neuroexcitatory activity, we evaluated its neurotoxic potential following both subcutaneous administration to immature rats and direct microinjection into the brain of the adult rat (73). The latter approach was employed to provide evidence that if CSS were to accumulate in brain, as might occur in cysteine oxidase deficiency, it could have toxic consequences for central neurons. CSS reproduced the Glu-type lesion (dendrosomatotoxic, but axon-sparing) following either subcutaneous administration or direct injection into the diencephalon (62,73).

Homocysteic, N-Methyl Aspartic, and Kainic Acids

As an extension of the CSS experiments, we microinjected three of the more potent Glu analogs—DL-homocysteic acid (HCA), N-methyl-DL-aspartic acid (NMA), and kainic acid (KA)—directly into the adult rat diencephalon and found (82) that each produced an acute axon-sparing neurotoxic reaction with the severity of the acute reaction being directly proportional to the excitatory potency of the injected compound. Small doses (for example, 3.5 nmoles KA) were sufficient to destroy large numbers of neurons. We proposed, therefore, that these analogs might be used as lesioning agents for removing nerve cell bodies from a given brain region without damaging axons terminating in or passing through the region. We also noted that although many neurons were sensitive to the toxic effects of these compounds, some appeared to be either relatively or completely resistant (82). The latter phenomenon has long been recognized as a characteristic of Glu neurotoxicity (56,62,73). (For a series of recent studies addressing the selectivity and axon-sparing characteristics of KA neurotoxicity, see ref. 45).

Comparison of Glutamate and the More Potent Analogs

Recently, we compared the toxic activities of several excitotoxins on striatal neurons using the size of the tissue zone devoid of neurons 1 week following

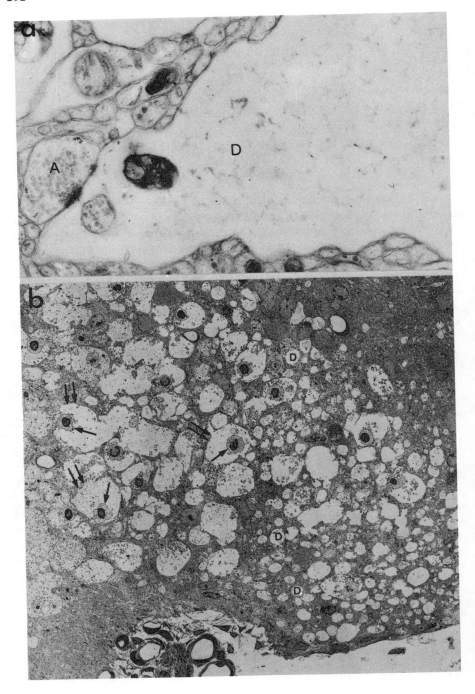

intrastriatal injection as the index of neurotoxic potency (62,66). The order of potency (Fig. 5) for the excitotoxins as evaluated in this manner was KA > NMA > HCA > Glu, which is the known order of potency for the excitatory activities of these compounds. It is to be noted that the toxic activity of Glu is much weaker than that of KA. Indeed, the large doses of Glu required to induce substantial neuronal loss might lead one to suspect that the damage results from nonspecific factors, such as the hypertonicity of the injected solution. The fact that equally hypertonic solutions of GABA or NaCl induced no damage, however, argues against this conclusion (Fig. 5). The loss of striatal neurons following Glu injection and the failure of equimolar doses of GABA to destroy striatal neurons is illustrated in Fig. 6. Comparing the doses of KA (< 2 nmoles) and Glu (1,000 nmoles) required to induce roughly the same amount of striatal damage, it appears that KA is more than 500 times more powerful. This is very similar to the relative potencies of KA and Glu as neurotoxins when given systemically (46,77). Although the remarkably greater excitotoxic potency of KA compared to Glu is not fully understood, most investigators agree that protective mechanisms that inactivate Glu more efficiently than KA play an important role (see chapters by Johnston and McGeer, *this volume*). McGeer et al. (44) and Coyle et al. (16) have presented data suggesting a cooperation mechanism whereby the potency of KA is explained in terms of an ability of KA to act in concert with endogenous Glu to enhance the excitotoxic action of Glu at its natural synaptic receptor sites.

Excitotoxin-Induced Axon-Sparing Lesions

In the acute stages of the neurotoxic reaction induced by Glu or its excitatory analogs, dendritic and somal components of the neuron undergo extreme swelling and degenerative changes, whereas axons passing through or terminating in the region exhibit no changes. It was on the basis of such ultrastructural evidence that the Glu-type lesion was characterized as dendrosomatotoxic but axon-sparing (57,71,73,79,81). More recently, as the direct toxic action of potent Glu analogs, such as kainic acid, on various brain regions has been examined, additional evidence for the axon-sparing nature of the excitotoxic type of lesion has accumulated. In the kainate-lesioned striatum, for example, Coyle et al. (17) and McGeer and McGeer (43) have reported a marked loss of choline acetyltransferase and glutamic acid decarboxylase, enzyme markers for the intrinsic neurons of the striatum, without loss of tyrosine hydroxylase, an enzyme marker for extrinsic dopaminergic

FIG. 3. **a:** An axodendritic synaptic scene from the AH region of an infant mouse 30 min after a 3 mg/g s.c. dose of Glu. The presynaptic axonal component (A) appears normal, but the postsynaptic dendritic process (D) is massively dilated and contains scattered particles of debris, a multivesicle body, and a condensed, vacuolated mitochondrion undergoing degeneration. × 36,000. (From Olney, ref. 57.) **b:** A survey electron micrograph depicting the lateral edge of the AH from a 10-day-old mouse 6 hr after a 1 mg/g oral dose of Glu administered by feeding tube. Necrotic-neurons are present throughout the AH region. The typical pyknotic nuclei (*arrows*) and swollen cytoplasm (*double arrows*) are clearly evident. The smaller vacuous profiles (D) are massively dilated dendrites. × 900. (From J. Olney, *unpublished*.)

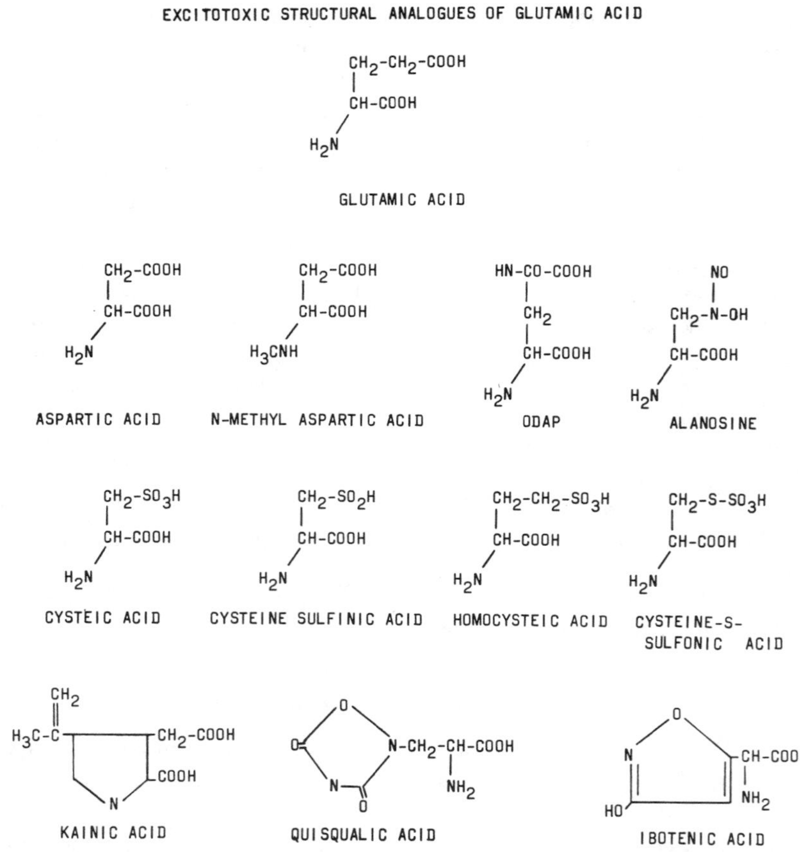

FIG. 4. Representative acidic amino acids known to have both neuroexcitatory and neurotoxic activity. ODAP (β-N-oxalyl-1-$\alpha\beta$-diaminopropionic acid) is an excitotoxin found naturally in chick peas; it is thought to be the neurotoxic factor responsible for the human crippling disease, neurolathyrism (74). Alanosine is an antibiotic and antileukemic agent recently demonstrated to be both a neurotoxin (65) and neuroexcitant (Curtis and Lodge, *personal communication*). Cysteine-S-sulfonic acid is an excitotoxin associated with the neurodegenerative metabolic disorder, sulfite, oxidase deficiency (73). For detailed information on other uncommon excitotoxins such as kainic, quisqualic, and ibotenic acids, please consult ref. 46. Thus far, we have not tested any excitatory analog of Glu and found it lacking in neurotoxic activity. The order of toxic potency as established in studies by us (58,60,65,69,71,73,74,77,80) or Coyle et al. (16) for these compounds is kainic > quisqualic = ibotenic > DL-N-methyl Asp > D-homocysteic = ODAP > alanosine = cysteine-S-sulfonic > L-homocysteic > cysteic = cysteine sulfinic = cysteic = L-Asp = D-Asp = D-Glu = L-Glu. Without significant exception, the same order has been described for the excitatory activities of these compounds (see chapters by Curtis and Johnston, this volume).

axons terminating in the striatum. In the hippocampus, Nadler et al. (48) have demonstrated histologically that KA deletes select intrinsic neuronal populations with subsequent degeneration of their axons while sparing axons of extrinsic origin terminating in the lesioned zone. We have examined several brain regions by

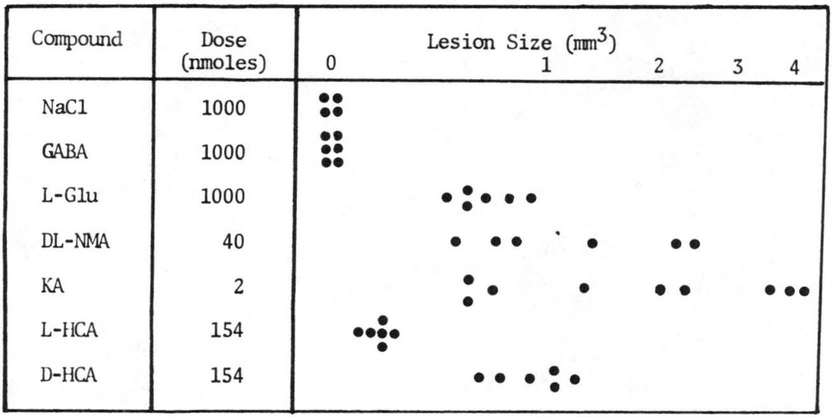

Compound	Dose (nmoles)	Lesion Size (mm³) 0 1 2 3 4
NaCl	1000	•• •• (at 0)
GABA	1000	•• •• • (at 0)
L-Glu	1000	• •• • • (around 1)
DL-NMA	40	• •• · • ••
KA	2	•• • • •• •••
L-HCA	154	••:•• (near 0)
D-HCA	154	•• • •• •

FIG. 5. NaCl, GABA, and several excitotoxins compared for striatal toxicity. All compounds were dissolved in sterile water and adjusted to neutral pH with NaOH. NaCl, GABA, and Glu were injected in 1-μl volume and the other compounds in 0.5-μl volume into the adult rat neostriatum. Animals were sacrificed by aldehyde perfusion fixation 1 week after injection and histological sections cut from araldite embedded blocks were evaluated for neuronal loss at the injection site. Since the shape of lesions tended to be eliptical, we established the margins of the lesion (area without neurons) at the site of injection and used a formula for the volume of an epipsoid ($v = \pi/6 ab^2$ where a is the widest and b the narrowest diameter) to obtain a rough three-dimensional estimate of the total tissue mass devoid of neurons. Each dot represents a single animal. There was no detectable loss of neurons in NaCl or GABA injected brains. Since 2 nmoles KA destroyed more neurons than 40 nmoles NMA or 1,000 nmoles Glu, KA would appear to be > 20 and > 500 times more powerful than NMA and Glu, respectively. Comparison of the D- and L-isomers of HCA at the same dose revealed the D-isomer to be substantially more effective than the L-isomer in destroying striatal neurons.

electron microscopy 1 to 3 weeks after a KA lesion and have found numerous well-preserved axon terminals in zones that have sustained a total loss of intrinsic neurons (67,68). Such axons often retain postsynaptic densities that represent receptor membranes relinquished by the dying neurons to their prior presynaptic contacts (Fig. 7). Herndon and Coyle have observed the same phenomenon in the KA-lesioned cerebellar cortex (28).

Localization of the Toxic Mechanism to the Extracellular Compartment

In view of the well-known property of brain tissue to concentrate Glu from an incubation medium, the question arises whether cellular uptake of Glu with consequent derangement in intracellular metabolism could be the basis for Glu-induced neuronal necrosis. We attempted to study this recently (69) by taking advantage of the observation by Cox et al. (18) that D-HCA, a potent excitatory analog of Glu, is not taken up intracellularly by brain tissue, whereas L-HCA, a weaker excitant, is taken up by both high- and low-affinity transport systems. We compared the neurotoxicity of these stereoisomers by systemic administration to infant mice or direct injection into the striatum of adult rats, reasoning that if the D-isomer proved to be a more potent neurotoxin, this would imply a link between the toxic and

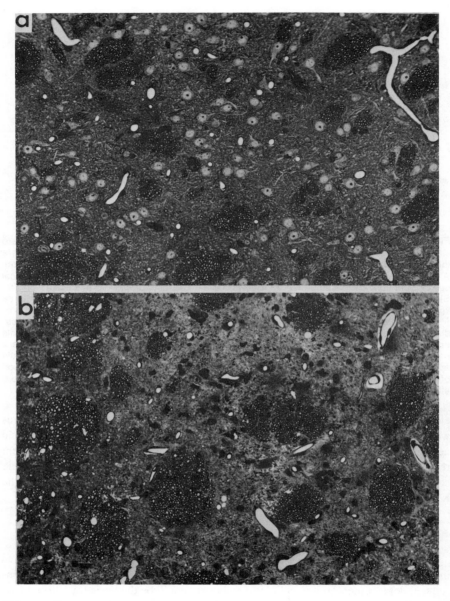

FIG. 6. **a:** The neostriatal scene appears entirely normal despite the injection of 1,000 nmoles GABA into this region 21 days previously. No neurons have been deleted. **b:** A comparable neostriatal scene near a site where 1,000 nmoles Glu were injected 21 days previously. All neurons have been eliminated from this striatal area, but axonal bundles appear quite normal. The numerous small dark bodies are glial cells, which typically proliferate at the scene of an excitotoxin-induced striatal lesion. × 266.

excitatory activities and identify both as actions exerted from the extracellular compartment. We found the D-isomer substantially more powerful than the L-isomer in necrosing AH neurons of the infant mouse (69) or striatal neurons of the adult rat (Fig. 5). This suggests that the toxic and excitatory activities of the HCA molecule are extracellularly mediated and argues against either phenomenon being dependent on intracellular uptake or the associated metabolic derangement. The same can be said for the KA molecule in view of recent evidence (Johnston, *this volume*) that this potent excitotoxin is not actively taken up into rat brain slices.

EXCITOTOXIC AMINO ACIDS AS NEUROENDOCRINE PROBES

The AH has long been recognized as an important, albeit poorly understood, neuroendocrine regulatory center. It is contiguous with the ME and the infundibular stalk that connect the pituitary gland to the brain. AH neurons have short axons that are thought to terminate in the external mantle of the median eminence (EMME) on or near portal vessels that convey various regulatory factors (hypophysiotrophic hormones) from the EMME zone to the adenohypophysis. Other hypothalamic and possibly extrahypothalamic centers also contribute axons to the EMME region, and this had made it difficult, in the absence of specific tools, to delineate the endocrine regulatory roles of individual centers such as AH. Fortunately, excitotoxic amino acids have properties that make them very promising tools for probing the neuroendocrine functions of AH neurons.

For an agent to serve as a useful systemic neuroendocrine probe, it must have access from blood to the endocrine hypothalamus. Perez and Olney (85), applying quantitative histochemical methods to obtain a microregional delineation of Glu uptake patterns, demonstrated a fourfold elevation of Glu in the AH-ME following a 2 g/kg (s.c.) dose of Glu administered to immature mice, but detected no changes in Glu concentrations in other hypothalamic or thalamic regions samples. Moreover, the time course of Glu accumulation in the AH-ME paralleled the time course of lesion formation in that region (55,57). Since it is well established that the systemic administration of high doses of Glu results in little or no change in whole brain Glu concentrations (36), it is quite clear that the homeostatic mechanisms regulating net Glu influx from blood into the AH-ME are very different from those pertaining to brain proper. Furthermore, since various acidic excitatory analogs of Glu mimic the selectivity of Glu in damaging the AH when administered subcutaneously (65,71,73,77), these compounds presumably share with Glu the ready access it has to AH-ME neurons. Selective (net) uptake into the AH coupled with the dual capacity to either stimulate the firing of AH neurons or destroy them makes these agents ideal for use as either provocative or ablative neuroendocrine investigational tools.

ABLATION APPROACH

The first report (55) of Glu-induced brain damage was accompanied by a brief description of the neuroendocrine disturbances that animals treated in infancy

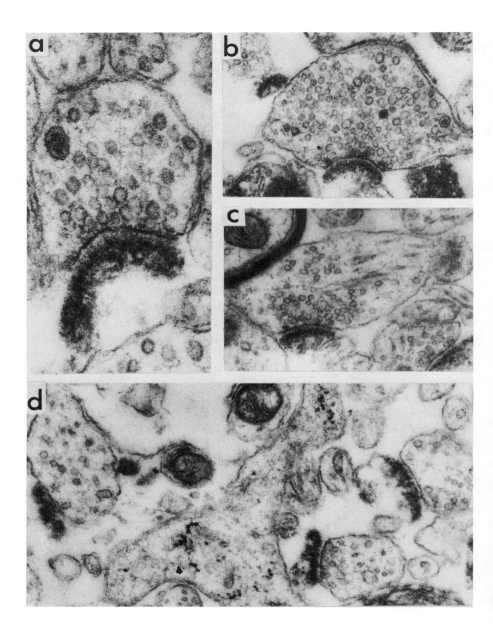

typically manifest as adults, and it was postulated that all or nearly all such disturbances stem from the loss of AH neurons (Fig. 2). This interpretation, if correct, suggests that a full characterization of the endocrine deficiency syndrome associated with Glu treatment should provide valuable clues to the endocrine regulatory functions subserved by AH neurons. The major features of the syndrome as originally described (55) were obesity, skeletal stunting, adenohypophyseal hypoplasia, female sterility, and pathomorphological changes in the reproductive organs of the female. Available evidence confirming and expanding on each feature will be summarized.

Pituitary Status

Olney noted (55) that although Glu-induced damage to the infant hypothalamus was not accompanied by acute pathological changes in the pituitary gland, the anterior lobe of the pituitary was characteristically very small when treated animals attained adulthood (Fig. 8). The severe hypoplasia of the anterior pituitary was ascribed to the removal of trophic influences ordinarily exerted on that organ by AH neurons. A markedly undersized anterior pituitary in both male and female adult animals (mice, rats, and hamsters) following neonatal Glu treatment has now been described by numerous researchers (14,30,37,49,51,53,55,87,92,96,103). The pituitary content of growth hormone (GH), luteinizing hormone (LH) and prolactin (Prl) in Glu-treated rodents is reduced either commensurate with or in excess of the reduction in mass of the anterior pituitary (Table 1) (49,92). When adult rats treated neonatally with Glu are challenged with appropriate releasing agents, however, normal or supranormal LH, TSH, or Prl responses are elicited (14,53), indicating unimpaired ability of the pituitary to release trophic hormones.

Obesity

In mice treated in infancy with either single or multiple subcutaneous injections of Glu, Olney (55) described an obesity syndrome in which treated animals, initially

FIG. 7. **a–c:** The synaptic complexes depicted are characteristic of those abundantly present in the striatum 1 to 3 weeks after a 10-nmole KA injection. They appear to be asymmetric synapses of the type that axon terminals ordinarily make with dendritic spines and shafts of striatal neurons. The presynaptic element appears healthy and contains numerous synaptic vesicles. The synaptic cleft remains distinct and has postsynaptic dense material extending from it that does not appear fundamentally different from the postsynaptic web of an intact synapse. **a,** × 96,000; **b,c,** × 45,000. (From Olney and de Gubareff, ref. 67.) **d:** The synaptic complexes characteristic of those seen in the olfactory cortex 1 week after KA administration. It was recently demonstrated (68) that olfactory cortical neurons are so sensitive to KA toxicity that they selectively degenerate following small doses of KA injected by any of several routes of administration (intradiencephalic, intrastriatal, intraventricular, or subcutaneous). The olfactory cortex is innervated by axons from the olfactory bulb, which putatively use Glu as transmitter. The terminals of these neurons are depicted here with postsynaptic receptor membranes remaining attached to them after KA-induced degeneration of the olfactory cortical neurons that previously housed these membranes. × 36,000. (From Olney and de Gubareff, ref. 68.)

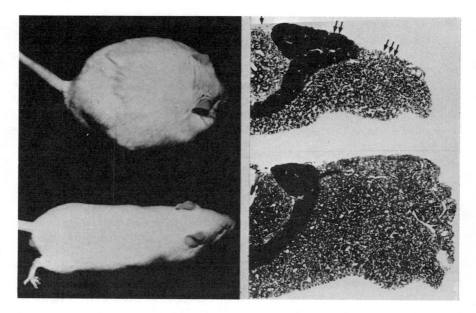

FIG. 8. Left: Two 9-month-old male litter mates. The experimental mouse (*left*) received daily subcutaneous Glu injections on postnatal days 1 to 10 and at 9 months weighed 84 g. The untreated control animal (*right*) weighed 44 g. In addition to obesity, note that the experimental animal is short and its body coat is not as sleek as that of the control. (From Olney, ref. 55.)
Right: Directly across from each animal is its pituitary gland. Only ½ of each pituitary is shown. The anterior lobe (*triple arrows*) of the Glu-treated animal is much smaller than normal. The intermediate lobes (*double arrows*) of the two pituitaries are the same size, although one looks larger because it was sectioned in a slightly different plane. The posterior lobes (*single arrow*) are the same size. Pituitaries, × 38. (From Olney and Price, ref. 75.)

lower in body weight, surpassed the weight of litter mate controls at about 45 days of age and thereafter continued to amass considerable carcass fat at a slow but steady pace throughout adulthood (Fig. 8). Measurement of food intake revealed the treated animals to be slightly hypophagic compared to controls and they appeared lethargic (55). Many researchers have now confirmed Glu-induced obesity in either mice (2,9,12,20,86,89) or rats (35,51,54,92). Bunyan et al. (9), finding Glu treatment of newborn mice to be almost 100% reliable in inducing a high degree of obesity, stressed the potential value of the Glu-obese mouse as a model for studying obesity. Cameron and colleagues (12,89) administered Glu to KK mice, an inbred strain with a high genetic susceptibility to diabetes, and found (89) that it unmasked diabetes; treated animals not only became markedly obese, but developed hyperglycemia accompanied by gross hyperinsulinemia, implying a state of insulin resistance. They considered the hyperglycemia and hyperinsulinemia to be a direct result of the hypothalamic abnormality induced by Glu in this diabetes-prone strain and suggested that further study of this model may shed light on the role of the hypothalamus in obesity and diabetes.

In Glu-treated rats, Knittle and Ginsberg-Fellner (35) established significant

TABLE 1. *Endocrine organ weights and hormone values in Glu-treated rodents*

Animal	Exp.	N	Pituitary (mg)	Testes/Ovaries (mg)	Adrenals (mg)	Pituitary GH (µg)	Plasma GH (ng/ml)	Plasma corticosterone (µg/100 ml)
Mice								
♂	Glu	9–12	0.54 ± 0.08[a]	167.5 ± 16.7[a]	8.1 ± 0.5[b]	8.7 ± 1.2[a]	5.3 ± 0.6[a]	37.0 ± 4.0[a]
	Control	10–16	2.04 ± 0.08	282.0 ± 7.4	9.0 ± 0.5	249.3 ± 14.9	122.7 ± 32.0	5.7 ± 1.3
♀	Glu	8–11	0.86 ± 0.07[a]	12.9 ± 2.3[a]	11.4 ± 1.1[b]	15.1 ± 2.6[a]	3.7 ± 0.9[a]	35.2 ± 5.3[c]
	Control	9	2.94 ± 0.14	39.1 ± 8.1	8.8 ± 0.4	269.2 ± 18.9	17.7 ± 2.5	12.4 ± 1.9
Rats								
♂	Glu	7–26	5.62 ± 0.23[a]	2.67 ± 0.10[a]	38.2 ± 4.0[b]	527.0 ± 77.0[a]	21.0 ± 4.5[a]	6.2 ± 1.2[b]
	Control	7–26	8.75 ± 0.43	3.35 ± 0.13	43.2 ± 3.3	909.0 ± 91.0	110.5 ± 18.9	6.2 ± 1.6
♀	Glu	6–18	7.74 ± 0.40[a]	93.9 ± 5.5[a]	38.0 ± 4.0[d]	327.0 ± 16.0[a]	13.2 ± 1.9[a]	31.0 ± 5.4[c]
	Control	8–19	12.28 ± 0.49	127.3 ± 4.8	59.0 ± 5.1	618.0 ± 56.0	34.9 ± 4.4	10.3 ± 2.7

Glu was given to all experimental animals s.c. at graduated doses from 2.2 to 4 mg/g in 5 daily treatments from days 2 through 7 postnatally. Animals were sacrificed at 5 months of age.
[a] $p < 0.001$; [b] Not significant; [c] $p < 0.01$; [d] $p < 0.02$.
From B. Massey, K. Kipnis, and J. Olney, *unpublished*.

obesity by measuring the weight of epididymal fat pads 5 months after treatment of rat pups. The increase in the weight of fat pads was traced to an increased lipid content per cell rather than an increased number of adipose cells. Also noted was a decreased responsiveness to the lipolytic effect of epinephrine and an increased responsiveness to the antilipolytic effects of insulin.

All researchers who have attempted to measure food intake in Glu-treated obese animals have reported them to be either hypophagic or normophagic, but never hyperphagic (2,9,20,51,54,55,92). Activity levels of Glu-obese rodents have been reported as increased (2), unchanged (54), and decreased (51,86) compared with controls, so that it remains unclear what role energy expenditure plays in Glu obesity.

Effects on Somatotrophin Axis

An approximate 10% reduction in linear bone growth in Glu-treated mice (Fig. 8), initially established by skeletal X-ray measurements (55), has been confirmed in both mice (2,9,30,31,86,89,50) and rats (51,54,92,102). Marked reductions in pituitary GH content and basal serum levels of GH (Table 1) have also been a consistent finding in both mice or rats treated neonatally with Glu (14,53,92,102). Terry et al. (102), by studying serum GH patterns over time in individual rats treated neonatally with Glu, have established that the pulsatile output of GH is markedly diminished. Reduced tissue levels of somatostatin (SRIF) in two specific brain regions, the mediobasal hypothalamus (14,53,102) and basolateral amygdala (102)—regions known to be involved in regulating GH secretion—have also been reported in Glu-treated rats.

Effects on Gonadotrophin Axis

Initial observations on Glu-treated mice (55) suggested disturbances in the neural-gonadal axis of only the female, but subsequent studies have extended the finding to include both female and male mice, rats, and hamsters (14,30,37,49, 51,87,92,103). Holzwarth-McBride et al. (30) and Pizzi et al. (87) treated large numbers of mice neonatally with Glu. In both studies, treated animals had significantly reduced weights of testes and ovaries, and reproductive capacity was significantly impaired in both sexes. Delayed vaginal opening and disturbed estrous cycles have also been reported in Glu-treated female mice (49,87).

In Glu-treated rats, the ovaries (14,51,92,103), uteri (51), testes (14,51,92,103), and seminal vesicles (14) were reportedly smaller than controls, although testicular weight differences were significant in some studies only when given as absolute weights, uncorrected for body weight. Estrous cycle disturbances have been observed in Glu-treated rats (14,103), but abnormal reproductive capacity has not been demonstrated. A marked decrease in the pituitary content of LH (92) and a decrease in basal serum LH (14,51) were found in both male and female Glu-treated rats, although the depressed serum LH levels were not considered statistically significant in one study (51).

Lamperti and Blaha (37) administered Glu subcutaneously to neonatal hamsters in doses of either 4 or 8 mg/g and observed a reduction in the weights of testes, seminal vesicles, ovaries, uteri, and anterior pituitaries in all treated animals. In hamsters given 8 mg/g, all females were acyclic, with ovaries containing only secondary-stage follicles, and males had markedly atrophic testes that were devoid of spermatids and spermatozoa in the seminiferous tubules. Whereas such extreme changes following 8 mg/g treatments might relate to the destruction of AH neurons, it must not be forgotten that exceedingly high doses of Glu, in the mouse and rat at least, destroy preoptic hypothalamic neurons that are believed to regulate the ovarian cycling in the female rodent (24).

Effects on Prl

Nagasawa et al. (49) administered a single dose of Glu (4 mg/g) to neonatal female mice and observed that treated animals in adulthood had a reduced pituitary content of Prl and suppressed development of mammary systems. In rats treated neonatally with Glu, Clemens et al. (14) elicited serum Prl peaks in response to 5-hydroxytryptophan challenge that were significantly higher than controls. In ovariectomized rats treated neonatally with Glu, estradiol benzoate treatments induced large Prl elevations in controls, but very slight Prl responses in Glu-treated animals.

Thyroid Status

The apparent lethargy and roughness of body coat noted by Olney (55) in Glu-obese mice suggested a thyroid-deficient state. Nemeroff et al. (53), who are the only researchers to report thyroid function data on Glu-treated animals, considered their Glu-treated rats hypothyroid on the basis of a significantly lower serum triiodothyronine and free thyroxine index compared to controls.

Adrenal Status

Several groups have reported that the adrenal weights of MSG-treated animals were smaller than in controls, but usually not significantly so. The adrenal weights given in Table 1 are fairly characteristic. Of greater potential interest, however, is the recent observation of B. Massey, D. Kipnis, and J. Olney (*unpublished*) that Glu-treated mice and female rats have exceedingly elevated basal corticosterone levels (Table 1).

Mediobasal Hypothalamus: Putative Transmitters and Hypophysiotropic Factors

Nemeroff et al. (51) have reported the loss of dopamine (DA) histofluorescence in AH cell bodies of rats treated neonatally with Glu, but no appreciable change in the intensity of DA histofluorescence in the EMME, where DA neurons of AH are thought to terminate. Helme et al. (27) described the loss of DA histofluorescence in the EMME of Glu-treated mice, but not rats. Holzwarth-McBride et al. (30) noted reduced DA histofluorescence in both AH and EMME of Glu-treated mice. Employing the Palkovits microdissection approach, Nemeroff et al. (53) found that the Glu-lesioned rat AH has normal tissue concentrations of luteinizing hormone-releasing hormone (LHRH), thyrotropin-releasing hormone (TRH), serotonin, and norepinephrine, but greatly reduced concentrations of DA and choline acetyltransferase, an enzyme marker for cholinergic neurons. Clemens et al. (14) recently confirmed a 60% reduction in the DA concentration of the mediobasal hypothalamus in Glu-treated rats. Lechan et al. (38) reported no reduction in LHRH concentrations and no loss of immunohistochemical staining for LHRH in Glu-lesioned mouse hypothalamus.

Holzwarth-McBride et al. (30) and Paull and Lechan (84) examined the EMME ultrastructurally after destroying an estimated 80% of the AH neurons by Glu treatment of immature mice. Neither group was able to find significant signs of axon terminal degeneration or change in the makeup, size, or configuration of the ME. Olney et al. (78) conducted a similar study in which the AH neurons were destroyed by Glu treatment of either infant or adult mice, and the EMME was examined at numerous posttreatment intervals as AH neurons were acutely degenerating and being phagocytized. At approximately 16 hr posttreatment, they detected axon terminal degeneration in the EMME region, but the numbers of degenerating processes were so few that they concluded that AH neurons are only a minor source of the fibers that project on the EMME (78).

Other Excitotoxins Used As Ablative Neuroendocrine Tools

Schainker and Olney (50) explored the potential of other excitotoxins (L-aspartate and L-cysteate) to reproduce the endocrine disturbances associated with Glu treatment in mice. Animals treated with cysteate and aspartate (Asp) developed obesity, skeletal stunting, and decreased weights of the adenohypophyses, ovaries, and testes, whereas NaCl-treated animals were free from such stigmata. Pizzi et al. (88) confirmed the above in Asp-treated mice and, in addition, established that Asp treatment results in impaired reproductive capacity of both sexes.

PROVOCATIVE APPROACH

If excitatory amino acids in toxic systemic doses destroy AH neurons by an excitatory mechanism, as proposed above, it follows that the same compounds in subtoxic doses—that is, doses that do not damage AH neurons—may excite them to fire at an accelerated rate, just as the firing of other central neurons is stimulated by the microelectrophoretic application of these compounds. Since bioelectric discharge is presumably the first step in the chain of events through which AH neurons influence pituitary hormonal outputs, increased AH discharge activity should give rise to altered hormonal outputs. Experiments recently undertaken to explore this proposition will be summarized.

Acute Effects of Glu on LH, GH, and Prl

In 1976 we reported (63,64) that the subcutaneous administration of Glu to adult male Holtzman rats at 1 mg/g, approximately 25% of the dose required to destroy AH neurons in the adult rat, results in significant elevations of serum LH and testosterone 15 min after Glu treatment. Subsequently, we observed that the peak LH response to Glu occurs even more rapidly—in the 5- to 10-min posttreatment interval (Fig. 9). Therefore, we have focused upon the 7½-min interval in our more recent stimulated LH output studies (see below). Terry et al. (102) administered Glu, 1 mg/g s.c., to adult male rats and observed an acute elevation of serum Prl 15 min after treatment and a sustained (for 5 hr) suppression of the pulsatile output of GH.

Acute Effects of Potent Glu Analogs on Serum LH

If the hypothesis is correct that Glu in subtoxic doses induces acute elevations in LH by stimulating bioelectric activity of AH neurons, it should be possible to elicit a similar LH response with subtoxic doses of other neuroexcitatory analogs of Glu. To test this proposal, we administered HCA, NMA, and KA, three potent excitatory analogs of Glu, to 25-day-old male rats in doses ranging from 1 to 100 mg/kg and determined serum LH levels 7½ min after subcutaneous injection of each excitant (91). Each excitant elicited a rapid and striking LH elevation, the lowest effective doses being 5, 16, and 50 mg/kg for KA, NMA, and HCA, respectively (Fig. 10).

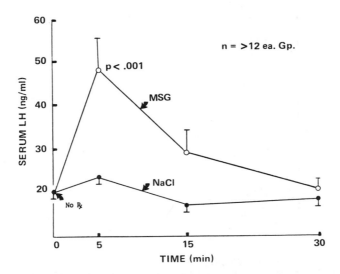

FIG. 9. Serum LH levels (mean ± SEM) of adult male rats that received 1 g/kg s.c. Glu 5 to 30 min before sacrifice. A rapid rise in serum LH levels occurred within 5 min of Glu administration. By 30 min, LH values were the same for experimental and untreated or NaCl-treated control animals. The latter received 0.35 g NaCl/kg, the molar dose equivalent of 1 g/kg Glu. (From Olney and Price, ref. 75.)

Thus, the order of potency observed for the LH-stimulating activities of these agents (KA > NMA > HCA) was the same as had previously been shown for both their excitatory and toxic activities. We also evaluated the brain-damaging potential of these compounds on 25-day-old male rats and found that the lowest doses of NMA and HCA effective in producing AH-ME damage were 3 to 4 times higher than the lowest doses required to induce LH elevations (91). For KA, however, there was little or no practical margin between a brain-damaging and LH-stimulating dose; moreover, we found that KA tended to damage several extrahypothalamic regions of brain more readily than it damaged AH. Thus, we judged NMA and HCA highly promising but KA relatively unsuitable for neuroendocrine investigational purposes.

GABA or Taurine Blockade of NMA-Stimulated LH Release

In the early work of Curtis and Watkins (19), it was observed that when applied iontophoretically, the neuroinhibitory amino acids such as β-alanine, GABA, and taurine were effective in blocking the excitatory responses of central neurons to excitants such as Glu, Asp, or cysteate. To test the hypothesis that excitatory amino acids induce LH release by excitation of AH neurons, we administered NMA to 25-day-old rats in sufficient dosage (25 mg/kg) to assure a reliable LH response and attempted to block this response by simultaneous administration of GABA or

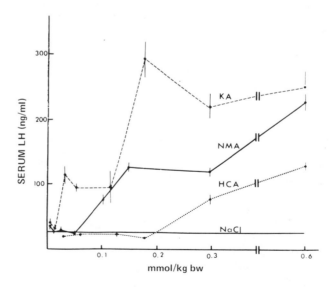

FIG. 10. A total of 255 male rats, 25 days old, were treated with various doses of HCA, NMA, or KA, and serum LH levels were determined 7½ min later. KA induced significant ($p < 0.001$) elevations beginning at a dose of 5 mg (0.03 nmoles)/kg, NMA at 16 mg (0.1 nmoles)/kg, and HCA at 50 mg (0.3 nmoles)/kg. The horizontal line above the *abscissa* represents pooled base-line values determined on untreated or NaCl-treated control rats. (Modified from Price et al., ref. 91.)

taurine. We also studied the effects of DA blocking agents (pimozide and chlorpromazine) in the same study. The results (Fig. 11) of these experiments were unequivocal. Neither GABA nor taurine by itself influenced serum LH concentrations in the interval tested (7½ min following injection), but when given with NMA, either compound completely blocked the striking 5- to 10-fold LH elevations typically observed in rats 7½ min after NMA treatment. Pimozide and chlorpromazine were ineffective in blocking the NMA-induced release of LH.

Blockade of NMA Activity by α-Aminoadipate

In very recent experiments we have administered α-amino-DL-adipate (α-AA) together with NMA in an effort to block either the LH-releasing activity or the

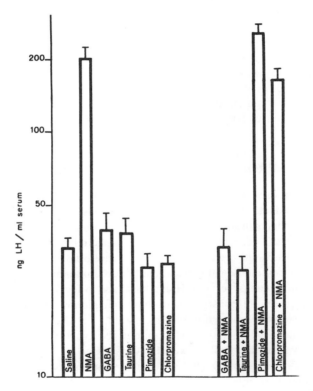

FIG. 11. Serum LH levels (mean ± SEM) 7½ min after subcutaneous treatment of 25-day-old male rats with various agents in the following doses: 25 mg/kg NMA, 1 g/kg GABA, 1 g/kg taurine, 0.6 mg/kg pimozide, 25 mg/kg chlorpromazine. When given alone, NMA induced a significant LH elevation ($p < 0.001$), which was effectively blocked by GABA or taurine, but not by pimozide or chlorpromazine. When GABA, taurine, pimozide, or chlorpromazine were administered without NMA, LH values were not significantly different from saline control values. In drug combination experiments, GABA or taurine were administered concurrently with NMA, but pimozide and chlorpromazine were given as multiple pretreatments 24, 14, and 2 hr prior to NMA. (From Olney and Price, ref. 75.)

neurotoxic effects of NMA. Our preliminary findings are that α-AA effectively antagonizes the LH-releasing action of NMA in 25-day-old rats (M. T. Price and J. W. Olney, *unpublished*) or the toxic action of NMA on AH neurons of the adult mouse (J. W. Olney, *unpublished*). The impetus for conducting such studies came from the recent microelectrophoretic experiments of Biscoe et al. (7) showing that α-AA specifically antagonizes the depolarizing action of NMA on central mammalian neurons. Although our findings with α-AA are very preliminary, we believe them worth mentioning because of the distinct possibility that α-AA, unlike GABA or taurine, specifically antagonizes the action of NMA at its excitatory receptor locus. Although GABA and taurine are very effective in blocking the LH-releasing action of NMA, they are totally ineffective in antagonizing the neurotoxic activity of NMA (J. W. Olney and T. Fuller, *unpublished*). We interpret this as evidence that GABA and taurine act at a different receptor site than NMA—a site that aborts the action potential of AH neurons and hence the LH-releasing action of NMA, without preventing its action at excitatory receptors on the dendritic and somal surfaces of AH neurons where the initial excitatory stimulus is translated into a toxic effect. Blockade of both the LH-releasing action and AH neurotoxic effects of NMA by α-AA, in light of evidence that α-AA is a specific antagonist of NMA-induced excitation, strongly suggests that NMA-induced excitation of AH neurons is the triggering mechanism underlying both the LH releasing and neurotoxic actions of NMA.

NEUROENDOCRINE DATA SUMMARIZED

For ablation purposes, it is possible to delete about 80 to 90% of the neurons of AH by treating rodents in infancy with subcutaneous Glu or certain excitatory analogs of Glu. Rodents thus treated, including mice, rats, and hamsters, manifest a complex syndrome of endocrine-type disturbances, including normophagic obesity, skeletal stunting, impaired reproductive capacity (not established in rats), reduced mass of the anterior pituitary and gonads, and reduced pituitary content of GH, Prl, and LH. Pulse amplitude of serum GH and basal serum levels of GH and probably LH are also depressed. Responsiveness of hypothalamic Prl release mechanisms to estrogenic and serotonergic stimuli are altered. In Glu-treated rats, serum triiodothyronine and free thyroxine indices are in the hypothyroid range and serum corticosterone levels are markedly elevated in Glu-treated male and female mice and female rats. The mechanism by which Glu treatment results in obesity remains a mystery, although the AH lesion is probably responsible since treatment with other Glu analogs that reproduce the AH lesion also cause obesity.

Analysis of the mediobasal hypothalamus following Glu treatment suggests that in terms of neurotransmitter content, the deleted AH neurons are composed of at least two subpopulations—dopaminergic and cholinergic. Whether any AH neurons contain LHRH remains unclear, but Glu ablation studies suggest that if there are LHRH-containing AH neurons, they do not supply more than a small percentage of the total LHRH-containing fibers that terminate in the EMME region. Although evidence also remains incomplete regarding the contribution by AH of DA-containing fibers to the EMME region, it appears likely that AH contributes only a portion of such fibers and that the remainder come from extra-AH sources that have yet to be identified.

Reduced tissue levels of SRIF in the mediobasal hypothalamus and basolateral amygdala have been reported in Glu-treated rats. Since Glu does not directly damage the amygdala, loss of SRIF in amygdala may reflect atrophy of the SRIF system through disuse, i.e., the AH lesion may break circuits that promote GH secretion, leaving the animal in a GH-deficient state in which SRIF inhibitory circuits are unneeded and unused. Alternatively, AH neurons may be the target cells upon which SRIF-containing fibers from the amygdala project. Removal of the target cells may cause retrograde degeneration of the SRIF system (or failure of the system to fully develop).

Evidence pertaining to the acute effects of systemically administered excitotoxins on neuroendocrine systems (provocative approach), although only preliminary, suggests that subtoxic doses of Glu stimulate a burst of LH and Prl output, but suppress GH pulsatile output in adult male rats. The most potent excitatory analogs of Glu (HCA, NMA, and KA) are effective in stimulating LH output in direct proportion to their excitatory potencies. NMA, being particularly reliable and potent in stimulating LH output and lacking the erratic toxicity of KA, may be an ideal probe for studying LH release mechanisms. It has been demonstrated that NMA-induced LH release is totally blocked by simultaneous subcutaneous administration of GABA or taurine, but is unaffected by pimozide or chlorpromazine (dopamine receptor blocking agents). Failure of the latter agents to influence NMA-induced LH release tends to exclude dopaminergic neurons as the subpopulation of AH neurons through which NMA induces LH release. Future research should include studies designed to determine whether this phenomenon is mediated by cholinergic AH neurons.

The demonstration by Clemens et al. (14) that rats with AH lesions have a weak Prl response to estrogenic stimulation suggests that AH neurons may be a major link in an estrogen feedback loop for Prl regulation. Whether this feedback loop involves the DA subpopulation of AH neurons remains to be clarified, as does the exaggerated Prl response of the Glu-lesioned rat to serotonergic stimuli (14) and the acute burst of Prl output reported following Glu administration to normal male rats (102).

PROSPECTS FOR PRIMATE NEUROENDOCRINE STUDIES

Others have reported failure to demonstrate brain damage from Glu administration in primates (see chapters by Worden and Reynolds, *this volume*); such evidence is subject to the interpretation that primates are not susceptible to the mechanism of Glu neurotoxicity. If this were so, it would limit the usefulness of excitatory amino acids as neuroendocrine probes. We believe primates are quite susceptible, however, since we were able to locate Glu-type lesions in the AH in all six infant rhesus monkeys treated either orally or subcutaneously with Glu, but found no cytopathol-

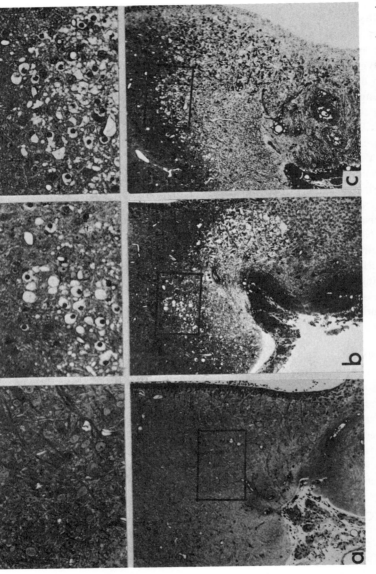

FIG. 12. The AH and infundibular region of 3 rhesus monkey hypothalami are depicted in the lower figures with magnified views (*boxed areas*) of AH presented above. **a:** From a control infant rhesus monkey given NaCl s.c. at 24 mmoles/kg (the molar equivalent of 4 g/kg Glu). The tiny dark profiles lining the ventricle are red blood cells, which reflect the hemorrhagic consequences of administering such a high solute load to an infant monkey. The hypothalamus is histologically normal. **b:** From an experimental infant monkey given Glu s.c. at 2.7 g/kg. The acute edema sweeping across the AH region and numerous acutely necrotic neurons (bull's eye profiles) in the middle of AH—the characteristic signs of a Glu-type lesion in any species—are clearly evident. **c:** From a fetal rhesus monkey following administration of a 2 g/kg *i.v.* dose of Glu to the pregnant mother (152nd day of gestation). The fetus was removed by cesarean section and sacrificed 5 hr following maternal Glu administration. The typical signs of a Glu lesion are unmistakably present in the AH region of this fetal hypothalamus. Lower panel, × 72; upper panel, × 216.

ogy in any of three rhesus monkey controls treated with NaCl (Figs. 12 and 13) (79,81). Furthermore, we recently completed a more extensive evaluation of these monkey brains and found typical Glu-type lesions in the AP of Glu-treated animals, but no such cytopathology in controls (Fig. 14). In addition, we administered (59) Glu intravenously to a pregnant rhesus monkey in late gestation (152 days), delivered a viable fetus by cesarian section 5 hr later, examined its brain by light and electron microscopy, and found sizeable acute Glu-type lesions in both the AH-ME and AP (Figs. 12 and 14); the mother also sustained similar damage from the high intravenous load of Glu (Fig. 14).

In light of the above, we believe excitatory amino acids may prove to be valuable neuroendocrine investigational tools for use with subhuman primates. Ablational studies could be undertaken with Glu treatments being given either pre- or postnatally. Since monkeys are scarce and costly, however, we believe that the emphasis should be on the provocative rather than ablative approach, the advantage of the former being that numerous experiments can be performed on a single animal. The observation that Glu destroys AP neurons in monkeys is of considerable interest in that Glu is known to have emetic properties in several species (42,81,104), including humans (40), and AP neurons are believed to subserve an emesis chemoreceptor trigger function (8). Glu may prove useful as either a provocative or ablative tool for studying the role of AP neurons and of neurotransmitter mechanisms in emesis chemoregulation in subhuman primates.

EXCITOTOXINS AS FOOD ADDITIVES: SAFETY IMPLICATIONS

Glu has long been used as a food-flavoring agent and currently remains among the additives listed by the Food and Drug Administration (FDA) as GRAS (generally regarded as safe). In the absence of regulatory restrictions, Glu was added liberally to processed infant foods for many years. In 1969, when the potential of Glu to induce brain damage in infant animals following oral administration was demonstrated, baby food manufacturers voluntarily stopped adding Glu to baby foods. About the same time, however, they began adding protein hydrolysates (rich in Glu and aspartate) in concentrations sufficient to maintain the free Glu content in baby foods at flavor levels to which the maternal palate had been conditioned. In addition, babies continued to be fed Glu-supplemented processed foods from the adult table. In 1973, at FDA request, the Federated American Societies for Experimental Biology (FASEB) formed an 11-member scientific advisory committee to review the safety of GRAS food additives. In 1976 this Committee advised FDA that neither Glu nor protein hydrolysates could be considered safe for use in baby or junior foods (98). The FASEB Committee is now finalizing a set of recommendations that presumably will guide the FDA in making a regulatory decision. Concerning the much-contested issue of primate susceptibility to Glu-induced brain damage, this Committee evaluated all relevant evidence and expressed the following opinion, "Despite the fact that the brain damage reported in neonatal mice, rats and monkeys by some research groups could not be confirmed by other groups, the Select

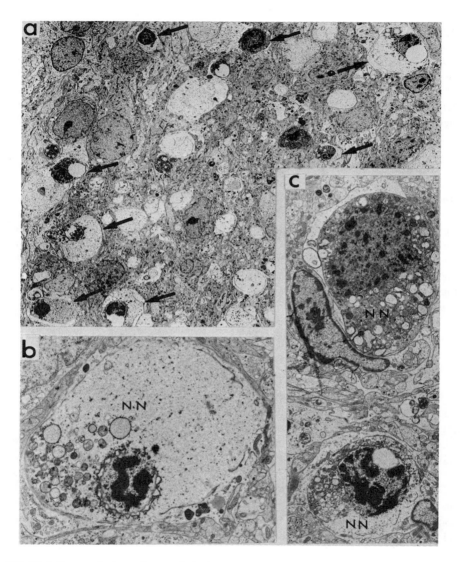

FIG. 13. **a:** Survey electron micrograph depicting a scene from an infant rhesus monkey that was tube fed a 2 g/kg dose of Glu mixed in milk 5 hr earlier. Note the numerous acutely necrotic neurons (*arrows*) throughout the field—a finding not readily explained in terms other than Glu neurotoxicity. Compare these acutely necrotic neurons and those in **b** and **c** of this figure with those from the Glu-treated infant mouse illustrated in Fig. 3b. × 950. **b,c:** Electron micrographs depicting acutely necrotic neurons (NN) from the AH region of an infant rhesus monkey that was tube fed a 1 g/kg dose of Glu mixed 5 hr earlier. In **c** there are two NN; the one above is being phagocytized by another cell having a flattened oblong nucleus. **b,** × 2,850; **c,** × 1,425.

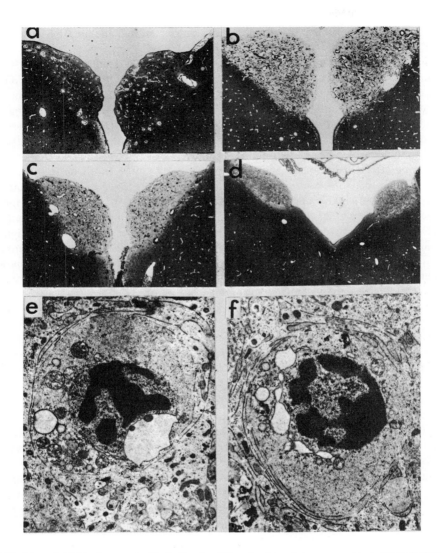

FIG. 14. **a–d:** Light micrographs illustrating the AP region from 4 rhesus monkeys, the 3 depicted in **b–d** being Glu treated and the 1 in **a** being an NaCl-treated control. The AP is a symmetrical organ in the primate, the margins of which are precisely outlined by the acute edema reaction in the Glu-treated monkeys (**b–d**), which gives the AP region an abnormally rarefied appearance not seen in the NaCl control **a**. The AP illustrated in **d** is from a pregnant rhesus monkey 5 hr after 2 g/kg Glu i.v. dose and the AP in **b** is from her fetus. The APs in **a** and **c** are from infants treated with NaCl (12 mmoles/kg) and Glu (16 mmoles/kg), respectively. **a–c,** × 27; **d,** × 18. **e,f:** Electron micrographs illustrating acutely necrotic neurons from the AP region of an infant rhesus monkey that was tube fed 2 g Glu/kg (12 mmoles/kg) mixed in milk 5 hr previously. × 1,800.

Committee concludes that the morphologic changes are real and are reproducible" (98).

I agree with the tentative conclusion of the FASEB Committee (98) that Glu cannot be considered safe for use in infant or junior foods. If the amounts of Glu added to foods ingested by immature humans were in the range of those found naturally in human milk, no safety issue would have arisen. However, it is quite predictable that the concentrations used for food flavoring will always be excessive compared to concentrations in human milk because no flavor effect is achieved from Glu in the concentration range found in milk. For example, to achieve the desired flavoring effect, baby food manufacturers have used Glu in levels up to 0.6% (> 750 mg Glu per 4.5-oz jar of strained baby food) (50). Since human breast milk contains free Glu in the range of 30 to 35 mg/150 ml, one jar of the above baby food would contain 20 to 25 times more free Glu than is found in one feeding of human milk and would provide a human infant with more than 125 mg Glu/kg body weight, which is 25% of the oral load (500 mg/kg) known to destroy hypothalamic neurons in infant animal brain (70,106). Although blood-brain barriers protect most central neurons from Glu, it must be recognized that (a) certain brain regions lack such protection; (b) it requires only a transient increase in blood Glu levels for neurons in such regions to be destroyed; (c) mechanisms for preventing transient blood Glu elevations may be ineffective in youth or disease; (d) the addition of Glu to foods ingested by immature humans entails risk without benefit (meets no health or nutritional needs). The interested reader will find a more detailed discussion of safety considerations elsewhere (60,61).

CONCLUDING REMARKS

Considerable interest has developed recently in the use of excitotoxic amino acids as research tools for producing "axon-sparing" lesions in the central nervous system. I have given an historical account revealing how this application of excitotoxins in neurobiological research developed as an outgrowth of earlier inquiries into the neurotoxic properties of Glu, particularly molecular specificity inquiries that led to the realization that the neuroexcitatory amino acids in general—not just Glu—are axon-sparing neurotoxins. I have not covered in detail the most recent research developments pertaining to the use of the more potent excitotoxins as lesioning tools because an entire book concerning this theme, with contributions from all leading researchers in the field, was recently published (45). In this review I have placed proportionally greater emphasis on the uses of Glu and its excitotoxic analogs as neuroendocrine investigational agents, a topic that has been relatively slighted in the review literature. I have attempted to convey an impression of the extent and variety of information referable to neuroendocrine mechanisms that can be obtained from simple experiments involving systemic administration of excitotoxins, either alone or in combination with other pharmacological agents, to experimental animals. Excitotoxins penetrate select portions of the endocrine hypothalamus from blood and have the dual capacity of either stimulating or destroying hypothalamic neurons, hence the versatility of serving alternatively as either provocative or ablative neuroendocrine research tools. Evidence presented here and elsewhere (59,81) documenting Glu-induced hypothalamic damage in rhesus monkeys suggests that the excitatory amino acids may be suitable agents for examining neuroendocrine regulatory mechanisms in primates as well as rodents. Risk factors associated with the use of Glu as an additive in foods ingested by human infants and children are briefly discussed.

ACKNOWLEDGMENTS

This work was supported by USPHS grants DA-00259 and NS-09156, and Research Career Development Award MH-38894.

REFERENCES

1. Abraham, R., Dougherty, W., Golberg, L., and Coulston, F. (1971): The response of the hypothalamus to high doses of monosodium glutamate in mice and monkeys: Cystochemistry and ultrastructural study of lysosomal changes. *Exp. Mol. Pathol.,* 15:43.
2. Araujo, P. E., and Mayer, J. (1973): Activity increase associated with obesity induced by monosodium glutamate in mice. *Am. J. Physiol.,* 225:764–765.
3. Arees, E., and Mayer, J. (1970): Monosodium glutamate-induced brain lesions: Electron microscopic examination. *Science,* 170:549–550.
4. Arees, E., and Mayer, J. (1972): Monosodium glutamate-induced brain lesions in mice. *J. Neuropathol. Exp. Neurol.,* 31:181.
5. Berry, H. K., Butcher, R. E., Elliot, L. A., and Brunner, R. L. (1974): The effect of monosodium glutamate on the early biochemical and behavioral development of the rat. *Dev. Psychobiol.,* 7:165–173.
6. Bhagavan, H. N., Coursin, D. B., and Stewart, C. N. (1971): Monosodium glutamate induced convulsive disorders in rats. *Nature,* 232:275–276.
7. Biscoe, T. J., Evans, R. H., Francis, A. A., Martin, M. R., Watkins, J. C., Davies, J., and Dray, A. (1977): D-α-Aminoadipate as a selective antagonist of amino acid-induced and synaptic excitation of mammalian spinal neurones. *Nature,* 270:743–745.
8. Borison, H. L., and Brizzee, K. R. (1951): Morphology of emetic chemoreceptor trigger zone in cat medulla oblongata. *Proc. Soc. Exp. Biol. Med.,* 77:38–42.
9. Bunyan, J., Murrell, E. A., and Shah, P. P. (1976): The induction of obesity in rodents by means of monosodium glutamate. *Br. J. Nutr.,* 35:25–39.
10. Burde, R. M., Schainker, B., and Kayes, J. (1971): Monosodium glutamate: Acute effect of oral and subcutaneous administration on the arcuate nucleus of the hypothalamus in mice and rats. *Nature,* 233:58.
11. Burde, R. M., Schainker, B., and Kayes, J. (1972): Monosodium glutamate: Necrosis of hypothalamic neurons in infant rats and mice following either oral or subcutaneous administration. *J. Neuropathol. Exp. Neurol.,* 31:181.
12. Cameron, D. P., Poon, T. K.-Y., and Smith G. C. (1976): Effects of monosodium glutamate administration in the neonatal period on the diabetic syndrome in KK mice. *Diabetologia,* 12:621–626.
13. Carson, K. A., Nemeroff, C. B., Rone, M. S., Youngblood, W. M., Prange, A. J., Jr., Hanker, J. S., and Kizer, J. S. (1977): Biochemical and histochemical evidence for the existence of a tuberoinfundibular cholinergic pathway in the rat. *Brain Res.,* 129:169–173.
14. Clemens, J. A., Roush, M. E., and Shaar, C. J. (1977): Effect of glutamate lesions of the arcuate nucleus on the neuroendocrine system in rats. *Neurosci Abstr.,* 3:341.
15. Cohen, A. I. (1967): An electron microscopic study of the modification by monosodium glutamate of the retinas of normal and "rodless" mice. *Am. J. Anat.,* 120:319.
16. Coyle, J. T., Biziere, K., and Schwarcz, R. (1978): Neurotoxicity of excitatory amino acids in the

neural retina. In: *Kainic Acid As A Tool In Neurobiology*, edited by E. G. McGeer, J. W. Olney, and P. L. McGeer, Raven Press, New York.
17. Coyle, J. T., McGeer, E. G., McGeer, P. L., and Schwarcz, R. (1978): Neostriatal injections: A model for Huntington's chorea. In: *Kainic Acid As A Tool In Neurobiology*, edited by E. G. McGeer, J. W. Olney, and P. L. McGeer, Raven Press, New York.
18. Cox, D. W. G., Headley, M. H., and Watkins, J. C. (1977): Actions of L- and D-homocysteate in rat CNS: A correlation between low affinity uptake and the time courses of excitation by microelectrophoretically applied L-glutamate analogs. *J. Neurochem.*, 29:579–588.
19. Curtis, D. R., and Watkins, J. C. (1960): The excitation and depression of spinal neurons by structurally related amino acids. *J. Neurochem.*, 6:117–141.
20. Djazayery, A., and Miller, D. S. (1973): The use of gold-thioglucose and monosodium glutamate to induce obesity in mice. *Proc. Nutr. Soc.*, 32:30–31A.
21. Everly, J. L. (1971): Light microscopic examination of MSG-induced lesions in brain of fetal and neonatal rats. *Anat. Rec.*, 169:312.
22. Freedman, J. K., and Potts, A. M. (1962): Regression of glutamine I in the rat retina by administration of sodium-L-glutamate II. *Invest. Ophthalmol.*, 1:118.
23. Goodman, L. S., Swinyard, E. A., and Tomans, J. E. P. (1946): Effects of L-glutamic acid and other agents on experimental seizures. *Arch. Neurol. Psychiatr.*, 56:20.
24. Gorski, R. A. (1971): Gonadal hormones and the perinatal development of neuroendocrine function. In: *Frontiers in Neuroendocrinology*, edited by L. Martini and W. F. Ganong, pp. 237–290. Oxford University Press, New York.
25. Hamatsu, T. (1964): Effect of sodium iodate and sodium L-glutamate on ERG and histological structure of retina of adult rabbits. *Nippon Ganka Gakkei Zasshi*, 68:1621.
26. Hansson, H. A. (1970): Ultrastructure studies on long-term effects of MSG on rat retina. *Virchows Arch. [Cell Pathol.]*, 6:1.
27. Helme, R. D., Lemkey-Johnston, N., and Smith, G. C. (1973): Effect of monosodium L-glutamate on the median eminence of rats and mice. *J. Anat.*, 116:466.
28. Herndon, R. M., and Coyle, J. T. (1978): Glutamergic innervation, kainic acid and selective vulnerability in the cerebellum. In: *Kainic Acid As A Tool In Neurobiology*, edited by E. G. McGeer, J. W. Olney, and P L. McGeer. Raven Press, New York.
29. Holzwarth, M. A., and Hurst, E. M. (1974): Manifestations of monosodium glutamate (MSG) induced lesions of the arcuate nucleus of the mouse. *Anat. Rec.*, 178:378.
30. Holzwarth-McBride, M. A., Hurst, E. M., and Knigge, K. M. (1976): Monosodium glutamate induced lesions of the arcuate nucleus. I. Endocrine deficiency and ultrastructure of the median eminence. *Anat. Rec.*, 186:185–196.
31. Inouye, M., and Murakami, U. (1974): Brain lesions and obesity in mouse offspring caused by maternal administration of monosodium glutamate during pregnancy. *Congenital Anomalies*, 14:77–83.
32. Johnston, G. A. R. (1973): Convulsions induced in 10-day-old rats by intraperitoneal injection of monosodium glutamate and related excitant amino acids. *Biochem. Pharmacol.*, 22:137–140.
33. Karlsen, R. L., and Fonnum, F. (1976): The toxic effect of sodium glutamate on rat retina: Changes in putative transmitters and their corresponding enzymes. *J. Neurochem.*, 27:1437–1441.
34. Kenney, R. H., and Tidball, C. S. (1972): Human susceptibility to oral monosodium glutamate. *Am. J. Nutr.*, 25:140–146.
35. Knittle, J. L., and Ginsberg-Fellner, F. (1970): Cellular and metabolic alterations in obese rats treated with monosodium glutamate during the neonatal period. *Bull. Am. Pediatr. Soc. Gen. Meeting*, Program Abstract, p. 6.
36. Lajtha, A., Berl, S., and Waelsch, H. (1959): Amino acid and protein metabolism of the brain. IV. The metabolism of glutamic acid. *J. Neurochem.*, 3:322–332.
37. Lamperti, A., and Blaha, G. (1976): The effects of neonatally-administered monosodium glutamate on the reproductive system of adult hamsters. *Biol. Reprod.*, 14:362–369.
38. Lechan, R. M., Alpert, L. C., and Jackson, I. M. D. (1976): Synthesis of luteinising hormone releasing factor and thyrotropin-releasing factor in glutamate-lesioned mice. *Nature*, 264:463–465.
39. Lemkey-Johnston, N., and Reynolds, W. A. (1974): Nature and extent of brain lesions in mice related to ingestion of monosodium glutamate. A light and electron microscope study. *J. Neuropathol. Exp. Neurol.*, 33:74–97.
40. Levey, S., Harroun, J. E., and Smyth, C. J. (1949): Serum glutamic acid levels and the occurrence of nausea and vomiting after the intravenous administration of amino acid mixtures. *J. Lab. Clin. Med.*, 34:1238–1249.

41. Lucas, D. R., and Newhouse, J. P. (1957): The toxic effect of sodium L-glutamate on the inner layers of the retina. *AMA Arch. Ophthalmol.*, 58:193.
42. Madden, S. C., Woods, R. R., Schull, F. W., Remington, J. H., and Whipple, G. H. (1945): Tolerance to amino acid mixtures and casein digests given intravenously. Glutamic acid responsible for reactions. *J. Exp. Med.*, 81:439–448.
43. McGeer, E. G., and McGeer, P. L. (1976): Duplication of biochemical changes of Huntington's chorea by intrastriatal injections of glutamic and kainic acids. *Nature*, 263:517–518.
44. McGeer, E. G., McGeer, P. L., and Singh, K. (1978): Kainate induced degeneration of neostriatal neurons: Dependency upon corticostriatal tract. *Brain Res.*, 139:381.
45. McGeer, E. G., Olney, J. W., and McGeer, P. L., ed. (1978): *Kainic Acid As A Tool In Neurobiology*. Raven Press, New York.
46. Mizukawa, K., Shimizu, K., Matsuura, T., Ibata, Y., and Sano, Y. (1976): The influence of kainic acid on the tuberoinfundibular dopaminergic tract of the rat: Fluorescence histochemistry and electron microscopic investigation. *Acta Histochem. Cytochem.*, 9:315–322.
47. Murakami, U., and Inouye, M. (1971): Brain lesions in the mouse fetus caused by maternal administration of monosodium glutamate (preliminary report) *Congenital Anomalies*, 11:171–177.
48. Nadler, J. W., Perry, B. W., and Cotman, C. W. (1978): Preferential vulnerability of hippocampus to intraventricular kainic acid. In: *Kainic Acid As A Tool In Neurobiology*, edited by E. G. McGeer, J. W. Olney, and P. L. McGeer. Raven Press, New York.
49. Nagasawa, H., Yanai, R., and Kikuyama, S. (1974): Irreversible inhibition of pituitary prolactin and growth hormone secretion and of mammary gland development in mice by monosodium glutamate administered neonatally. *Acta Endocrinol.* 75:249–259.
50. National Research Council (1970): *Safety and Suitability of MSG and Other Substances in Baby Foods.* (Report of Subcommittee.) National Academy of Sciences, Washington, D.C.
51. Nemeroff, C. B., Grant, L. D., Bissette, G., Ervin, G. N., Harrell, L. E., and Prange, A. J., Jr. (1977): Growth, endocrinological and behavioral deficits after monosodium L-glutamate in the neonatal rat: Possible involvement of arcuate dopamine neuron damage. *Psychoneuroendocrinology*, 2:179–196.
52. Nemeroff, C. B., Grant, L. D., Harrell, L. E., Bissette, G. Ervin, G. N., and Prange, A. J., Jr. (1975): Histochemical evidence for the permanent destruction of arcuate dopamine neurons by neonatal monosodium L-glutamate in the rat. *Neurosci. Abstr.*, 1:434.
53. Nemeroff, C. B., Konkol, R. J., Bissette, G., Youngblood, W., Martin, J. B., Brazeau, P., Rone, M. S., Prange, A. J., Jr., Breese, G. R., and Kizer, J. S. (1977): Analysis of the disruption in hypothalamic-pituitary regulation in rats treated neonatally with monosodium L-glutamate (MSG): Evidence for the involvement of tuberoinfundibular cholinergic and dopaminergic systems in neuroendocrine regulation. *Endocrinology*, 101:613–622.
54. Nikoletseas, N. M. (1978): Obesity in exercising, hypophagic rats treated with monosodium glutamate. *Physiol. Behavior*, 19:767–773.
55. Olney, J. W. (1969): Brain lesions, obesity and other disturbances in mice treated with monosodium glutamate. *Science*, 164:719–721.
56. Olney, J. W. (1969): Glutamate-induced retinal degeneration in neonatal mice. Electron microscopy of the acutely evolving lesion. *J. Neuropathol. Exp. Neurol.*, 28:455–474.
57. Olney, J. W. (1971): Glutamate-induced neuronal necrosis in the infant mouse hypothalamus. An electron microscopic study. *J. Neuropathol. Exp. Neurol.*, 30:75–90.
58. Olney, J. W. (1972): A model for manipulating CNS cytotoxic-effects between neuronal and non-neuronal compartments. *J. Neuropath. Exp. Neurol.*, 31:181.
59. Olney, J. W. (1974): Toxic effects of glutamate and related amino acids on the developing central nervous system. In: *Heritable Disorders of Amino Acid Metabolism*, edited by W. L. Nyhan, pp. 501–512. John Wiley & Sons, Inc., New York.
60. Olney, J. W. (1975): Another view of Aspartame. In: *Sweeteners: Issues and Uncertainties.* National Academy of Science Forum, G.P.O., Washington, D.C.
61. Olney, J. W. (1976): Brain damage and oral intake of certain amino acids. In: *Transport Phenomena in the Nervous System: Physiological and Pathological Aspects* (Advances in Experimental Medicine & Biology 69:497–506), edited by G. Levi, L. Battistin and A. Lajtha, Plenum Press, New York.
62. Olney, J. W. (1978): Neurotoxicity of excitatory amino acids. In: *Kainic Acid As A Tool In Neurobiology*, edited by E. G. McGeer, J. W. Olney, and P. L. McGeer. Raven Press, New York.
63. Olney, J. W., Cicero, T. J., Meyer, E. R., and de Gubareff, T. (1976): Acute glutamate-induced elevations in serum testosterone and luteinizing hormone. *Neurosci. Abstr.*, 2:677.

64. Olney, J. W., Cicero, T. J., Meyer, E. R., and de Gubareff, T. (1976): Acute glutamate induced elevations in serum testosterone and luteinizing hormone. *Brain Res.*, 112:420–424.
65. Olney, J. W., Fuller, T., and de Gubareff, T. (1977): Alanosine—an antileukemic agent with glutamate-like neurotoxic properties. *J. Neuropathol. Exp. Neurol.*, 36:619.
66. Olney, J. W., and de Gubareff, T. (1978): Glutamate neurotoxicity and Huntington's chorea. *Nature*, 271:557–559.
67. Olney, J. W., and de Gubareff, T. (1978): The fate of synaptic receptors in the kainate-lesioned striatum. *Brain Res.*, 140:340–343.
68. Olney, J. W., and de Gubareff, T. (1978): Extreme sensitivity of olfactory cortical neurons to kainic acid toxicity. In: *Kainic Acid As A Tool In Neurobiology*, edited by E. G. McGeer, J. W. Olney, and P. L. McGeer. Raven Press, New York.
69. Olney, J. W., de Gubareff, T., and Misra, C. H. (1977): D and L Isomers of homocysteic acid compared for neurotoxicity. *Neurosci. Abstr.*, 3:518.
70. Olney, J. W., and Ho, O. L. (1970): Brain damage in infant mice following oral intake of glutamate, aspartate or cysteine. *Nature*, 227:609–610.
71. Olney, J. W., Ho, O. L., and Rhee, V. (1971): Cytotoxic effects of acidic and sulphur containing amino acids on the infant mouse central nervous system. *Exp. Brain Res.*, 14:61–76.
72. Olney, J. W., Ho, O. L., Rhee, V., and de Gubareff, T. (1973): Neurotoxic effects of glutamate. *N. Engl. J. Med.*, 289:1374–1375.
73. Olney, J. W., Misra, C. H., and de Gubareff, T. (1975): Cysteine-S-sulfate: Brain damaging metabolite in sulfite oxidase deficiency. *J. Neuropathol. Exp. Neurol.*, 34:167.
74. Olney, J. W., Misra, C. H., and Rhee, V. (1976): Brain and retinal damage from the lathyrus excitotoxin, β-N-oxalyl-L-α,β-diaminopropionic acid (ODAP). *Nature*, 264:659–661.
75. Olney, J. W., and Price, M. T. (1978): Excitotoxic amino acids as neuroendocrine probes. In: *Kainic Acid As A Tool In Neurobiology*, edited by E. G. McGeer, J. W. Olney, and P. L. McGeer. Raven Press, New York.
76. Olney, J. W., Rhee, V., and de Gubareff, T. (1977): Neurotoxic effects of glutamate on mouse area postrema. *Brain Res.*, 120:151–157.
77. Olney, J. W., Rhee, V., and Ho, O. L. (1974): Kainic acid: A powerful neurotoxic analogue of glutamate. *Brain Res.*, 77:507–512.
78. Olney, J. W., Schainker, B., and Rhee, V. (1975): Chemical lesioning of the hypothalamus as a means of studying neuroendocrine function. In: *Hormones Behavior and Psychopathology*, edited by E. Sachar, pp. 153–158. Raven Press, New York.
79. Olney, J. W., and Sharpe, L. G. (1969): Brain lesions in an infant rhesus monkey treated with monosodium glutamate. *Science*, 166:386–388.
80. Olney, J. W., and Sharpe, L. G. (1970): Monosodium glutamate: Specific brain lesion questioned. *Science*, 167:1016–1017.
81. Olney, J. W., Sharpe, L. G., and Feigin, R. D. (1972): Glutamate-induced brain damage in infant primates. *J. Neuropathol. Exp. Neurol.*, 31:464–488.
82. Olney, J. W., Sharpe, L. G., and de Gubareff, T. (1975): Excitotoxic amino acids. *Neurosci. Abstr.*, 1:371.
83. Paull, W. K. (1975): An autoradiographic analysis of arcuate neuron sensitivity to monosodium glutamate. *Anat. Rec.*, 181:445.
84. Paull, W. K., and Lechan, R. (1974): The median eminence of mice with a MSG induced arcuate lesion. *Anat. Rec.*, 178:436.
85. Perez, V. J., and Olney, J. W. (1972): Accumulation of glutamic acid in the arcuate nucleus of the hypothalamus of the infant mouse following subcutaneous administration of monosodium glutamate. *J. Neurochem.*, 19:1777–1782.
86. Pizzi, W.J., and Barnhart, J. E. (1976): Effects of monosodium glutamate on somatic development, obesity and activity in the mouse. *Pharmacol. Biochem. Behav.*, 5:551–557.
87. Pizzi, W. J., Barnhart, J. E., and Fanslow, D. J. (1977): Monosodium glutamate administration to the newborn reduces reproductive ability in female and male mice. *Science*, 196:452–454.
88. Pizzi, W. J., Tabor, J. M., and Barnhart, J. E. (1977): Somatic, behavioral, and reproductive disturbances in mice following early administration of sodium L-aspartate. *Neurosci. Abstr.*, 3:355.
89. Poon, T. K.-Y., and Cameron, D. P. (1976): Effects of monosodium glutamate (MSG) on diabetes and obesity in KK mice. *Aust. N.Z. J. Med.*, 6:247.
90. Potts, A. M., Modrell, K. W., and Kingsbury, C. (1960): Permanent fractionation of the ERG by sodium glutamate. *Am. J. Ophthalmol.*, 50:900.

91. Price, M. T., Olney, J. W., and Cicero, T. J. (1978): Acute elevation of serum luteinizing hormone induced by kainic acid, N-methyl aspartic acid or homocysteic acid. *Neuroendocrinology* 26:352–358.
92. Redding, T. W., Schally, A. V., Arimura, A., and Wakabayashi, I. (1971): Effect of monosodium glutamate on some endocrine functions. *Neuroendocrinology*, 8:245–255.
93. Reif-Lehrer, L. (1976): Possible significance of adverse reactions to glutamate in humans. *Fed. Proc.*, 35:2205–2211.
94. Reif-Lehrer, L. (1977): A questionnaire study of the prevalence of Chinese Restaurant Syndrome. *Fed. Proc.*, 36:1617–1623.
95. Reif-Lehrer, L., and Stemmermann, M. C. (1975): Monosodium glutamate intolerance in children. *N. Engl. J. Med.*, 293:1204–1205.
96. Schainker, B., and Olney, J. W. (1974): Glutamate-type hypothalamic-pituitary syndrome in mice treated with aspartate or cysteate in infancy. *J. Neurotrans.*, 35:207–215.
97. Schaumburg, H. H., Byck, R., Gerstl, R., and Mashman, J. H. (1969): Monosodium glutamate: Its pharmacology and role in the Chinese Restaurant Syndrome. *Science*, 163:826–828.
98. Select Committee on GRAS Substances (1976): *Health Aspects of Certain Glutamates as Food Ingredients, Tentative Report Prepared for the Food and Drug Administration*. Federated American Societies for Experimental Biology, Washington, D.C.
99. Snapir, N., Robinzon, B., and Perek, M. (1973): Development of brain damage in the male domestic fowl injected with monosodium glutamate at five days of age. *Pathol. Eur.*, 8:265–277.
100. Stewart, C. N., Coursin, D. B., and Bhagavan, H. N. (1972): Electroencephalographic study of L-glutamate induced seizures in rats. *Toxicol. Appl. Pharmacol.*, 23:635–639.
101. Tafelski, T. J. (1976): Effects of monosodium glutamate on the neuroendocrine axis of the hamster. *Anat. Rec.*, 184:543–544.
102. Terry, L. C., Epelbaum, J., Brazeau, P., and Martin, J. B. (1977): Monosodium glutamate: Acute and chronic effects on growth hormone (GH), prolactin (Prl) and somatostatin (SRIF) in the rat. *Fed. Proc.*, 36:500.
103. Trentini, G. P., Botticelli, A., and Botticelli, C. S. (1974): Effects of monosodium glutamate on the endocrine glands and on the reproductive function of the rat. *Fertil. Steril.* 25:478–483.
104. Unna, K., and Howe, E. E. (1945): Toxic effects of glutamic and aspartic acid. *Fed. Proc.*, 4:138.
105. Weindl, A. (1973): Neuroendocrine aspects of circumventricular organs. In: *Frontiers in Neuroendocrinology*, edited by L. Martini and W. F. Ganong, pp. 1–32. Oxford University Press, London.
106. Worden, A. N. (1977): Paper presented at Hearings of Select Committee on GRAS Substances of the Federated American Societies for Experimental Biology. Washington, D.C., July 1977.

Glutamic Acid: Advances in Biochemistry and Physiology, edited by L. J. Filer, Jr., et al.
Raven Press, New York © 1979.

Attempts to Establish the Safety Margin for Neurotoxicity of Monosodium Glutamate

L. Airoldi, A. Bizzi, M. Salmona, and S. Garattini

Istituto di Ricerche Farmacologiche "Mario Negri," 20157 Milan, Italy

For every chemical, with practically no exceptions, it is possible to find a route of administration and a dose that will induce a toxic effect in some animal species. Once this toxic effect has been found, the problem is to establish its relevance to man. Usually, a safety ratio is calculated by dividing the dose that induces a toxic effect in a given animal species by the dose utilized in man; the larger the ratio, the safer the chemical. More recently, however, there has been a trend to use not the dose, but the plasma, or tissue level, of the chemical under study as a more reliable parameter in making calculations (1,3). It is, in fact, known that different animal species dispose of chemicals in a different manners, thus making the plasma or tissue concentrations a better parameter than the dose for extrapolating biological activity across animal species.

This is the approach utilized here in an attempt to interpret the relevance for man of some interesting findings concerning the neurotoxicity of an ubiquitous amino acid, glutamic acid (Glu).

Necrotic lesions of the hypothalamus (arcuate nucleus) have been found as a result of the administration of monosodium glutamate (MSG) to newborn animals (12,14–17). The mouse appears to be the most sensitive species, followed by rats and guinea pigs (Heywood and Worden, *this volume*). At variance with initial observations (17; Olney, *this volume*) in a variety of experimental conditions, monkeys are apparently insensitive to this toxic effect of MSG (11,21,24; Reynolds et al., *this volume*).

BASAL PLASMA AND BRAIN LEVELS IN DIFFERENT ANIMAL SPECIES

Preliminary investigations indicated that the basal plasma concentration of Glu did not substantially change in rats as a result of overnight fasting, so no reference will be made to the nutritional conditions of control animals. Similarly, no significant differences were found between the plasma or brain Glu levels in male and female rats. Furthermore, no difference was found between venous (jugular vein) and arterial (carotid artery) concentrations of Glu, either in the basal condition or after a load of Glu in rats. The ratio of plasma to red cells also showed no change in arterial and venous blood following administration of Glu.

TABLE 1. *Glu in plasma and brain of several animal species at different ages*

Animal species	Age (days)	Plasma (μmoles/ml ± SE)	Brain (μmoles/g ± SE)
Mouse	1	0.15 ± 0.01	—
Mouse	7	0.15 ± 0.01	4.42 ± 0.18[a]
Mouse	Adults	0.17 ± 0.01	8.21 ± 0.21
Rat	1	0.25 ± 0.02	3.21 ± 0.14[a]
Rat	7	0.20 ± 0.01	4.83 ± 0.14[a]
Rat	15	0.17 ± 0.01	6.50 ± 0.06[a]
Rat	Adults	0.15 ± 0.02	8.05 ± 0.25
Guinea pig	1	0.19 ± 0.01	8.01 ± 0.19
Guinea pig	7	0.17 ± 0.01	8.04 ± 0.15
Guinea pig	15	0.23 ± 0.01	9.15 ± 0.41
Guinea pig	Adults	0.20 ± 0.01	7.56 ± 0.15
Rabbit	Adults	0.12 ± 0.01	10.65 ± 0.40
Dog (beagle)	Adults	0.050 ± 0.003(81)	5.80 ± 0.02
Monkey (rhesus)	Adults	0.13 ± 0.01(22)	9.14 ± 1.97
Human	Premature newborns	0.50 ± 0.07(21)	—
Human	Adults	0.06 ± 0.003(109)	—

Glu was assayed by an enzymatic method according to Bernt and Bergmeyer (2). Results represent the average of at least five determinations unless otherwise indicated in parentheses.
[a] $p < 0.01$ compared to values in adults of the same species (Student's t- and Dunnett tests).

Table 1 indicates that the plasma Glu levels in mice, rats, guinea pigs, rabbits, and rhesus monkeys are not markedly different. In contrast, dog and man show Glu plasma values that are about 50% lower than the other species considered. Furthermore, they are not influenced by growth, the levels in newborns and adults being similar in mice, rats, and guinea pigs, with the possible exception of the high plasma Glu levels of newborn rats the first day of life. Conversely, human premature newborns show much higher Glu plasma levels than adults. The brain levels of Glu are comparable in adult animals with the exception of dogs, which have the lowest level among the species investigated.

In mice and rats, brain Glu was low at birth and increased with age, whereas in guinea pigs, brain levels were similar at and after birth. These results are in agreement with the finding that guinea pigs are born with a more mature brain than mice and rats (6). It should be recalled that human newborns are also considered relatively more mature in their brain myelinization than rats and mice.

KINETICS OF Glu IN NEWBORNS AND ADULTS

Administration by gavage of a standard dose of MSG (1 g/kg, 10% w/v) results in a marked increase of plasma Glu in all the animal species studied (Table 2). However, the extent of the increase is different. Peak plasma Glu levels are about 12 times higher than the basal concentration in mice, 13 in rats, 11 in guinea pigs, 2 in rabbits, 35 in dogs, and 4 in monkeys. The area under the curve (AUC) is also different, being lowest in rabbits and highest in guinea pigs; the plasma half-life

TABLE 2. *Kinetic parameters for Glu in different animal species after administration by gavage of 1 g/kg of MSG (10% w/v)*

Animal species	Age (days)	Plasma peak (μmoles/ml ± SE)	Plasma AUC (μmoles/ml × min)	Plasma $T_{1/2}$ (min)
Mouse	7	2.10 ± 0.06	375	111
Mouse	90	2.08 ± 0.02	309	98
Rat	7	2.89 ± 0.11	1,391	237
Rat	90	1.91 ± 0.09	288	88
Guinea pig	7	1.91 ± 0.02	321	101
Guinea pig	90	2.28 ± 0.03	487	173
Rabbit	Adults	0.29 ± 0.14	13	53
Dog (beagle)	Adults	1.78 ± 0.31[a]	143	49
Monkey (rhesus)	Adults	0.58 ± 0.33[a]	53	99

Results represent the average of at least five animals.
[a] Vomiting occurred.

TABLE 3. *Kinetic parameters of plasma Glu in 7-day-old mice after administration by gavage of different doses of MSG (10% w/v)*

Dose (g/kg)	Plasma peak (μmoles/ml ± SE)	Plasma AUC (μmoles/ml × min)	Plasma $T_{1/2}$ (min)
0.25	0.72 ± 0.04	103	79
0.50	1.07 ± 0.02	200	71
1.00	2.10 ± 0.05	375	111

Results represent the average of 5 animals.

$T_{1/2}$ ranges from 49 min in dogs to 173 min in guinea pigs. There are also, as summarized in Table 2, quantitative age-dependent differences. For rats, and to a lesser extent for mice too, plasma Glu levels expressed as AUC are higher for newborns than adults, whereas for guinea pigs the opposite holds true. This confirms the importance of maturity at birth as a factor in the ability to dispose of Glu. Table 3 shows that the plasma peak level, the plasma AUC, but not the $T_{1/2}$, increase linearly for newborn mice as a function of the MSG dose given by tube.

The concentration at which MSG is given is also an important factor in the kinetics of Glu. At equal doses the MSG concentration determines the volume in which it is given and hence the time required to administer the dose orally to newborn animals. Table 4 confirms that the plasma peak rises for newborn mice and rats as the concentration at which MSG is given increases. When the same dose of MSG (1 g/kg) is given to newborn rats by tube, the plasma AUC is five times higher when the concentration is increased from 2 to 10%. Similarly in mice, given an oral dose of 0.5 g/kg at concentrations from 2 to 20%, the AUC increases by a factor of about 2.5. Similar results were obtained in adult animals. However, the concentration at which MSG is given becomes less important when the dose is increased (7).

TABLE 4. *Kinetic parameters of plasma Glu in 7-day-old mice and rats after administration by gavage of MSG at different concentrations*

Animal species	Dose (g/kg)	Concentration (% w/v)	Plasma peak (μmoles/ml ± SE)	Plasma AUC (μmoles/ml × min)	Plasma $T_{1/2}$ (min)
Mouse	0.5	2	0.82 ± 0.03	126	88
Mouse	0.5	5	0.86 ± 0.04	147	117
Mouse	0.5	10	1.08 ± 0.02	200	71
Mouse	0.5	20	1.43 ± 0.02	318	106
Rat	1.0	2	0.97 ± 0.05	275	183
Rat	1.0	5	2.26 ± 0.36	1,085	237
Rat	1.0	10	2.89 ± 0.11	1,391	241

Different times were required for administering MSG by gavage at different concentrations. Results represent the average of five animals.

TABLE 5. *Plasma peak, AUC, and $T_{1/2}$ of MSG given to adult mice either by gavage or mixed with the diet*

MSG (g/kg)	Percent	Plasma peak (μmoles/ml ± SE)	Plasma AUC (μmoles/ml × min)	Plasma $T_{1/2}$ (min)
1.00	10[a]	2.08 ± 0.02	309	98
1.00	Meal[b]	0.27 ± 0.01	4.8	—

Results represent the average of five animals.
[a] By gavage (1 min).
[b] Mixed with food (30 min).

PLASMA LEVELS OF Glu WHEN MSG IS GIVEN WITH FOOD

Usually, Glu is not given to humans by tube feeding, but is added in various amounts to food eaten in a given period of time. It therefore seems reasonable to investigate differences in plasma Glu levels in laboratory animals when the substance is given by tube feeding or mixed with food.

To this end, adult mice were trained to eat their daily food intake during 1 hr in the morning. Glu (as MSG) was added to a commercial diet in such a way as to give 1 g/kg body weight during an eating period of 30 min; mice were then killed at various times after the meal for determination of plasma Glu and for calculation of the kinetic parameters (Table 5).

It is interesting to note that 1 g/kg of MSG given by tube feeding induces a high peak level of plasma Glu and a large AUC, whereas when given with the meal results in only very slight increases in plasma Glu. Hence, when MSG is given with a meal, it is much more slowly absorbed and/or much more rapidly disposed than when given by gavage. The reasons for this difference are not yet clear, but may include decreased absorption of MSG due to competition with other amino acids,

TABLE 6. *Relation between plasma Glu peak and brain concentrations in different species (newborn and adult animals)*

Animal species	Age (days)	Oral dose MSG (g/kg)	Glu factor of increase	
			Plasma	Brain
Mouse	3	0.5 (5%)	8	NS
	7	1 (10%)	13	NS
	90	1 (10%)	12	NS
Rat	3	1 (10%)	14	NS
	7	1 (10%)	18	NS
	7	2 (5%)	12	NS
	7	2 (20%)	19.6	1.6[a]
	90	1 (10%)	10	NS
Guinea pig	7	2 (20%)	12	NS
	90	1 (10%)	12	NS

NS, not significant.
[a] $p < 0.05$.

increased metabolism in the intestine due to the slower rate of absorption, interaction with other nutrients, and others.

LEVELS OF Glu IN BRAIN

Throughout the variety of conditions described above, the levels of brain Glu remained unchanged in both mice and rats, adults and newborns, even when plasma Glu levels were up to 15 times the basal concentrations. Rat brain Glu concentration significantly increased, by about 60%, only when the oral dose of MSG was raised to 2 g/kg (20% w/v) and peak plasma Glu exceeded the basal level by about 19 times (Table 6).

The fact that the level of whole brain Glu was not changed even by doses of MSG that induce some brain damage (15) does not exclude the possibility, as shown by Perez and Olney (19), that Glu may increase in a small brain area such as the arcuate nucleus of the hypothalamus, where the amino acid has been seen to produce toxic effects (necrosis). However, recent data (8) indicate that in adult rats Glu does not substantially increase in the retina, another tissue where, nevertheless, it causes damage (4).

In relation to these data, it was of interest to study whether Glu was capable of accumulating in discrete brain areas, particularly the nucleus arcuatus. Experiments were carried out in rats. The nucleus arcuatus was obtained by the punching technique of Palkovitz et al. (18). The thalamus lateralis was selected as a control area. The results obtained are reported in Table 7 for adult rats and in Table 8 for newborn rats. The levels of Glu are expressed in nmoles/mg protein. Doses of MSG that increase plasma levels by about 11 times in both adult and newborn rats do not affect the level of Glu in the arcuate nucleus or in the lateral thalamus (L. Airoldi et

TABLE 7. *Relation between Glu levels in plasma and in the nucleus arcuatus and lateral thalamus in adult male rats*

Min. after treatment	Plasma Glu (μmoles/ml ± SE)	Glu (nmoles/mg protein ± SE)	
		N. arcuatus	L. thalamus
0	0.21 ± 0.02	132 ± 5	110 ± 4
15	1.56 ± 0.21	136 ± 5	113 ± 9
30	2.30 ± 0.11	136 ± 7	118 ± 2
60	1.63 ± 0.08	125 ± 6	117 ± 5
180	1.43 ± 0.09	131 ± 9	109 ± 2
360	0.62 ± 0.08	125 ± 6	112 ± 6

MSG was given by gavage at the dose of 4 g/kg (20% w/v). Glu was measured according to the method of Young et al. (25). Results represent the average of five animals.

TABLE 8. *Relation between Glu levels in plasma and in the nucleus arcuatus and lateral thalamus in 4-day-old rats*

Min after treatment	Plasma Glu (μmoles/ml ± SE)	Glu (nmoles/mg protein ± SE)	
		N. arcuatus	L. thalamus
0	0.23 ± 0.02	114 ± 10	124 ± 14
60	2.76 ± 0.37	115 ± 3	126 ± 23
180	0.86 ± 0.28	123 ± 10	128 ± 4

MSG was given by gavage at the dose of 2 g/kg (20% w/v). Glu was measured according to the method of Young and Lowry (25). Results represent the average of five animals.

al., *unpublished*). The reason for the discrepancy between these results in rats and the data obtained by Perez et al. (20) in mice is at present unknown. However, it seems that the explanation given for the selective neurotoxic effect of MSG in mice cannot be generalized to all animal species.

SIGNIFICANCE OF KINETIC DATA FOR THE NEUROTOXICITY OF Glu

The kinetic data previously summarized might help us interpret the incidence of hypothalamic lesions observed in newborn and adult animals (12,14). Table 9 summarizes some of the data reported in the literature or in this book concerning the incidence of hypothalamic lesions in various animal species, both newborns and adults. The newborn mouse is the species most sensitive to the neurotoxic effect of MSG given by gavage, followed by rats and guinea pigs. For monkeys there are contradictory reports, but most of the authors could not find the lesion. Dogs are also apparently insensitive (Heywood and Worden, *this volume*).

The effect of chemicals always depends on two factors: the concentrations present at the site of action and the sensitivity of the target organ. If we try to correlate the

TABLE 9. *Correlation between the incidence of hypothalamic lesions and Glu plasma AUC after administration of MSG by stomach tube to several animal species*

Dose (g/kg)	Incidence of hypothalamic lesions (and AUC)	
	Newborns	Adults
0.5	0 (—)[a] Rat 0 (200) Mouse	
1.0	0 (813) Rat 0 (321) Guinea pig 100 (375) Mouse	0 (288) Rat 0 (309) Mouse 0 (487) Guinea pig 0 (143) Dog 0 (53) Rhesus monkey
2.0	60 (—)[a] Rat 20 (516) Guinea pig	30 (—)[a] Rat 0 (179) Dog 0 (86) Rhesus monkey
4.0	100 (—)[a] Rat 80 (—)[a] Guinea pig	60 (494) Rat 0 (187) Dog 0 (150) Rhesus monkey

[a] Data not available.

exposure to MSG, expressed as plasma AUC (Table 9) with hypothalamic lesions, we note that no lesions are observed unless the AUC is more than 200 μmoles/ml $\times$ min. However, this value may not be enough to induce a lesion, as shown for instance in newborn rats and guinea pigs. It should be stressed that this conclusion is tentative, since so little data is available and most of the reports used for this correlation were from different laboratories.

The discussion can be more elaborate concerning mice, since this species is the most sensitive to the neurotoxicity of MSG and has therefore been widely tested. Table 10 summarizes the percentage of mice with hypothalamic lesions according to several authors who used various doses and concentrations of MSG. The toxicological data compared with the plasma AUC obtained under similar conditions show that the percentage of lesions is related more to the plasma AUC than to the dose of MSG administered. For instance, the same dose of MSG (0.5 g/kg) can give 50% lesions if used at a 20% concentration or no lesions when employed at 2, 5, or 10%. The total exposure to MSG, therefore, expressed in this study by the AUC value, may be a better index of toxicity than the administered dose. Within the limits of the present results, it looks as though plasma exposure to MSG over about 200 μmoles/ml $\times$ min (or about 33.330 μg/ml $\times$ min) may result in hypothalamic damage.

PLASMA Glu LEVELS IN HUMAN PREMATURE NEWBORNS

Plasma Glu levels measured in 21 premature newborns (average weight 2,008 $\pm$ 17 g, age 19 $\pm$ 3 days) in blood obtained from the umbilical vein were considerably higher than in human adults (0.50 $\pm$ 0.07 μmoles/ml $\pm$ SE as opposed to 0.06

TABLE 10. *Correlation between hypothalamic lesions and plasma AUC in newborn mice after administration of MSG at various doses and concentrations*

Oral dose (g/kg)	Concentration (% w/v)	Hypothalamic lesions (%)	Plasma AUC (μmoles/ml × min)
0.25	10	0[a]	103
0.50	2	0[b]	126
0.50	5	0[b]	147
0.50	10	0[c]	200
0.50	20	50[a]	318
0.75	20	75[a]	—
1.00	2	50[b]	—
1.00	5	80[b,d]	—
1.00	10	100[b]	375
1.00	20	100[a]	—

[a] From Olney, ref. 13.
[b] R. Heywood et al., *personal communication*.
[c] From Takasaki, ref. 23.
[d] From Lemkey-Johnston and Reynolds, ref. 9.

± 0.003 μmoles/ml ± SE). There was no correlation between Glu plasma level and weight ($r^2 = 0.25$) or age ($r^2 = 0.30$).

Human milk is known to contain relatively high levels of free Glu (22). By calculating the amount of milk given with a meal, it is possible to work out that premature newborns receive between 1 and 11 mg/kg (2.76 ± 0.51) of Glu in a single intake.

In blood samples taken for medical reasons from premature newborns, it was possible to measure the levels of Glu at various times after a meal. The results obtained show that Glu plasma levels were 0.54 ± 0.09 μmoles/ml (± SE) when measured between 5 and 90 min after the milk meal. This value is not statistically different from the basal value. These studies indicate that premature newborns can metabolize Glu after a milk meal so that plasma Glu levels are kept within the range of normal values (B. Assael et al., *unpublished*).

PLASMA Glu LEVELS IN MAN

As previously mentioned, the levels of plasma Glu in man are somewhat lower than in the rodents examined in this study. A total of 109 volunteers (49 females and 60 males, 26.04 ± 0.53 years of age and 62.03 ± 0.99 kg body weight) received a dose of 60 mg/kg of MSG with a bouillon (2% solution), which is a large dose for any food enriched with MSG. The subjects were fasted for 5 hr and were then required to drink the bouillon in 3 min, after which they received no further food for another 5 hr. Blood samples were withdrawn before MSG and 15, 30, 60, and 120 min after MSG.

The average peak level was reached 30 min after MSG administration (0.194 ± 0.09 μmoles/ml). The apparent plasma $T_{1/2}$ was 68 ± 4.7 min, while the AUC was

TABLE 11. *AUC of plasma Glu after oral doses of MSG in man*

No. subjects	Dose MSG		AUC
	mg/kg	Percent	(μmoles/min × min)
7	30	2	1.69
109	60	2	5.56
6	120	4	8.00
5	60	Tomato juice	2.89
5	—	Meal	0.22

different ($p < 0.05$) for females (4.74 ± 0.53 μmoles/ml × min) and males (6.23 ± 0.49 μmoles/ml × min). The distribution of basal Glu levels and of the AUC was normal (0.06 and 0.08 by Kolmogorov-Smiznov's test); there was no correlation between basal Glu and AUCs (linear regression) or between basal plasma levels or AUCs and age, weight, or cigarette, coffee, or tea consumption (multiple regression analysis). The AUCs of subjects with postprandial symptoms (headache, gastric acidity, etc.) were not statistically different from subjects free of symptoms. Previous studies performed according to a double-blind design by Morselli et al. (10) did not show any specific effect of MSG in terms of side effects (26).

Table 11 summarizes the plasma AUC in adults after the administration of 30, 60, and 120 mg/kg of MSG. Plasma AUCs are clearly proportional to the dose given. Table 11 also gives the plasma AUC values obtained after adding 60 mg/kg of MSG to tomato juice at the concentration of 2%. MSG added to tomato juice results in lower Glu plasma levels than when it is given with a bouillon. Therefore, in man, as in mice, MSG taken with nutrients gives considerably lower plasma Glu levels.

RELEVANCE OF Glu NEUROTOXICITY FOR MAN

The data from the literature summarized above (see also Table 9) indicate that the newborn mouse appears to be the most sensitive animal species to the neurotoxicity of MSG. In the absence of better criteria, it seems safest to take the kinetic data of newborn mice as a basis for extrapolating safety ratios from animals to man.

As reported for mice in Table 10, an ED_{50} of MSG (dose producing hypothalamic lesions in 50% of the subjects; i.e., 500 mg/kg by gavage in solution at 20%) results in a plasma AUC of 318 μmoles of Glu/ml × min. If this kinetic parameter is compared with the AUC obtained in man after different doses of MSG, it is possible to obtain the safety ratios reported in Table 12. A ratio of 188 is calculated for human ingestion of a single dose 14 to 15 times the U.S. average daily intake of MSG (5). This figure becomes 57 when a dose of 60 mg/kg is utilized. It should be remembered that this is the largest single dose still palatable, at least according to Western taste preferences. Even at unpalatable doses (120 mg/kg of MSG), unlikely to be ingested under normal conditions, the safety ratio is still 40.

TABLE 12. Safety ratios for MSG calculated assuming that the sensitivity of man to the neurotoxic effect of MSG is similar to that of newborn mice

Animal species	Oral dose (mg/kg)	AUC (μmoles/ml × min)	Safety ratio according to Dose	Safety ratio according to AUC
Mouse	500	318	—	—
Man	30[a]	1.69	16.6	188
	60	5.56	8.3	57
	120[b]	8.00	1.6	40
	60[c]	2.89	8.3	110
	Meal	0.22	—	1,445

[a] This is 14–15 times the average daily intake of MSG in U.S.
[b] Unpalatable.
[c] Dissolved in tomato juice.

It should, however, be clearly stated that these calculations have only theoretical value because they have no direct bearing on the utilization of MSG as a food additive. In fact, when 60 mg/kg of MSG are given in tomato juice instead of in aqueous solution, the safety ratio increases to 110.

Furthermore, when MSG was given to sensitive animal species with a meal even at very high doses and for prolonged periods of time brain lesions or any other toxic effect were not reported (Takasaki, *this volume*), reinforcing the view that gavage of aqueous solutions is a route of administration that gives rise to high Glu plasma levels not achievable by regular feeding.

The data utilized for establishing safety ratios were obtained in adult humans. It is, however, believed that these conclusions can also be applied to infants in view of the fact that newborn prematures can metabolize free Glu at the dose of 2.7 ± 0.51 mg/kg (see above) and that within the range of doses at which MSG can be consumed, Filer et al. (*this volume*) have not observed any difference between infants and adults in absorption and/or disposal of MSG.

In conclusion, therefore, although Olney's findings are extremely interesting in terms of utilizing MSG as a toxicological tool, it is fortunate that they were obtained under a number of experimental conditions (animal species, age, maturity, route of administration, dose, and concentration of MSG) that all tend to induce high, long-lasting plasma Glu levels not likely to occur in man, infant, or adult, even under extreme conditions of MSG intake.

REFERENCES

1. Baker, S. B. de C., and Foulkes, D. M. (1973): Blood concentrations of compounds in animal toxicity tests. In: *Biological Effects of Drugs in Relation to Their Plasma Concentrations*, edited by D. S. Davies and B. N. C. Prichard, pp. 41–50. Macmillan, London.
2. Bernt, E., and Bergmeyer, H. U. (1974): UV assay with glutamate dehydrogenase and NAD. In: *Methods of Enzymatic Analysis, Vol. 4*, edited by H. U. Bergmeyer, pp. 1704–1708. Academic Press, New York.
3. Bizzi, A., Veneroni, E., Salmona, M., and Garattini, S. (1977): Kinetics of monosodium glutamate in relation to its neurotoxicity. *Toxicol. Lett.*, 1:123–130.

4. Cohen, A. I. (1967): An electron microscopic study of the modification by monosodium glutamate of the retinas of normal and "rodless' mice. *Am. J. Anat.*, 120:319–326.
5. Committee on GRAS List Survey—Phase III (1976): *Application of a probabilistic method in estimating daily intakes of certain GRAS substances.* Food and Nutrition Board, Division of Biological Sciences, Assembly of Life Sciences, National Research Council, National Academy of Sciences, Washington, D.C. *(In preparation).*
6. Folch-Pi, J. (1955): Composition of the brain in relation to maturation. In: *Biochemistry of the Developing Nervous System*, edited by H. Waelsch, pp. 121–136. Academic Press, New York.
7. James, R. W., Heywood, R., Worden, A. N., Garattini, S., and Salmona, M. (1978): The oral administration of MSG at varying concentrations to male mice. *Toxicol. Lett.*, 1:195–199.
8. Liebschutz, J., Airoldi, L., Brownstein, M. J., Chinn, N. G., and Wurtman, R. J. (1977): Regional distribution of endogenous and parental glutamate, aspartate and glutamine in rat brain. *Biochem. Pharmacol.*, 26:443–446.
9. Lemkey-Johnston, N., and Reynolds, W. A. (1972): Incidence and extent of brain lesions in mice following ingestion of monosodium glutamate (MSG). *Anat. Rec.*, 172:354–359.
10. Morselli, P. L., and Garattini, S. (1970): Monosodium glutamate and the Chinese restaurant syndrome. *Nature*, 227:611–612.
11. Newman, A. J., Heywood, R., Palmer, A. K., Barry, D. H., Edwards, F. P., and Worden, A. N. (1973): The administration of monosodium L-glutamate to neonatal and pregnant rhesus monkeys. *Toxicology*, 1:197–204.
12. Olney, J. W. (1969): Brain lesions, obesity and other disturbances in mice treated with monosodium glutamate. *Science*, 164:719–721.
13. Olney, J. W. (1969): Glutamate-induced retinal degeneration in neonatal mice. Electron microscopy of the acutely evolving lesion. *J. Neuropathol. Exp. Neurol.*, 28:455–474.
14. Olney, J. W. (1971): Glutamate-induced neuronal necrosis in the infant mouse hypothalamus. An electron microscopic study. *J. Neuropathol. Exp. Neurol.*, 30:75–90.
15. Olney, J. W., and Ho, O.-L. (1970): Brain damage in infant mice following oral intake of glutamate, aspartate or cysteine. *Nature*, 227:609–611.
16. Olney, J. W., Ho, O.-L., and Rhee, V. (1971): Cytotoxic effects of acidic and sulphur containing amino acids on the infant mouse central nervous system. *Exp. Brain Res.*, 14:61–76.
17. Olney, J. W., Sharpe, L. G., and Feigin, R. D. (1972): Glutamate induced brain damage in infant primates. *J. Neuropathol. Exp. Neurol.*, 31:464–488.
18. Palkovits, M., Brownstein, M., Saavedra, J. M., and Axelrod, J. (1974): Norepinephrine and dopamine content of hypothalamic nuclei of the rat. *Brain Res.*, 77:137–149.
19. Perez, V. J., and Olney, J. W. (1972): Accumulation of glutamic acid in the arcuate nucleus of the hypothalamus of the infant mouse following subcutaneous administration of monosodium glutamate. *J. Neurochem.*, 19:1777–1782.
20. Perez, V. J., Olney, J. W., and Robin, S. J. (1973): Glutamate accumulation in infant mouse hypothalamus: Influence of temperature. *Brain Res.*, 59:181–189.
21. Reynolds, W. A., Lemkey-Johnston, N., Filer, L. J. Jr., and Pitkin, R. M. (1971): Monosodium glutamate: Absence of hypothalamic lesions after ingestion by newborn primates. *Science*, 172:1342–1344.
22. Stegink, L. D., Filer, L. J. Jr., and Baker, G. L. (1972): Monosodium glutamate: Effect on plasma and breast milk amino acid levels in lactating women. *Proc. Soc. Exp. Biol. Med.*, 140:836–841.
23. Takasaki, Y. (1978): Studies on brain lesion by administration of monosodium L-glutamate to mice. I. Brain lesions in infant mice caused by administration of monosodium L-glutamate. *Toxicology*, 9:293–305.
24. Wenn, C., Kenneth, C. H., and Gershoff, S. N. (1973): Effects of dietary supplementation of monosodium glutamate on infant monkeys, weanling rats and suckling mice. *Am. J. Clin. Nutr.* 26:803–813.
25. Young, R. L., and Lowry, O. H. (1966): Quantitative methods for measuring the histochemical distribution of alanine, glutamate and glutamine in brain. *J. Neurochem.*, 13:785–793.
26. Zanda, G., Franciosi, P., Tognoni, G., Rizzo, M., Standen, S. M., Morselli, P. L., and Garattini, S. (1973): A double blind study on the effects of monosodium glutamate in man. *Biomedicine*, 19:202–204.

Glutamic Acid: Advances in Biochemistry and Physiology, edited by L. J. Filer, Jr., et al.
Raven Press, New York © 1979.

Factors Affecting Plasma Glutamate Levels in Normal Adult Subjects

Lewis D. Stegink, L. J. Filer, Jr., G. L. Baker, S. M. Mueller, and M. Y-C. Wu-Rideout

Departments of Pediatrics, Biochemistry, and Neurology, The University of Iowa College of Medicine, Iowa City, Iowa 52242

The dicarboxylic amino acids, glutamate and aspartate, occupy unique positions in intermediary metabolism. Since they are important in energy production, movement of reducing equivalents into the mitochondrial matrix, urea synthesis, glutathione synthesis, and as neurotransmitters (52), it is not surprising that cells contain considerable quantities of free glutamate and aspartate. In particular, these amino acids are the major amino acids found in the mitochondria of the cell, where they may comprise 50 to 70% of the total free amino acids (21). Considerable quantities are normally found in human brain and liver (44,45,50).

Like all chemical compounds, glutamate and aspartate exert toxic effects when administered at high doses to susceptible animal species. In the case of glutamate salts, toxic effects in animals are associated with two factors: (a) high blood glutamate levels and (b) a species of animal susceptible to glutamate toxicity.

The neonatal rodent is acutely sensitive to large doses of glutamate administered either orally or intravenously. Administration of large quantities of glutamate to the newborn rodent produces a variety of neurotoxic effects (see review in ref. 52), the most marked of which is hypothalamic neuronal necrosis (22,34,35,38). The neurotoxic effects of glutamate in the neonatal primate, however, are highly controversial. The initial reports from the St. Louis group indicating that high glutamate doses cause hypothalamic lesions in the neonatal primate (39,40) have not been confirmed by four other independent laboratories (1,2,31,49,58,60). The latter research groups, however, had no difficulty in producing the rodent lesion. It has been suggested that the failure of research groups other than the St. Louis group to produce a lesion in the neonatal primate reflected a failure to elevate plasma glutamate levels (40). However, this is not the case, since our research group has studied animals in which plasma glutamate levels were grossly elevated without finding any evidence of hypothalamic neuronal necrosis (58).

Our data indicate that grossly elevated plasma glutamate levels are associated with neuronal necrosis in the neonatal mouse (59), but equivalent elevation of plasma glutamate in neonatal primates is not associated with neuronal necrosis (58). However, the continuing controversy over the advisability of glutamate ingestion by man and the finding that glutamate neurotoxicity in sensitive animal species is always associated with grossly elevated plasma glutamate levels have led us to evaluate the factors that affect plasma glutamate levels in adult humans after ingestion of glutamate in either free or protein-bound form.

We will divide our discussion into three general areas: (a) normal biological variation in plasma glutamate levels, (b) the dose-response effect of glutamate ingested in water on plasma glutamate concentration, and (c) the dose-response effect of glutamate ingested with meals on these levels.

BIOLOGICAL VARIATION

Glutamate and aspartate account for 20 to 25% of the total amino acid composition of dietary proteins, including those found in human milk (23). In addition, human milk contains considerable quantities of these amino acids in free form (5,52,54). Thus, humans normally ingest a considerable portion of their protein-amino acid intake in the form of these amino acids.

In view of the large daily intake of glutamate in the diet, it is not surprising that glutamate and aspartate are rapidly metabolized by both the gut and the liver. Similarly, it is not surprising that individual differences in either absorption or metabolism of glutamate would affect plasma levels.

As shown in Fig. 1, glutamate and aspartate are absorbed from the intestinal lumen both as peptides and free amino acids (18,24). Protein-meal studies in man (32,33) strongly suggest that only the neutral amino acids and the dibasic amino acids are quantitatively taken in as free amino acids (18). The imino acids and glycine, as well as the dicarboxylic amino acids, all appear to enter the mucosal cells as constituents of small peptides, where specific intracellular peptidases hydrolyze them to component amino acids (24).

Despite the fact that most of the glutamate and aspartate peptides produced by

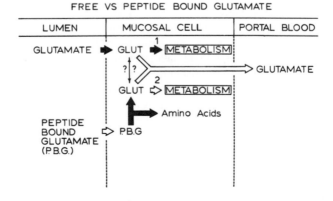

FIG. 1. Intestinal absorption of free vs peptide-bound glutamate.

pancreatic digestion of dietary protein are absorbed intact, the gut also has specific transport sites for free glutamate and aspartate, with maximal values for their transport noted in the terminal ileum (47,51).

A considerable fraction of glutamate is metabolized during the absorptive process for both free (25,29,30,41,43,48) and protein-bound (8,11,13,14,42,46,62) dicarboxylic amino acids. Much of the α-amino nitrogen initially present in these amino acids appears in the portal blood as alanine. As shown in Fig. 2, this alanine undoubtedly results from the transamination of pyruvate by the dicarboxylic amino acids, producing α-ketoglutarate or oxaloacetate as the other product. This process appreciably reduces the quantity of dicarboxylic amino acids released into the portal blood. Glutamate and aspartate escaping mucosal metabolism are carried by the portal vein to the liver, which controls the composition of the amino acid mixture released to the peripheral circulation. Our data in the neonatal pig and monkey (53,55,56,58) indicate a considerable conversion of glutamate and aspartate by the gut and liver into glucose and lactate, which then appear in the peripheral blood. One important question in considering sensitivity to glutamate is whether subgroups of the population metabolize glutamate less efficiently than the general population. Indeed, several children have been reported with apparent defects in glutamate metabolism and transport (26–28,63). Our data suggest that the neonatal rodent and nonhuman primate metabolize glutamate less readily than adult animals (57). The human infant has been shown to be able to metabolize protein-bound glutamate at the same rate as the human adult (17).

To evaluate the question of variable rates of glutamate absorption and metabolism, we reviewed our data to determine whether there was a population subset metabolizing glutamate poorly. In 1972, we reported apparent variations in the metabolism and absorption of glutamate during studies in lactating women

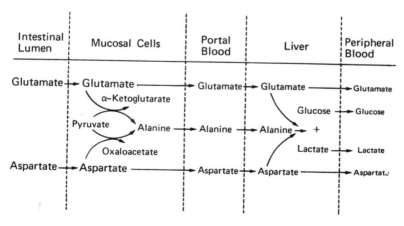

FIG. 2. Dicarboxylic amino acid absorption from the intestinal lumen, showing mucosal cell transamination to yield alanine and hepatic conversion of these amino acids to glucose and lactate as factors modulating peripheral plasma levels. (Reprinted with permission from Stegink, ref. 52.)

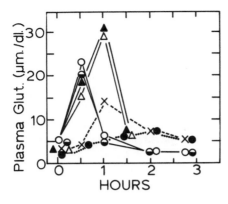

FIG. 3. Plasma glutamate levels in 8 subjects given 6-g loads of MSG in capsules with water. All symbols show values in individual subjects.

(5,54). In the original studies, four subjects were administered 6-g loads of monosodium glutamate (MSG) in capsules with water. We have since studied two additional nonlactating subjects under similar conditions. Figure 3 shows plasma levels in these subjects. Four of the subjects showed significant increases in plasma glutamate levels, with concomitant increases in plasma aspartate, whereas two subjects showed much smaller increases. These variations could have several causative factors: (a) biological variation in absorption and metabolism; (b) a differing rate of capsule dissolution in the gut; or (c) the possibility that two of the subjects had ingested a small amount of food prior to the study, contrary to instructions. Biological variation would be consistent with the earlier data of Himwich et al. (19), who noted considerable variation in plasma glutamate levels after MSG loading.

Recently, we have studied the variation in normal subjects given MSG dissolved in water. Six normal subjects (3 male, 3 female) drank a water solution (4.2 ml/kg) containing MSG at a level providing 100 mg/kg body weight. Peak plasma glutamate levels in these subjects are shown in Table 1. A substantial variation in peak plasma glutamate response was noted. However, all subjects showed substantial increases in plasma glutamate, in contrast to the study described earlier (54) where the equivalent load of MSG was given in capsule form. The variable rates of absorption and metabolism are consistent for each subject. Figure 4 shows plasma

TABLE 1. *Peak plasma glutamate levels in normal subjects given MSG in water at 100 mg/kg body weight*

Subject	Sex	Peak plasma glutamate levels (μmoles/dl)
1	F	87
2	F	56
3	F	30
4	M	61
5	M	62
6	M	51

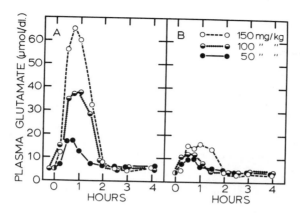

FIG. 4. Plasma glutamate levels in 2 adult subjects (**A** & **B**) administered MSG at 50, 100, and 150 mg/kg body weight dissolved in tomato juice.

glutamate levels in two subjects studied with varying doses of MSG administered in tomato juice. One subject metabolizes glutamate much more rapidly than the other, and this difference holds throughout the dose range studied. Moreover, the results are highly reproducible.

This variation in plasma glutamate response is not limited to subjects ingesting free glutamate dissolved in water. We have noted similar variation in normal adult subjects administered high protein meals where the major load comes from peptide-bound glutamate. Plasma glutamate levels were measured in six normal subjects (3 male, 3 female) ingesting a high protein meal consisting of an egg-milk custard (16). This meal, whose composition is shown in Table 2, was fed to the subjects at a level providing 1 g protein/kg body weight. As shown in Fig. 5, plasma glutamate levels in these subjects fall into three different patterns with two subjects in each group. Two subjects had essentially no change in plasma glutamate levels after ingestion of the meal. This result was reproducible. These same six subjects later participated in a meal study in which they were fed a high protein meal (1 g/kg)

TABLE 2. Composition of custard meal

Component	Weight (g)	Water (g)	Protein (g)	Fat (g)	CHO (g)
Egg	150	110	20	17	1
NFDM[a]	150	6	54	1	78
Fructose	30	0	—	—	30
Water	200	200	—	—	—
Total	530	316	74	18	109

Calculated protein content: 13.96 g protein/100 g custard. Actual protein content of each custard batch confirmed by analysis. Custard administered at a level providing 1 g protein/kg body weight.
[a] NFDM, nonfat dry milk.

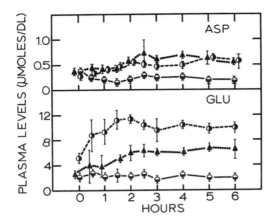

FIG. 5. Mean plasma glutamate (GLU) and aspartate (ASP) levels in 6 normal subjects administered a high protein meal (1 g protein/kg). The curves for these subjects appear to break down into three sets, with 2 subjects in each set. All symbols show values in the three sets of 2 subjects each.

consisting of a hamburger and a milk shake (4). Each subject showed a plasma glutamate response identical to that shown in the custard study. It was clear that the amino acids from the protein meal were absorbed by all subjects, since plasma levels of the branched-chain amino acids increased.

In summary, our data, like those of Himwich et al. (19), but in contrast to those of Bizzi et al. (7), indicate a considerable variation in the absorption and metabolism of both free and peptide-bound glutamate by normal adults.

PLASMA GLUTAMATE RESPONSE TO MSG ADMINISTERED IN WATER

Glutamate absorption and clearance following increasing doses of glutamate administered in water were studied in normal adult subjects. As expected, plasma glutamate levels in adult man increased sharply after ingestion of glutamate salts in water, with peak levels showing a high correlation to the administered dose.

Six normal adult subjects (3 male, 3 female) were studied after ingestion of MSG dissolved in water. The MSG was administered at levels of 100 and 150 mg/kg body weight. The 150 mg/kg body weight dose was chosen since it represents the Acceptable Daily Intake (ADI) for MSG as set by the WHO/FAO. The MSG was dissolved in water to provide either a 2.4 or 3.6% solution, and the subjects ingested these solutions at a level of 4.2 ml/kg body weight. Each subject was tested at both doses, with a 1-week interval between tests. Administration of doses was randomized.

Plasma glutamate levels in these subjects are shown in Table 3. Mean peak plasma glutamate levels were 50 μmoles/dl at a dose level of 100 mg/kg and 70 μmoles/dl at a dose level of 150 mg/kg. However, as noted previously (Table 1), considerable individual variation in peak plasma glutamate levels was observed. Subjects showing highest levels at the 100 mg/kg dose showed the highest values at the 150 mg/kg dose.

TABLE 3. *Plasma glutamate levels in normal adult subjects administered MSG at 100 and 150 mg/kg body weight dissolved in water*

Time (min)	Plasma levels (μmoles/dl)	
	100 mg/kg	150 mg/kg
0	2.69 ± 0.88	3.66 ± 1.00
15	14.4 ± 15.0	15.7 ± 11.8
30	47.1 ± 25.4	66.5 ± 50.0
45	50.1 ± 23.9	71.8 ± 35.7
60	24.9 ± 11.3	56.2 ± 29.8
90	8.19 ± 5.31	27.5 ± 13.4
120	4.28 ± 2.57	7.93 ± 3.72
150	3.72 ± 2.24	6.04 ± 4.18
180	4.19 ± 2.84	4.19 ± 1.39
240	3.05 ± 1.61	4.17 ± 2.28

Values listed as mean ± SD.

The magnitude of the observed plasma glutamate levels was surprising. Our previous studies with normal adult subjects given a load of MSG approximating 100 mg/kg in capsules had shown mean peak values of 16 μmoles/dl. Figure 6 compares plasma glutamate levels in the subjects given MSG at 100 mg/kg level dissolved in water, or given in capsules. The data suggest that administration of MSG in capsules affects plasma glutamate levels. However, it is also possible that these differences result from individual variations in the population groups studied.

The peak plasma glutamate levels observed after glutamate administration in water are consistent with those reported by other investigators in man. Figure 7 compares peak plasma glutamate levels obtained in our studies at 100 and 150 mg/kg body weight with values in other laboratories for human subjects adminis-

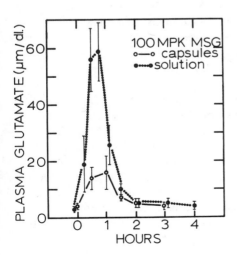

FIG. 6. Plasma glutamate levels in normal subjects after administration of 100 mg/kg loads either in capsules or in 4.2 ml water/kg body weight. Data are shown as the mean ± SEM.

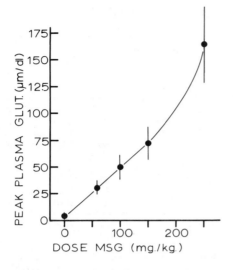

FIG. 7. Mean (± SEM) peak plasma glutamate levels in normal adults administered MSG in water. Data at 60 mg/kg from Bizzi et al. (7), data at 100 and 150 mg/kg from our studies, and data at 240 mg/kg from Himwich et al. (19).

tered glutamate at 60 (7) or 250 mg/kg body weight (19). These data indicate a high correlation of MSG dose to plasma glutamate levels in the 60 to 250 mg/kg body weight range when MSG is dissolved in water. The slope of this curve is significantly greater than that observed for adult mice and monkeys given equivalent doses (57).

Since tomato juice is often used as a vehicle for MSG administration during studies of Chinese restaurant syndrome, plasma glutamate levels were also measured in eight normal subjects administered MSG dissolved in tomato juice. Eight normal adult subjects (4 male, 4 female) were studied. The study was carried out double blind, with the subjects receiving either MSG (150 mg/kg) or NaCl (a dose equivalent to the sodium content of MSG) dissolved in unsalted tomato juice (Diet Delight, California Canners and Growers, San Francisco). The order in which the subjects received MSG or NaCl was randomized. The tomato juice was administered at a level of 4.2 ml/kg body weight, with sufficient MSG to give a 3.6% solution.

Plasma glutamate levels in these subjects are shown in Fig. 8. A marked elevation in plasma glutamate was noted after ingestion of MSG in tomato juice, whereas no change was seen after ingestion of the tomato juice containing NaCl. The free glutamate content of the unsalted tomato juice was 1 mg/4.2 ml of juice. Thus, subjects received approximately 1 mg MSG/kg body weight when ingesting tomato juice with NaCl. The data demonstrate a marked increase in plasma glutamate levels after ingestion of 150 mg MSG/kg body weight dissolved in tomato juice. Mean peak plasma levels were slightly lower than those noted in a different group of subjects ingesting the equivalent dose of MSG (150 mg/kg) in water (Table 3). It is not clear whether this difference reflects an effect of the other components of the tomato juice or a difference in the subject population studied.

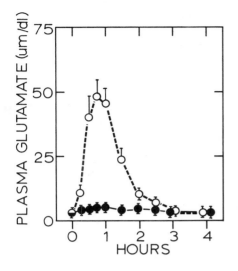

FIG. 8. Mean (± SEM) plasma glutamate levels in normal subjects administered glutamate (○) at 150 mg/kg body weight or NaCl (●) at a sodium content equal to that of the MSG dissolved in tomato juice.

EFFECT OF MSG INGESTION WITH MEALS ON PLASMA GLUTAMATE LEVELS

The critical questions about glutamate concern its use as a food additive. Thus, it is important to evaluate the effect of MSG addition to food systems on plasma amino acid levels. It is generally conceded that the large quantities of glutamate contained in dietary protein are readily metabolized. However, it has been suggested that the addition of free glutamate to a meal as MSG might produce a very rapid early increase in plasma glutamate and aspartate levels (36,37). It was suggested that the added free glutamate is metabolized less readily than the peptide-bound glutamate, producing a rapid early rise in plasma glutamate levels. We evaluated the plasma amino acid response in normal subjects ingesting MSG with a variety of meal systems to test the hypothesis that such an addition significantly affects plasma glutamate levels.

The data in Table 4 are taken from Appendix E of the Committee on GRAS List Survey—Phase III report (10). These data indicate a mean expected daily intake of 6.8 mg/kg body weight in the age group with the highest ingestion level. In this age group, a mean level of 30 mg/kg body weight represents the 90th percentile for the total expected *daily* ingestion of added MSG.

For our first study (4), a level of 34 mg MSG/kg body weight was chosen, slightly above the 90th percentile of daily ingestion. This dose was fed to normal adult volunteers in conjunction with a high protein meal. The meal chosen was a hamburger-milk shake system having the composition shown in Table 5. This meal was prepared with and without MSG. Normal adult volunteers were fed each meal in a quantity sufficient to provide a protein load of 1 g protein/kg body weight and a total glutamate load of 171 to 198 mg/kg body weight (Table 6). Figure 9 shows plasma glutamate and aspartate levels in these subjects. No significant

TABLE 4. *Expected daily intake of MSG based on person-days*

	Intakes (mg/kg/day)			
		Percentile		
Age	Mean	90th	99th	99.9th
0–5 months	0.3	0	11	25
6–11 months	1.9	1.9	36	46
12–23 months	6.8	30	43	61
2–5 years	5.5	23	37	56
6–17 years	2.7	10	25	40
18+ years	1.5	7	12	19

From Committee on GRAS List Survey—Phase III, ref. 10.

TABLE 5. *Composition of the hamburger—milk shake meal for a 70-kg adult*

Component	Quantity (g)	Protein (g)	Fat (g)	CHO (g)	Energy (kcal)
Hamburger	222	61	25.5	0	346
Bun	50	4.5	1.5	25.5	133
Milk	100	3.5	3.5	5	66
Ice cream	50	2	5	11	95
Total	422	71	35.5	72	640

For the 70-kg person, the meal supplies about 1 g/kg body weight as protein (=38% of total energy). Quantity of the hamburger in each meal was varied with each individual so as to provide protein at 1 g/kg body weight.

TABLE 6. *Estimated intake of protein, aspartic acid, and glutamic acid in meal studies*

Study	Protein (g/kg)	Total aspartate (g/kg)	Total glutamate (g/kg)
Hamburger-shake	1.0	90	171
Hamburger-shake with MSG	1.0	90	198[a]

[a] Corrected for the sodium content and water of hydration of MSG (78% of MSG is glutamate).

differences in plasma glutamate and aspartate levels were noted between groups. In particular, plasma glutamate and aspartate levels did not increase in the early part of the absorption-metabolism curve. A large number of blood samples were taken in the early postprandial state to detect such a rise if it did occur.

It has been suggested that under some circumstances, erythrocytes might carry amino acids to an extent greater than plasma (3,12–15). Accordingly, erythrocyte glutamate levels were also measured. As shown in Fig. 10, erythrocyte glutamate

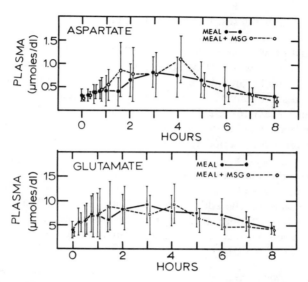

FIG. 9. Plasma glutamate and aspartate levels (mean ± SD) in normal adult subjects after ingestion of a high protein meal (1 g/kg) with and without added MSG (34 mg/kg body weight). (From Baker, Filer, Jr., and Stegink, ref. 4.)

and aspartate levels did not differ between the two groups. Thus, we concluded that the addition of MSG to a high-protein meal at a level of 34 mg/kg body weight had no significant effect on either plasma or erythrocyte glutamate and aspartate levels (4).

In view of the rapid metabolism of MSG when added to meals, additional studies

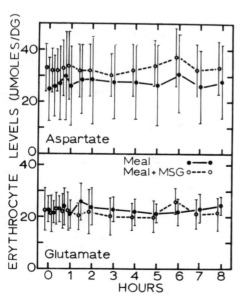

FIG. 10. Erythrocyte aspartate and glutamate levels (mean ± SD) in normal adults ingesting a high protein meal with and without added MSG (34 mg/kg). (From Baker, Filer, Jr., and Stegink, ref. 4.)

were carried out at higher doses. In these studies, glutamate was added to meals at levels of 0, 100, and 150 mg/kg body weight. The 150 mg/kg body weight load represents the ADI of MSG as published by the WHO/FAO.

The meal system used was Sustagen (composition shown in Table 7). Sustagen administered at 4.2 ml/kg provides 0.4 g protein, 1.12 g carbohydrate, and 6.61 kcal/kg. Six fasted adult humans (3 male, 3 female) were fed Sustagen meals with and without added MSG (6). Subjects were tested at 10-day intervals with the sequence randomized according to a Latin square design (9). Serial blood samples were obtained over a 6-hr period for determination of plasma concentrations of amino acids.

Table 8 shows the plasma glutamate levels in these subjects. Ingestion of Sustagen alone increased plasma glutamate levels from 4 μmoles/dl to approximately 6 to 7 μmoles/dl. The addition of MSG at 100 and 150 mg/kg increased plasma glutamate levels further, reaching mean values approximating 11 μmoles/dl. However, the levels noted at 150 mg/kg body weight in these studies are no higher

TABLE 7. *Composition of the Sustagen meal system fed was 4.2 ml/kg body weight*

Component	Quantity (gm/kg)	Energy (kcal/kg)
Protein	0.40	1.6
Fat	0.059	0.53
Carbohydrate	1.12	4.48
Water	4.2	0
Total	5.78	6.61

Sustagen (Mead-Johnson) also contains appropriate vitamins and minerals.

TABLE 8. *Plasma glutamate levels in normal subjects fed Sustagen with and without added MSG*

Time (min)	Plasma levels (μmoles/dl)		
	No MSG	100 mg MSG/kg	150 mg MSG/kg
0	4.21 ± 1.39	4.06 ± 0.63	4.59 ± 2.03
15	5.48 ± 2.03	6.90 ± 1.87	6.66 ± 2.62
30	6.43 ± 3.03	8.55 ± 3.30	8.89 ± 3.42
45	6.64 ± 1.99	11.2 ± 4.89	9.44 ± 2.58
60	6.37 ± 2.35	10.2 ± 4.38	10.7 ± 2.41
90	5.25 ± 1.67	8.61 ± 2.84	10.8 ± 3.10
120	5.78 ± 1.21	8.88 ± 2.49	9.27 ± 4.20
150	5.87 ± 2.15	7.78 ± 2.41	7.99 ± 2.46
180	6.68 ± 2.10	6.65 ± 2.99	6.59 ± 1.68
240	5.04 ± 1.24	5.11 ± 2.39	5.57 ± 0.48
300	4.21 ± 1.39	4.06 ± 0.63	4.59 ± 2.03
360	5.48 ± 2.03	6.90 ± 1.87	6.66 ± 2.62

Values listed as mean ± SD.

TABLE 9. *Correlation of plasma glutamate levels with glutamate ingested in a meal system*

Meal	Protein intake (g/kg)	MSG added (mg/kg)	Total glutamate (mg/kg)	Plasma glutamate levels (μm/dl)		
				Fasting[a]	Peak[a]	Range[b]
Custard (adults)	1.0	0	207	3.3 ± 1.6	6.3 ± 3.4	3–12
Hamburger— milk shake	1.0	0	171	4.1 ± 1.8	7.1 ± 3.9	4–15
Hamburger— milk shake + MSG	1.0	34	198	3.4 ± 1.0	8.8 ± 5.0	4–13
Sustagen	0.40	0	80	4.6 ± 1.6	7.6 ± 1.6	7–10
Sustagen + MSG	0.40	150	197	4.1 ± 1.2	10.5 ± 2.7	9–14

$N = 6$.
[a] Mean ± SD.
[b] Peak values.

than those noted postprandially in some normal adults ingesting a high protein meal alone (Table 9).

Our data indicate that MSG added to meals is metabolized much more rapidly than when ingested in water or tomato juice. Figure 11 summarizes the differences noted in plasma glutamate levels after ingestion of MSG (150 mg/kg) in either water or Sustagen. These data demonstrate a marked difference in the metabolism of glutamate depending on whether it is ingested in water or with a meal.

It is not clear how ingestion of the meal modulates the absorption of glutamate and its metabolism. A slower rate of absorption would permit greater catabolism of

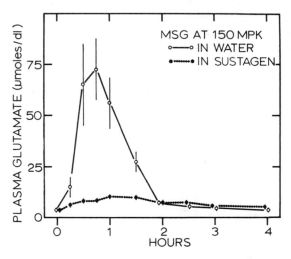

FIG. 11. Comparison of plasma glutamate levels (mean ± SEM) in normal adult subjects ingesting 150 mg MSG/kg body weight either dissolved in water or as part of a Sustagen meal.

glutamate by the intestinal mucosa, resulting in a decreased release of glutamate to portal blood. Alternatively, the carbohydrate present in Sustagen could serve as a source of pyruvate, facilitating the transamination of glutamate to α-ketoglutarate and its metabolism in the intestinal mucosa (Fig. 2). This would increase glutamate catabolism and decrease glutamate release to the peripheral circulation. The latter possibility was intriguing, since the Sustagen meal provided carbohydrate (as corn syrup solids) at 1.12 g/kg body weight.

To test this hypothesis, a normal subject was given 100 mg MSG/kg body weight, either dissolved in water or dissolved in water with sufficient carbohydrate to provide 1.12 g glucose/kg body weight. The carbohydrate was administered as Polycose (Ross Laboratories, Columbus, Ohio), a partially hydrolyzed corn starch preparation. The data in Fig. 12 compare glutamate and glucose levels in a typical individual after ingestion of 100 mg MSG/kg body weight in either water or Polycose. It is clear that the carbohydrate has a striking effect on the metabolism of glutamate. Changes in blood glucose and glutamate levels indicate that gastric emptying has occurred. It seems likely that the rapid metabolism of MSG noted after ingestion with meals reflects in part the carbohydrate content of the meal. Presumably, carbohydrate is absorbed into the intestinal mucosa and converted to glucose and pyruvate, ultimately facilitating the transamination and metabolism of glutamate by the mucosal cells (Fig. 13).

Studies of the Chinese restaurant syndrome described by Kenney (20) led us to examine the effect of the beverage mixture utilized in his studies on plasma glutamate levels. This study was prompted by the studies shown in Fig. 12 demonstrating the profound effect of added carbohydrate on the plasma glutamate levels after MSG loading. Since Kenney's beverage mixture contains 30 g sucrose, we wondered whether plasma glutamate levels would be depressed by this quantity of carbohydrate. Plasma glutamate levels were measured in one subject given MSG at 100 mg/kg body weight dissolved in water or the beverage mix. In both cases the

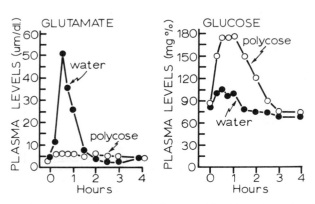

FIG. 12. Plasma glutamate and glucose levels in a normal adult subject administered MSG at 100 mg/kg body weight either in water or in a water solution containing 1.12 g Polycose/kg body weight.

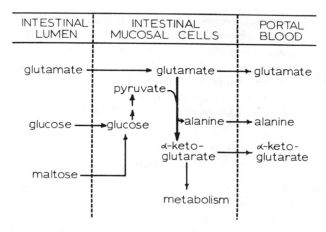

FIG. 13. Interrelationship between carbohydrate and glutamate metabolism in intestinal mucosal cells.

MSG concentration in the solution was 2.4%, and the mixture was administered at a level of 4.2 ml/kg body weight. In the first experiment, the subject ingested the MSG in water. In the second experiment, the basic beverage mixture described by Kenney was utilized, with the following exceptions: (a) additional MSG was added to bring the total dose to 100 mg/kg body weight and (b) the volume of the solution administered was increased so that the MSG concentration would be held constant at 2.4%. Figure 14 shows plasma glutamate and glucose levels in this study. The presence of sucrose (0.42 g/kg) had effected peak levels, but significant elevations of plasma glutamate occurred. This presumably reflects both the lower carbohydrate content of the beverage versus the Polycose feeding (0.42 versus 1.12 g/kg) and the

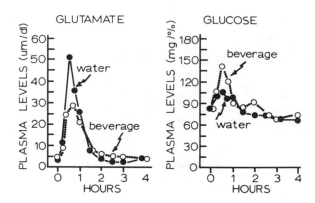

FIG. 14. Plasma glutamate and glucose levels in a single normal male subject administered MSG at 100 mg/kg body weight dissolved either in water or in a beverage mixture containing 30 g sucrose.

carbohydrate composition. Sucrose is composed of glucose and fructose. Fructose is not appreciably metabolized by the intestinal mucosa and probably does not exert a significant effect on mucosal metabolism of glutamate. Thus, the load of glucose available to the mucosal cells would be 0.21 g/kg. This suggestion is consistent with the data we have observed after ingestion of MSG with tomato juice (Fig. 8). The tomato juice used contained a significant quantity of carbohydrate (0.16 g/4.2 ml), largely as fructose (61). This level of carbohydrate had little effect on plasma glutamate levels.

SUMMARY

1. Our data indicate considerable variation in the absorption and metabolism of both free and peptide-bound glutamate by normal adults.

2. Plasma glutamate levels rapidly increase after MSG ingestion in water, with peak values showing a strong correlation with the dose administered.

3. Ingestion of MSG at 150 mg/kg with a single meal did not significantly elevate plasma glutamate levels above those noted after ingestion of a high protein meal alone (1 g protein/kg body weight). However, administration of this dose in water produced significant elevations of plasma glutamate levels.

4. The increased plasma glutamate levels noted after ingestion of MSG in water are markedly diminished by the presence of carbohydrates in the solution.

ACKNOWLEDGMENTS

These studies were supported in part by grants-in-aid from the Gerber Products Company, Searle Laboratories, and the International Glutamate Technical Committee.

REFERENCES

1. Abraham, R. W., Dougherty, W., Golberg, L., and Coulston, F. (1971): The response of the hypothalamus to high doses of monosodium glutamate in mice and monkeys. Cystochemistry and ultrastructural study of lysosomal changes. *Exp. Mol. Pathol.*, 15:43–60.
2. Abraham, R., Swart, J., Golberg, L., and Coulston, F. (1975): Electron microscopic observations of hypothalami in the neonatal rhesus monkeys (*Macaca mulatta*) after administration of monosodium-L-glutamate. *Exp. Mol. Pathol.*, 23:203–213.
3. Aoki, T. T., Brennan, M. F., Muller, W. A., Moore, F. D., and Cahill, G. F. Jr. (1972): The effect of insulin on muscle glutamate uptake. Whole blood versus plasma glutamate analysis. *J. Clin. Invest.*, 51:2889–2894.
4. Baker, G. L, Filer, L. J., Jr., and Stegink, L. D. (1977): Plasma and erythrocyte amino acid levels in normal adults fed high protein meals: Effect of added monosodium glutamate (MSG) or monosodium glutamate plus Aspartame (APM). *Fed. Proc.*, 36:1154.
5. Baker, G. L., Filer, L. J., Jr., and Stegink, L. D. (1979): Factors influencing dicarboxylic amino acid content of human milk (*this volume*).
6. Baker, G. L., Filer, L. J., Jr., and Stegink, L. D. (1978): Plasma amino acid levels in normal adults after ingestion of high doses of monosodium glutamate (MSG) with a meal. *Fed. Proc.*, 37:752.
7. Bizzi, A., Veneroni, E., Salmona, M., and Garattini, S. (1977): Kinetics of monosodium glutamate in relation to its neurotoxicity. *Toxicol. Lett.*, 1:123–130.

8. Christensen, H. N. (1949): Conjugated amino acids in portal plasma of dogs after protein feeding. *Biochem. J.*, 44:333–335.
9. Cochran, W. G., and Cox, G. M. (1950): *Experimental Design*, John Wiley & Sons, New York, p. 86.
10. Committee on GRAS List Survey—Phase III (1976): Estimating distribution of daily intake of monosodium glutamate (MSG), Appendix E. In: *Estimating Distribution of Daily Intake of Certain GRAS Substances*. Food and Nutrition Board, Division of Biological Sciences, Assembly of Life Sciences, National Research Council, National Academy of Sciences, Washington, D.C.
11. Dent, C. E., and Schilling, J. A. (1949): Studies on the absorption of proteins: The amino acid pattern in the portal blood. *Biochem. J.*, 44:318–333.
12. Elwyn, D. H. (1966): Distribution of amino acids between plasma and red blood cells in the dog. *Fed. Proc.*, 25:854–861.
13. Elwyn, D. H., Launder, W. J., Parikh, H. C., and Wise, E. M., Jr. (1972): Roles of plasma and erythrocytes in interorgan transport of amino acids in dogs. *Am. J. Physiol.*, 222:1333–1342.
14. Elwyn, D. H., Parikh, H. C., and Shoemaker, W. C. (1968): Amino acid movements between gut, liver, and periphery in unanesthetized dogs. *Am. J. Physiol.*, 215:1260–1275.
15. Felig, P., Wahren, J., and Raf, L. (1973): Evidence of interorgan amino acid transport by blood cells in humans. *Proc. Natl. Acad. Sci. USA* 70:1775–1779.
16. Filer, L. J., Jr., Baker, G. L., and Stegink, L. D. (1977): Plasma aminograms in infants and adults fed an identical high protein meal. *Fed. Proc.*, 36:1181.
17. Filer, L. J., Jr., Baker, G. L., and Stegink, L. D. (1979): Metabolism of free glutamate in clinical products fed human infants (*this volume*).
18. Gray, G. M., and Cooper, H. L. (1971): Protein digestion and absorption. *Gastroenterology*, 61:535–544.
19. Himwich, W. A., Petersen, I. M., and Graves, J. P. (1954): Ingested sodium glutamate and plasma levels of glutamic acid. *J. Appl. Physiol.*, 7:196–199.
20. Kenney, R. A. (1979): Placebo-controlled studies of human reaction to oral monosodium L-glutamate (MSG) (*this volume*).
21. King, M. J., and Diwan, J. J. (1972): Transport of glutamate and aspartate across membranes of rat liver mitochondria. *Arch. Biochem. Biophys.*, 152:670–676.
22. Lemkey-Johnston, N., and Reynolds, W. A. (1974): Nature and extent of brain lesions in mice related to ingestion of monosodium glutamate. *J. Neuropathol. Exp. Neurol.*, 33:74–97.
23. Macy, I. G., Kelly, H. J., and Sloan, R. E. (1953): The Composition of Milks, Publication 254, National Research Council, National Academy of Sciences, Washington, D.C.
24. Matthews, D. M. (1975): Intestinal absorption of peptides. *Physiol. Rev.*, 55:537–608.
25. Matthews, D. M., and Wiseman, G. (1953): Transamination by the small intestine of the rat. *J. Physiol.*, 120:55P.
26. Menkes, J. H., Alter, M., Steigleder, G. K., Weakley, D. R., and Sung, J. H. (1962): A sex-linked recessive disorder with retardation of growth, peculiar hair, and focal cerebral and cerebellar degeneration. *Pediatrics*, 29:764–779.
27. Melancon, S. B., Dallaire, L., Lemieux, B., Robitaille, P., and Potier, M. (1977): Dicarboxylic aminoaciduria: An inborn error of amino acid conservation. *J. Pediatr.*, 91:422–427.
28. Mueller, S. M., Stegink, L. D., and Reynolds, W. A. (1977): Elevated plasma glutamate levels without hypothalamic lesions. *Pediatr. Res.*, 11:564.
29. Neame, K. D., and Wiseman, G. (1957): The transamination of glutamic and aspartic acids during absorption by the small intestine of the dog in vivo. *J. Physiol.*, 135:442–450.
30. Neame, K. D., and Wiseman, G. (1958): The alanine and oxo acid concentrations in mesenteric blood during the absorption of L-glutamic acid by the small intestine of the dog, cat and rabbit in vivo. *J. Physiol.*, 140:148–155.
31. Newman, A. J., Heywood, R., Plamer, A. K., Barry, D. H., Edwards, F. P., and Worden, A. N. (1973): The administration of monosodium L-glutamate to neonatal and pregnant rhesus monkeys. *Toxicology*, 1:197–204.
32. Nixon, S. E., and Mawer, G. E. (1970): The digestion and absorption of protein in man. 1. The site of absorption. *Br. J. Nutr.*, 24:227–240.
33. Nixon, S. E., and Mawer, G. E. (1970): The digestion and absorption of protein in man. 2. The form in which digested protein is absorbed. *Br. J. Nutr.*, 24:241–258.
34. Olney, J. W. (1969): Brain lesions, obesity, and other disturbances in mice treated with monosodium glutamate. *Science*, 164:719–721.

35. Olney, J. W. (1969): Glutamate induced retinal degeneration in neonatal mice. Electron microscopy of the acutely evolving lesion. *J. Neuropathol. Exp. Neurol.*, 28:455–474.
36. Olney, J. W. (1975): Another view of Aspartame. In: *Sweeteners, Issues and Uncertainties*, pp. 189–195. Academy Forum, National Academy of Sciences, Washington, D.C.
37. Olney, J. W. (1975): L-Glutamic and L-aspartic acids—A question of hazard? *Food Cosmet. Toxicol.*, 13:595–600.
38. Olney, J. W., and Ho, O. L. (1970): Brain damage in infant mice following oral intake of glutamate, aspartate and cysteine. *Nature*, 227:609–611.
39. Olney, J. W., and Sharpe, L. G. (1969): Brain lesions in an infant rhesus monkey treated with monosodium glutamate. *Science*, 166:386–388.
40. Olney, J. W., Sharpe, L. G., and Feigin, R. D. (1972): Glutamate-induced brain damage in infant primates. *J. Neuropathol. Exp. Neurol.*, 31:464–488.
41. Parsons, D. S., and Volman-Mitchell, H. (1974): The transamination of glutamate and aspartate during absorption in vitro by small intestine of chicken, guinea pig and rat. *J. Physiol.*, 239:677–694.
42. Peraino, C., and Harper, A. E. (1961): Effect of diet on blood amino acid concentrations. *Fed. Proc.*, 20:245.
43. Peraino, C., and Harper, A. E. (1962): Concentrations of free amino acids in blood plasma of rats force-fed L-glutamic acid, L-glutamine and L-alanine. *Arch. Biochem. Biophys.*, 97:442–448.
44. Perry, T. L., Berry, K., Hansen, S., Diamond, S., and Mok, C. (1971): Regional distribution of amino acids in human brain obtained at autopsy. *J. Neurochem.*, 18:513–519.
45. Perry, T. L., Hansen, S., Berry, K., Mok, C., and Lesk, D. (1971): Free amino acids and related compounds in biopsies of human brain. *J. Neurochem.*, 18:521–528.
46. Pion, R., Fauconneau, G., and Rerat, A. (1964): Variation de la composition en acides amines du sang porte au cours de la digestion chez le porc. *Ann. Biol. Anim. Biochim. Biophys.*, 4:383–401.
47. Ramaswamy, K., and Radhakrishnan, A. N. (1966): Patterns of intestinal uptake and transport of amino acids in the rat. *Indian J. Biochem.*, 3:138–143.
48. Ramaswamy, K., and Radhakrishnan, A. N. (1970): Labeling patterns using C^{14}-labeled glutamic acid, aspartic acid and alanine in transport studies with everted sacs of rat intestine. *Indian J. Biochem.*, 7:50–54.
49. Reynolds, W. A., Lemkey-Johnston, N., Filer, L. J., Jr., and Pitkin, R. M. (1971): Monosodium glutamic: Absence of hypothalamic lesions after ingestion by newborn primates. *Science*, 172:1342–1344.
50. Ryan, W. L., and Carver, M. J. (1966): Free amino acids of human foetal and adult liver. *Nature*, 212:292–293.
51. Schultz, S. G., Yu-Tu, L., Alvarez, O. O., and Curran, P. E. (1970): Dicarboxylic amino acid influx across brush border of rabbit ileum. *J. Gen. Physiol.*, 56:621–639.
52. Stegink, L. D. (1976): Absorption, utilization and safety of aspartic acid. *J. Toxicol. Environ. Health*, 2:215–242.
53. Stegink, L. D., Brummel, M. C., Boaz, D. P., and Filer, L. J., Jr. (1973): Monosodium glutamate metabolism in the neonatal pig: Conversion of administered glutamate into other metabolites in vivo. *J. Nutr.*, 103:1146–1154.
54. Stegink, L. D., Filer, L. J., Jr., and Baker, G. L. (1972): Monosodium glutamate: Effect on plasma and breast milk amino acid levels in lactating women. *Proc. Soc. Exp. Biol. Med.*, 140:836–841.
55. Stegink, L. D., Filer, L. J., Jr., and Baker, G. L. (1973): Monosodium glutamate metabolism in the neonatal pig: Effect of load on plasma, brain, muscle, and spinal fluid free amino acid levels. *J. Nutr.*, 103:1135–1145.
56. Stegink, L. D., Pitkin, R. M., Reynolds, W. A., Boaz, D. P., Filer, L. J., Jr., and Brummel, M. C. (1975): Placental transfer of glutamate and its metabolites in the primate. *Am. J. Obstet. Gynecol.*, 122:70–78.
57. Stegink, L. D., Reynolds, W. A., Filer, L. J., Jr., Baker, G. L., Daabees, T. T., and Pitkin, R. M. (1979): Comparative metabolism of glutamate in mouse, monkey, and man (*this volume*).
58. Stegink, L. D., Reynolds, W. A., Filer, L. J., Jr., Pitkin, R. M., Boaz, D. P., and Brummel, M. C. (1975): Monosodium glutamate metabolism in the neonatal primate. *Am. J. Physiol.*, 229:246–250.
59. Stegink, L. D., Shepherd, J. A., Brummel, M. C., and Murray, L. M. (1974): Toxicity of protein hydrolysate solutions: Correlation of glutamate dose and neuronal necrosis to plasma amino acid levels in young mice. *Toxicology*, 2:285–299.
60. Wen, C., Hayes, K. C., and Gershoff, S. M. (1973): Effects of dietary supplementation of

monosodium glutamate on infant monkeys, weanling rats and suckling mice. *Am. J. Clin. Nutr.*, 26:803–813.
61. Williams, K. T., and Bevenue, A. (1954): Some carbohydrate components of tomato. *Agric. Food Chem.*, 2:472–474.
62. Wolff, J. E., Bergman, E. N., and Williams, H. H. (1972): Net metabolism of plasma amino acids by liver and portal-drained viscera of fed sheep. *Am. J. Physiol.*, 223:438–446.
63. Yoshida, T., Tada, K., Mizuno, T., Wada, Y., Akabane, J., Ogasawara, J., Minagwawa, A., Morikawa, T., and Okamura, T. (1964): A sex-linked disorder with mental and physical retardation characterized by cerebrocortical atrophy and increase of glutamic acid in cerebrospinal fluid. *Tokoku J. Exp. Med.*, 83:261–269.

Metabolism of Free Glutamate in Clinical Products Fed Infants

L. J. Filer, Jr., G. L. Baker, and L. D. Stegink

Departments of Pediatrics and Biochemistry, The University of Iowa College of Medicine, Iowa City, Iowa 52242

One of the tentative conclusions advanced by the Select Committee on GRAS Substances (SCOGS) (11) regarding the health aspects of glutamate salts as food ingredients was as follows:

> The evidence is insufficient to determine that the adverse effects reported are not deleterious to infants should glutamic acid, L-glutamic acid hydrochloride, monoammonium L-glutamate, monopotassium L-glutamate or monosodium L-glutamate be added to infant formulas and/or commercially prepared strained and junior foods.

It is difficult to understand the rationale for this conclusion if it is based on a knowledge of infant feeding practice and the statement by the SCOGS Committee (12) that "there appears to be no hazard of brain damage in the use of casein hydrolysate formulas."

During the neonatorum, infants are fed at intervals of 3 to 4 hr. Because of reduced gastric capacity, low-birth-weight infants are usually fed at 3-hr intervals. Many infants of less than 1,500 g birth weight are provided a source of nitrogen in the form of an amino acid mixture or a protein hydrolysate as part of their parenteral nutrition program. Since infants are fed frequently, they are in a constant postprandial state. If human young were unable to metabolize dicarboxylic amino acids effectively, survival of the species would be jeopardized. In fact, the nature and content of the free amino acids found in human milk and the amino acid content of the proteins of human milk would undoubtedly differ from reported values (1).

Measurement of plasma and erythrocyte concentrations of free amino acids has been used as an index of the ability of infants to metabolize amino acids. Biochemical immaturity and inborn metabolic defects in amino acid metabolism are associated with hyperaminoacidemia. Substrate overload in persons otherwise normal may potentially produce the same results.

In 1970 we initiated an ongoing study of plasma and erythrocyte aminograms of infants fed a variety of nitrogen sources either enterally or parenterally. These observations, collated in this report, show that irrespective of birth weight or age, the infant effectively metabolizes glutamic and aspartic acids provided either as the amino acid, peptides, or intact protein.

ENTERAL FEEDINGS

Term Infants

Healthy, normal term infants, breast or formula fed, are constantly enrolled in studies of normal growth and development within the University of Iowa Pediatric Metabolism Unit (6,7). At monthly intervals a biochemical profile is determined on small blood samples obtained 2 hr postprandially from a majority of all infants. Serum or plasma aminograms and erythrocyte aminograms are among the determined biochemical indices.

In 1971 Stegink and Schmitt (18) reported that healthy term infants, 28 to 33 days of age, showed no statistically significant difference in serum glutamate or aspartate when fed a conventional milk-based formula or Nutramigen (Mead Johnson Laboratories, Evansville, Ind.). The nitrogen source in the latter formula is an enzymatic hydrolysate of casein with approximately equal parts of free amino acids and polypeptides. Despite the relatively large amount of free glutamate (22% of total amino acids) present in the Nutramigen feeding, there was no evidence in the serum aminogram of glutamate overload.

Stegink and co-workers (19) extended their studies to include observations of infants fed a soy protein isolate formula (Isomil, Ross Laboratories, Columbus, Ohio) and older infants, 6 and 11 months of age. While these studies were designed to evaluate the response of infants to formulas fortified with DL-methionine, there was no evidence of any aberrant response in plasma concentrations of glutamate or aspartate.

The responses of term infants to enteral loads of glutamate and aspartate are summarized in Table 1. Two-hour postprandial concentrations of plasma or serum glutamate and aspartate are given for breast-fed infants, infants fed conventional milk-based formula, a soy isolate-based formula, and a formula prepared from an enzymatic hydrolysis of casein. From these data it is quite evident that the term infant responds to a formula containing dicarboxylic amino acids in free or peptide form in a manner comparable to that of the breast-fed infant. Plasma or serum concentrations of glutamate or aspartate are not elevated by the feeding of hydrolyzed versus milk protein or the ingestion of human milk with its relatively high content of free glutamic acid (1).

Concentrations of glutamate and aspartate in erythrocytes of breast-fed and formula-fed infants are summarized in Table 2. These observations, like those reported by Stegink and co-workers (16,17) in adults, do not show preferential accumulation of the dicarboxylic amino acids within the erythrocyte.

Low-Birth-Weight Infants

Since biochemical immaturity is a major problem for low-birth-weight (LBW) infants, we have investigated the plasma aminogram response of healthy, growing LBW infants, 11 to 35 days of age, to four formulas (5). Eight infants weighing 1.3 to 1.8 kg at birth were studied in a Latin square design so that each infant was fed all

TABLE 1. *Plasma glutamate and aspartate concentrations in term infants*

Feeding	Protein source	Intake (mg/kg/feeding)		Plasma concentrations (μmoles/dl)	
		Glutamate	Aspartate	Glutamate	Aspartate
Human milk	Human milk	65	33	12.3 ± 3.1	0.7 ± 0.4
Nutramigen[a]	Casein hydrolysate	80	28	10.1 ± 2.1	3.0 ± 2.5
Isomil	Soy isolate	95	57	13.0 ± 3.1	3.1 ± 2.1
Enfamil[a]	Cow milk	62	15	9.8 ± 2.8	2.4 ± 2.2

[a] Serum samples.

TABLE 2. *Erythrocyte glutamate and aspartate concentrations in term infants*

Feeding	N	Intake (mg/kg/feeding)		Erythrocyte concentration (μmoles/dl)	
		Glutamate	Aspartate	Glutamate	Aspartate
Human milk	13	65	33	42.4	13.3
Soy protein isolate	12	95	57	43.0	8.8

of the formulas. This approach enabled us to determine age effects on amino acid metabolism if they existed. All formulas were isocaloric and isonitrogenous at 67 Kcal and 1.5 g protein/100 ml. The data summarized in Table 3 show no statistically significant differences in 2-hr postprandial plasma concentrations of glutamate or aspartate for the four formulas studied. These observations are very important to the issue of glutamate and aspartate safety for infants because the nitrogen source in the experimental formula, identified as modified Pregestimil (Mead Johnson Laboratories, Evansville, Ind.), was an enzymatic hydrolysate of casein. Infants fed this formula received their dicarboxylic amino acids in both free and peptide form. Even under these circumstances plasma levels of glutamate or aspartate were not excessively elevated.

The data summarized in Table 3 are comparable to the data summarized in Table 1 and are indicative of the fact that LBW infants are as effective as term infants in metabolizing glutamic and aspartic acids.

One-Year-Old Infants

We recently tested the concept that infants may not metabolize a high protein meal as effectively as an adult (3). To test this hypothesis, it was necessary to have a high protein meal that could be eaten by both infants and adults. A custard whose composition is shown in Table 4 was prepared and fed to normal, healthy fasted adults and 1-year-olds. The custard contained 14% protein from milk and eggs and

TABLE 3. Postprandial plasma glutamate and aspartate concentrations in LBW infants

Feeding	Protein source	Intake (mg/kg/feeding)		Plasma concentrations (μmoles/dl)	
		Glutamate	Aspartate	Glutamate	Aspartate
SMA	Cow milk	69	10	12.4 ± 3.1	3.1 ± 2.8
Enfamil	Cow milk	82	20	13.5 ± 5.8	3.4 ± 1.3
Modified Pregestimil	Casein hydrolysate	94	32	11.0 ± 2.9	1.7 ± 0.8
5031A	Cow milk	90	24	10.3 ± 3.3	2.3 ± 1.0

TABLE 4. Custard composition

Component	Weight (g)	Protein (g)	Fat (g)	CHO (g)
Egg	150	20	17	1
Nonfat dried milk	150	54	1	78
Fructose	30			30
Water	200			
Total	530	74(14%)	18(3%)	109(21%)

TABLE 5. Custard study on adults and 1-year-olds

Subjects	N	Intake (mg/kg/feeding)		Plasma concentrations[a] (μmoles/dl)			
		Glutamate	Aspartate	Glutamate		Aspartate	
				Fast.	Post.	Fast.	Post.
Adults	6	220	77	3.3	6.4	0.4	0.6
Infants	24	230	80	5.8	11.0	0.6	1.0

[a] Fast., fasting; Post., postprandial.

was fed at a level to provide 1 g protein/kg body weight. A 10-kg infant had to eat 70 g of custard. In contrast, an 80-kg male had to eat 570 g. Serial blood samples were collected from each group of subjects. Adult subjects were bled at frequent intervals over a period of 6 hr using a heparinized indwelling catheter. Blood samples from infants were obtained by heel stick, with each infant providing four specimens at staggered intervals over a period of 4 hr.

Plasma concentrations of free amino acids doubled in both adults and infants as the result of the protein load. Plasma values peaked in approximately 2 hr and remained elevated during the entire study period. Protein intakes of 1 g/kg body weight at one feeding are not unusual for the infant (4) and many adults frequently achieve intakes of this magnitude or more.

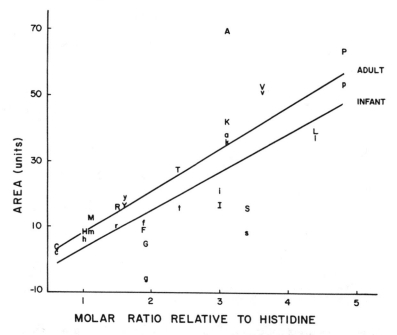

FIG. 1 Area under the curve for specific amino acids as a function of their molar ratio to histidine following ingestion of a high protein meal. Capital letters = adult data, lower-case letters = infant data.

As shown in Table 5, the concentration of plasma glutamate and aspartate in the fasting state is greater for infants than adults. When given identical protein loads, the percentage change in plasma glutamate and aspartate concentration is identical for the two age groups. Plasma concentrations of glutamate and aspartate in the postprandial state of 1-year-olds fed custard are similar to those observed for formula-fed LBW infants or formula-fed and breast-fed term infants.

Marrs and co-workers (8) have suggested a method of comparing the absorption of protein hydrolysates to amino acid mixtures. From timed plasma aminograms an absorption curve is plotted for each amino acid. The area under the curve is determined and this value is expressed as a function of the molar ratio of each amino acid relative to histidine. If protein hydrolysates and amino acid mixtures are similar, the plot will fit a common regression line.

This method of analysis has been applied to the aminogram data obtained on the infant and adult subjects fed custard. The molar ratio for 15 amino acids relative to histidine (indexed at 1) was calculated from the amino acid composition of the custard. Area under the curve in arbitrary units was calculated for these amino acids and plotted as a function of relative molar ratio (Fig. 1). Individual amino acids in this figure are identified according to the one-letter notation employed in the *Atlas of Protein Sequence and Structure* (2). If the infant and adult digest, transport, and

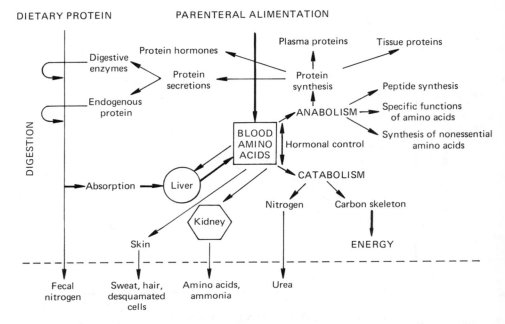

FIG. 2. Interrelationships of amino acid metabolism during enteral and parenteral feeding.

metabolize the egg-milk protein load in a similar manner, the data should fit a common regression line. As shown in Fig. 1, adults and infants have similar regression lines. Calculated coefficients of correlation are $r = 0.73$ for adults and $r = 0.81$ for infants.

PARENTERAL FEEDINGS

During enteral feeding, dietary or exogenous protein entering the gut is diluted by a considerable quantity of endogenous protein secreted into the gut (10) (Fig. 2). Proteins present in the gut are digested to free amino acids and small peptides, both of which enter the intestinal mucosal cell. Mucosal cells metabolize some amino acids, notably glutamic and aspartic acids, thereby offering some protection against excessive intake of these amino acids. Peptides are hydrolyzed to free amino acids during transit through the mucosal cells so that, in general, only free amino acids reach the liver via the portal circulation. Amino acids circulating in peripheral blood and available for tissue uptake are subject to complex hormonal controls that determine their precise anabolic or catabolic fate (13).

During parenteral feeding, both the gut and liver are bypassed since amino acids and peptides are administered directly into the peripheral circulation. From studies of the daily dietary intake and flux of phenylalanine and tyrosine for a 70-kg man, Munro (9) has postulated that man has a considerable capacity to remove via normal metabolic processes amino acids given parenterally.

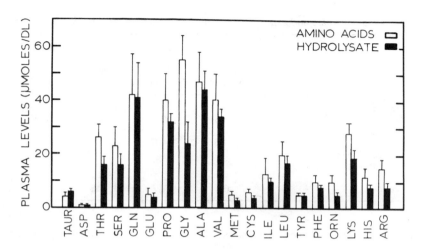

FIG. 3. Plasma amino acid concentrations in normal adult subjects fed parenterally using either a 5% casein hydrolysate-25% glucose solution or a 5% amino acid-25% glucose solution.

Adult Subjects

Stegink and co-workers (14) have studied glutamate and aspartate metabolism in a group of normal, healthy adult volunteers on total parenteral feedings. Subjects were assigned to one of two parenteral solutions. One group of subjects received nitrogen in the form of a casein hydrolysate (Amigen, Baxter Laboratories, Morton Grove, Ill.) containing large amounts of glutamic and aspartic acids. The other group of subjects was given a mixture of crystalline amino acids (Aminosyn, Abbott Laboratories, North Chicago, Ill.) free of glutamic and aspartic acids. Energy requirements were met with a 25% (w/v) solution of glucose. No significant differences in plasma concentrations of glutamate and aspartate were noted, indicative of the rapid metabolism of these amino acids when administered intravenously (Fig. 3). In a parallel study of a group of postsurgical (stressed) patients, no differences were observed in the capacity to metabolize the glutamate or aspartate present in the infused casein hydrolysate.

Term Infants

In 1971 Stegink and Baker (15) reported plasma aminograms on a series of six sick infants receiving total parenteral nutrition. The nitrogen source for these feedings was either an enzymatically digested casein (Amigen) or an enzymatically digested beef fibrin (Aminosol, Abbott Laboratories, North Chicago, Ill.).

Plasma glutamate and aspartate concentrations in such infants infused with protein hydrolysate solutions providing 0.2 g/kg body weight of glutamate and 0.05 g/kg body weight of aspartate were well within the normal postprandial limits observed for enterally fed infants (Table 6). These authors reported that the amino

TABLE 6. *Plasma glutamate and aspartate concentrations in infants infused with hydrolysates*

Infusion	N	Glutamate		Aspartate	
		Intake (mg/kg)	Plasma (μmoles/dl)	Intake (mg/kg)	Plasma (μmoles/dl)
Casein hydrolysate	6	195	6.1	50	1.1
Fibrin hydrolysate	2	35	4.5	65	0.4
Oral feedings of milk-based formula					
Fasting	15	—	2.7	—	0.5
Postprandial	15	62	7.7	15	2.1

TABLE 7. *Plasma glutamate and aspartate concentrations in LBW infants on total parenteral feedings*

Patient	Weight (kg)	Intake (mg/kg)		Plasma concentrations (μmoles/dl)	
		Glutamate	Aspartate	Glutamate	Aspartate
McC	1.38	165	39	3.6	1.6
MA	1.51	201	48	3.9	1.7

Solution: 2.5% casein hydrolysate, 24% glucose.

acid composition of the hydrolysate had a marked effect on plasma amino acid levels, with some degree of amino acid imbalance observed with both products. In general, plasma amino acid concentrations reflected concentrations of specific amino acids in the hydrolysate. For example, the low levels of cystine and tyrosine in the casein hydrolysate caused a reduction in plasma cystine and tyrosine. The beef fibrin hydrolysate, rich in glycine, produced an increase in plasma glycine concentration.

If a population with decreased ability to metabolize dicarboxylic amino acids existed, it might well be that of small premature infants on parenteral feedings. Such infants are biochemically immature and are further compromised by a feeding technique that bypasses the gut. Recently, we have measured amino acid levels in two LBW infants weighing 1.4 and 1.5 kg. The nitrogen source for these parenterally fed infants was a casein hydrolysate. Plasma concentrations of glutamate and aspartate were less than the postprandial concentrations for these amino acids observed in LBW infants following formula feedings (Table 7).

COMMENT

Winters and co-workers [20] have recently concluded that amino acid intakes resulting in plasma concentrations of individual amino acids within the normal range of postprandial values are both safe and efficacious. Normal postprandial values were defined as those found in breast-fed or formula-fed infants. Intakes of nitrogen as protein, peptides, or free amino acids that yield plasma values significantly above the postprandial range should be considered suspect for safety. Intakes of nitrogen yielding plasma values below this range should be considered suspect for nutritional efficacy.

In this report we have established postprandial values for plasma concentrations of glutamate and aspartate in both breast-fed and formula-fed infants. Like Winters and co-workers, we regard these as normative values.

Relative to these normative data we have demonstrated the following:

1. Term infants fed a casein hydrolysate formula (Nutramigen) have plasma glutamate and aspartate levels that agree with normative data.

2. LBW infants fed conventional milk-based formulas and a special formula prepared from a casein hydrolysate have plasma glutamate and aspartate levels that agree with normative data.

3. Fasting 1-year-old infants fed a high protein meal have plasma glutamate and aspartate levels that agree with normative data. In addition, analysis of plasma aminograms shows that these infants are as efficient as normal adult subjects in metabolizing a protein load.

4. Surgically stressed or ill term infants and LBW infants weighing 1,500 g or less given a casein hydrolysate solution parenterally have plasma glutamate and aspartate levels that agree with normative data.

Although glutamate and aspartate are both metabolically highly reactive amino acids requiring careful study, one cannot escape the fact that abnormally high plasma glutamate and aspartate levels have not been seen following enteral or parenteral administration of formula or intravenous solutions containing high levels of free glutamate and aspartate. On the basis of these clinical observations in man, it is difficult to understand the tentative conclusion advanced by the SCOGS Committee suggesting that insufficient evidence exists to determine whether glutamate affects infants adversely. Available clinical evidence strongly supports the conclusion that glutamate and aspartate ingested in association with an adequate source of carbohydrate are safe for infants.

ACKNOWLEDGMENTS

These studies have been supported in part by grants-in-aid from the Gerber Products Company, Fremont, Michigan; G. D. Searle & Company, Skokie, Illinois; and the International Glutamate Technical Committee, Tokyo, Japan.

REFERENCES

1. Baker, G. L., Filer, L. J., Jr., and Stegink, L. D. (1979): Factors influencing dicarboxylic amino acid content of human milk (*this volume*).
2. Dayhoff, Margaret O. (1972): Protein data introduction. In: *Atlas of Protein Sequence and*

Structure, Vol. 5, D-1. National Biomedical Research Foundation, Georgetown University Medical Center, Washington, D.C.
3. Filer, L. J., Jr., Baker, G. L., and Stegink, L. D. (1977): Plasma aminograms in infants and adults fed an identical high protein meal. *Fed. Proc.*, 36:1181.
4. Filer, L. J., Jr., and Martinez, Gilbert, A. (1964): Intake of selected nutrients by infants in the United States. *Clin. Pediatr.*, 3:633–645.
5. Filer, L. J., Jr., Stegink, L. D., and Chandramouli, B. (1977): Effect of diet on plasma aminograms of low birth weight infants. *Am. J. Clin. Nutr.*, 30:1036–1043.
6. Fomon, Samuel, J., Filer, L. J., Jr., Thomas, Lora, N., and Rogers, Ronald R. (1970): Growth and serum chemical values of normal breast-fed infants. *Acta Paediatr. Scand. (Suppl.)* 202:1–200.
7. Fomon, Samuel, J., Thomas, Lora, N., Filer, L. J., Jr., Ziegler, Ekhard, E., and Leonard, Michael, T. (1971): Food consumption and growth of normal infants fed milk-based formulas. *Acta Paediatr. Scand. (Suppl.)* 223:1–36.
8. Marrs, T. C., Addison, J. M., Burston, D., and Matthews, D. M. (1975): Changes in plasma amino acid concentrations in man after ingestion of an amino acid mixture simulating casein and a tryptic hydrolysate of casein. *Br. J. Nutr.*, 34:259–265.
9. Munro, H. N. (1976): Absorption and metabolism of amino acids with special emphasis on phenylalanine. *J. Toxicol. Environ. Health*, 2:189–206.
10. Munro, H. N. (1977): Parenteral nutrition: Metabolic consequences of bypassing the gut and liver. In: *Clinical Nutrition Update—Amino Acids*, edited by H. N. Munro, M. A. Holliday, and H. L. Greene, pp. 141–146. American Medical Association, Chicago.
11. Select Committee on GRAS Substances (1976): *Tentative Evaluation of the Health Aspects of Certain Glutamates as Food Ingredients.* Life Sciences Research Office, Federation of American Societies for Experimental Biology, Bethesda, Md.
12. Select Committee on GRAS Substances (1976): *Tentative Evaluation of the Health Aspects of Protein Hydrolysates as Food Ingredients.* Life Sciences Research Office, Federation of American Societies for Experimental Biology, Bethesda, Md.
13. Stegink, L. D. (1975): Amino acid metabolism. In: *Intravenous Nutrition for High Risk Infants*, edited by R. W. Winters and E. G. Hasselmeyer, p. 181. John Wiley & Sons, New York.
14. Stegink, L. D. (1977): Peptides in parenteral nutrition. In: *Clinical Nutrition Update—Amino Acids*, edited by H. N. Munro, M. A. Holliday, and H. L. Greene, pp. 192–198. American Medical Association, Chicago.
15. Stegink, L. D., and Baker, G. L. (1971): Infusion of protein hydrolysates in the newborn infant. Plasma amino acid concentrations. *J. Pediatr.*, 78:595.
16. Stegink, L. D., Filer, L. J., Jr., Baker, G. L., Mueller, S. M., and Wu-Rideout, M.Y-C. (1979): Factors affecting plasma glutamate levels in normal adult subjects *(this volume)*.
17. Stegink, L. D., Reynolds, W. A., Filer, L. J., Jr., Baker, G. L., and Daabees, T. T. (1979): Comparative metabolism of glutamate in the mouse, monkey, and man *(this volume)*.
18. Stegink, L. D., and Schmitt, J. L. (1971): Postprandial serum amino acid levels in young infants fed casein-hydrolysate based formulas. *Nutr. Rep. Int.*, 3:93.
19. Stegink, L. D., Schmitt, J. L., Meyer, P. D., and Kain, P. (1971): Effect of DL-methionine fortified diets on urinary and plasma methionine levels in young infants. *J. Pediatr.* 79:648–659.
20. Winters, Robert W., Heird, William C., Dell, Ralph B., and Nicholson, J. F. (1977): Plasma amino acids in infants receiving parenteral nutrition. In: *Clinical Nutrition Update—Amino Acids*, edited by H. N. Munro, M. A. Holliday, and H. L. Greene, pp. 147–154. American Medical Association, Chicago.

Glutamic Acid: Advances in Biochemistry and Physiology, edited by L. J. Filer, Jr., et al.
Raven Press, New York © 1979.

Placebo-Controlled Studies of Human Reaction to Oral Monosodium L-Glutamate

Richard A. Kenney

Department of Physiology, George Washington University Medical Center, Washington, D.C. 20037

In 1972 Kenney and Tidball (3) reported an investigation of human susceptibility to monosodium L-glutamate (MSG) administered orally. This study used two samples of tomato juice: one as the vehicle for administering the MSG and the other as a "placebo" in which an equivalent degree of saltiness was obtained by the addition of NaCl. They reported that 32% of the 77 subjects tested at the level of 5 g MSG dissolved in 150 ml of juice reported one or more of the following sensations: warmth or burning, stiffness or tightness, weakness in the limbs, pressure, tingling, headache, light-headedness, or heartburn or gastric discomfort. These subjects were then studied in greater depth. This study led the authors to conclusions summarized as follows:

1. When MSG was administered as a 1.32% solution in 150 ml of tomato juice, reports of symptoms were no more common than after ingesting a seasoned juice containing no MSG.
2. When administered in heavy concentration (3.33%), symptoms were provoked in 32% of the test population.
3. Symptom experience bore no apparent relation to the level of plasma glutamic acid before the ingestion of MSG or to the extent of its rise after the ingestion.
4. With increasing dosage of MSG there was an increasing frequency of reports of sensations of warmth or burning, stiffness or tightness, weakness in the limbs, and tingling.
5. In a limited number of cases tested, no objective correlate of the reported symptom could be discerned.

This present chapter reports two further placebo-controlled trials designed to examine this problem further.

TRIAL A

Fifty-one volunteers were recruited within the Medical Center by the local publication of a notice containing a description of the project objective and procedure used in the form of informed consent.

A schedule of administration of unique sets of three juices from a total of seven samples was prepared in which each set was identified by number. The juices were designated A–G. Of these, C, F, and G were placebo juices consisting of tomato juice with 0.8 g common salt added to each 150 ml. Samples A, B, D, and E contained in 150 ml of tomato juice with 5 g MSG, purchased in the form of Accent. Each set of three juices consisted of one placebo juice and two juices containing MSG. The administration sequences were selected so that placebo juice appeared an equal number of times in each of the three positions. The technician in charge was given the only available copy of the list of juice administrations. On enrolling in the experiment, the procedure was explained to each volunteer and each subject was given three packets of Granola bars (either "Cinnamon" or "Honey and Oats") and three 6-oz cans of either grapefruit or orange juice with the instructions that on each of the test days one packet of bars and one can of juice was to be taken as breakfast as near to 7:00 a.m. as possible. At this time the subjects also signed their consent forms and drew a number assigning their sequence of juices. The technicians kept sole charge of the subject identifications and was unaware of the composition of the juices.

The subjects reported to the laboratory at 10:00 a.m. on their assigned days and were issued the appropriate glass of juice and given a copy of a standard questionnaire. Completed questionnaires were returned at noon. The questionnaires were collected and were read and classified at the completion of the series. The classification system used was as follows:

A questionnaire specifying no unusual sensation was classified as 0.

A questionnaire reporting only headache, thirst, light-headedness, or gastric discomfort was classified as type A.

A questionnaire reporting one or more of the sensations of warmth or burning, stiffness or tightness, weakness in the limbs, or tingling was classified as type B.

A box was provided on the form for the specification of sensations other than those listed. This was seldom used and when used was essentially a paraphrase of one of the listed items, in other words no "new" sensations were reported, save for two reports of tremor and four of difficulty in reading fine print. The questionnaires were then identified with the juice administered on a particular day.

TABLE 1. *Response pattern to test and placebo tomato juice*

	Reaction to placebo		
	0	A	B
Reaction to test juice			
0/0	8	3	1
0/A	3	2	
A/A	2		
0/B	6		1
A/B	3	1	1
B/B	5	6	5

TABLE 2. *Average symptom scores (symptoms/day/subject)*

	Juice sample								
	A	B	C	D	E	F	G	H	I
MSG content	1 g	0	2 g	0	3 g	0	4 g	0	5 g
Reactors ($N = 7$)	0.36	0.28	0.59	0.43	1.59	0.28	1.71	0.36	2.17
Nonreactors ($N = 9$)	0.22	0.33	0	0.44	0	0	0.22	0.11	0.34

Although 51 subjects began the study, only 47 completed the series. The reasons for this dropout included a major fracture of a wrist requiring high levels of analgesics, intercurrent illness, and, one suspects, boredom.

Of the 47 subjects completing the series, 24 were men and 23 were women. In Table 1 are recorded the numbers of subjects showing various patterns of reaction classified by reaction types O, A, or B.

Accepting an A-type reaction to placebo juice as a nonspecific observation, two groups can be clearly identified, the 11 subjects (5 males and 6 females) in the upper pair of cells who may be regarded as confirmed *nonreactors* and the 11 subjects (5 males and 6 females) in the lower pair of cells who may be regarded as *confirmed reactors*. These individuals were invited to participate in further studies. Nine nonreactors and 7 reactors completed these studies.

As before, subjects ate the standard breakfast at 7:00 a.m. and reported to the laboratory at 10:00 a.m. The study involved nine attendances, and on each occasion the subject was given 150 ml of either placebo or juice containing MSG in the range of 1 to 5 g. The nine juices administered were designated A through I; their contents were as follows:

A = 1 g MSG
B = 0.8 g NaCl
C = 2 g MSG
D = 0.8 g NaCl
E = 3 g MSG
F = 0.8 g NaCl
G = 4 g MSG
H = 0.8 g NaCl
I = 5 g MSG

On the first attendance of the subject, a number from 1 to 10 was drawn by lot, this number serving to identify the sequence of juices for that subject for the 9 days. Only the technician was aware of which juice a subject received on a particular day. After drinking the juice under supervision, the subject was given a questionnaire to be completed over the course of the following 2 hr.

Before drinking the juice on 2 of the 9 days, each subject was tested for fine tremor of the hand and visual acuity was measured. These tests were repeated on subjects reporting symptoms at the time of symptom experience. The results were negative with respect to both tremor and visual acuity.

The questionnaires were classified on the same basis as in the earlier trial, save that a numerical value of 0.5 was assigned to the type A responses. Type B responses were quantitated in terms of the number of significant symptoms reported. When all questionnaires had been graded, the juice code was broken and mean symptom scores calculated for each juice for the reactor and nonreactor groups (Table 2).

These results essentially confirm the reliability of the two groups and confirm the earlier observation of the relationship in the reactor group between the number of symptoms reported and the quantity of MSG administered.

As a part of this phase of the study, seven reactor subjects in the reactor group and nine subjects in the nonreactor group volunteered to provide blood samples for analysis by Dr. Karl Folkers (University of Texas) for erythrocyte GOT estimation. The subjects reported to the laboratory at or about 10:00 a.m. on each of 3 days. On each occasion 10 to 15 ml blood were drawn by venipuncture and stored in ice prior to shipment to Dr. Folkers. In addition, on 1 of the 3 days a further blood sample was taken and submitted for a standard SMA 12/60 analysis by the Division of Laboratory Medicine. There was excellent coincidence between the groups of the mean values for the basal level of EGOT, SGOT, and all of the standard blood chemistry items save for glucose and LDH. The blood glucose in the reactor subjects was 71 ± 10 mg/dl compared with 81 ± 6 in the nonreactor subjects. The reactors showed a mean LDH value of 188 ± 18 U/liter; the nonreactors, a mean value of 166 ± 10 U/liter. Neither of these differences is statistically significant. This trial appeared to have accomplished the following:

1. Confirmed the existence in a random population of subjects of a group who react reliably and predictably to a heavy dose of MSG and of a group who are reliably symptom free after such a dose.

2. Confirmed previous observations that at the dosage level used 33% of the men and 50% of the women will report symptoms following MSG administration and, further, that reaction to placebo juice will occur in one out of six trials.

3. Confirmed that in the reactor subjects, reports of symptoms appear to be related to the quantity of MSG administered.

4. Failed to demonstrate any tremor or visual change accompanying the reaction.

5. Failed to demonstrate any significant difference in blood chemistry between the two groups.

TRIAL B

At this point, it was felt that elucidation of the action of MSG in producing symptoms called for a closer pursuit of the objective correlates of symptom experience. The series up to this point had been clouded in its analysis by the significant rate of placebo response. It was apparent that the subjects received symptom suggestion by the information provided when their consent to experimentation was obtained and further suggestion was provided by the questionnaire that the subjects had completed. In addition, tomato juice was perhaps not the ideal vehicle of administration. There being no ethical or legal way in which symptom suggestion at the time of informed consent could be avoided, it was decided to avoid reinforcement of the suggestion by questioning each subject after an administration either personally or over the telephone rather than by employing a check-list-type ques-

TABLE 3. Results of triangle tests

Test material	Subjective response				
	All different	All the same	Wrongly paired	Correctly paired	Total
Tomato juice	1	4	9	4	18
Soft drink	5	3	12	7	27

tionnaire. Pairs of samples of a soft-drink mix in powder form were made available for use in place of the tomato juice samples. The compositions of each pair are as given below:

Placebo (sample S)
Sucrose
Citric acid (monohydrate)
Trisodium-Citrate (2-hydrate)
Lemon flavor (natural)
Caramel color
Naringin[1]

Sample with MSG (sample G)
Sucrose
Citric acid (monohydrate)
MSG

Lemon flavor (natural)
Caramel color
Naringin[1]

Each pair of samples thus consisted of a placebo sample and a mixture containing 6 g MSG. The materials as supplied to the investigator were identified in a coded fashion so that the trial could be conducted in a double-blind format. The material was prepared for administration by dissolving the powder in precisely 200 ml of cool (15° C) distilled water.

Triangle tests of the soft-drink materials were made by a taste panel. The same panel was used to triangle test the tomato juice samples used in the earlier studies. The outcome of these tests is given in Table 3.

It would appear, therefore, that both vehicles are effective for concealment of MSG, but in the case of tomato juice, one is dealing with a substance the flavor of which is well known to most subjects and any addition, whether it be of salt or of MSG, produces a "spoiled tomato juice" flavor. The soft drink, on the other hand, provides a flavor never previously experienced, and since there exists no expectation of basic flavor, no clue is provided as to the presence of an additive. This material has therefore been used in the latest trial.

Since some months had elapsed from the earlier series of trials, the majority of the previous subject population was dispersed and the study had to be reinitiated. The protocol was further modified from that of the earlier studies by administering the trial solutions to the subjects in a fasting state.

Fifty-seven volunteer subjects were recruited from among Medical Center personnel. Of these, 9 subjects were black, 4 oriental, and 44 white. The age range was

---

[1] Naringenin-7-rhamnosidio-glycoside: grapefruit bitter principle (natural).

20 to 56 years; 22 were male and 35 female. The subjects read a statement of the purpose and protocol of the trial and, after asking any questions and receiving answers, signed the statement to acknowledge their informed consent to serve as a subject.

At this time each subject was registered with a number and assigned two pairs of samples selected at random from the supply. On each of the first 2 test days, one member of the assigned pair was administered; on the second 2 test days, the remaining assigned pair was used. The subjects reported to the laboratory kitchen in a fasting state within a few minutes of 8:00 a.m. on each test day. They were given their assigned juice and asked to drink it at a leisurely pace and to follow it with a small drink of cool water. This the subjects did in the presence of the investigator and then returned to their normal workplace with the instruction that they were not to eat or drink with the exception that if they experienced a severe thirst they might assuage this with water from the cooler.

An hour ($\pm$ 10 min) after drinking the juice the subjects whose workplace was within the laboratory building were visited and questioned. Those situated in other buildings were questioned by telephone. The questioning was stereotyped and followed the following plan:

1. What did you think of your juice this morning? Did you find it pleasant or unpleasant?
2. What made it pleasant/unpleasant?
3. Could you describe the taste?
4. What was the aftertaste?
5. How long did the aftertaste last? (At this point check on drinking of water relative to aftertaste.)
6. Did you experience any other sensations? (If the answer here is positive, details of nature, location, and time of onset and offset are sought.)
7. Are you still experiencing any of these sensations? (If the answer is positive, a second contact is made later.)

Answers to the questions were recorded as obtained. When the administration of the trial samples was completed, the decoding document was obtained and the sheets recording the subject's daily responses sorted and classified in terms of referring to an S (placebo) or G (MSG) juice.

Of the 57 subjects in the study, 48 received two pairs of samples and 9, for any one of several reasons, received one pair of samples. S and G juices were each therefore administered on a total of 105 occasions.

The sensations reported are recorded in Table 4. In only a few cases has it been necessary to paraphrase a subject's own terminology, probably because the statement of informed consent had in effect provided a vocabulary of description. Where a subject reported more than one sensation, these have been included in the two or more categories. These sensations clearly fall into two groups. One group of sensations common to exposure to both S and G samples and the other group almost exclusively related to exposure to G samples. The first group coincides with what in the

TABLE 4. *Sensations reported for soft drink*

| | Frequency of report | |
Description	S sample	G sample
Nausea or heartburn	4	3
Thirst or dryness	8	3
Light-headedness or headache	6	10
Tightness	1	9
Pressure	0	7
Tingling	0	5
Weakness	0	3
Warmth	0	1
Burning	0	2

Omitted from this record are 6 reports of tingling of the tongue that occurred as a regular response to either S or G samples in two subjects.

earlier trial was termed a type A response; the second with the type B response. Experience of type B sensations was confined to 16 of the 57 subjects, a frequency that agrees well with earlier experience. Of these 16 subjects, 15 were female and 1 male. The 15 women include 1 black subject and 1 oriental subject. The response pattern is given in Table 5.

TABLE 5. *Response pattern to soft drink*

| | Reaction to S | | | | | |
	0/0	0/A	0/B	A/A	A/B	B/B
Reaction to G						
0/0	19 (7)	3	(1)			
0/A	4	1		1		
0/B	4	2				
A/A	2			1		
A/B	4					
B/B	6 (2)					

Figures in parentheses refer to individuals who received only one pair of samples.

The loci of the type B sensations are confined to a very sharply defined and restricted part of the body surface, i.e., the face or head, in the shoulders, on the upper arms, or on the upper part of the chest (Fig. 1). The frequency of location of sensation to the various body areas is as given below; where more than one locus was mentioned, all the areas were included in the accounting:

Face or head   19 reports of tightness or pressure
Shoulders   8 reports of warmth, tingling, or tightness

Upper arms	4 reports of tingling or fatigue
Upper chest	4 reports of burning or pressure
Neck	1 report of warmth

Review of the reports in the earlier series confirmed this restricted location of B-type sensation.

The 16 subjects identified in this screening procedure formed the population for further studies. To better define the sensitivity of these individuals, trials were made with pairs of samples, with one sample containing MSG at either the 3- or 1.5-g level. Four subjects were lost to the study at this point, 2 being reluctant to experience further sensations and 2 whose duties made them unavailable. A double-blind trial with the remaining 12 subjects was made at the 3-g level, and 3 subjects, 2 female and 1 male, reported type B sensations. In each case the modality of the sensation and its locus was the same as that experienced at the 6-g level but at a lower perceived intensity in all cases. These 3 subjects were then further tested using the samples containing 1.5 g MSG. Two of these subjects reported mild, brief, transient type B sensations, and the third experienced nothing. It was apparent that these sensations would have gone without notice had not attention been directed to them by prior experience. One of the sensitive subjects reported the following:

6-g level: First experience—"Burning sensation in upper chest and arms—began 15 min after drinking juice—lasted 15 min."

Second experience—"Burning sensation began in chest and upper arms about 15 min after drinking juice—lasted 20 min."

3-g level: "Slight burning sensation in left shoulder off and on for 5 min—started 15 min after juice."

1.5-g level: "Little tingling in shoulders 20 min after drinking juice—lasted 2 min."

Three members of the reactor group who had consistent and well-defined symptoms were studied in some detail during symptom experience. These subjects

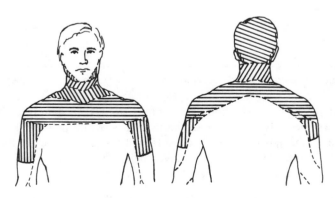

FIG. 1. The hatched areas, together with the face, define the body area in which type B sensations are experienced.

reported to the laboratory on each of 2 days in a fasted state. They were fitted with chest electrodes to record the electrocardiogram, with adhesive surface electrodes to record the electrical activity of muscles or muscle groups where these subjects had previously reported tightness or weakness. Temperature sensors were taped to the body surface in areas where sensations of warmth or burning had been reported. The subjects were seated comfortably in a room with background music, and the transducers were connected via appropriate signal-conditioning devices to a photographic recording galvanometer. The outputs were monitored continuously on a cathode ray tube and photographic hard-copy records taken from time to time. When the subjects were at ease, they were given 200 ml of one or other of a pair of the soft-drink solutions. This administration was double blind. The subjects were asked to volunteer statements about the sensations they experienced. All three reported sensations on the day the G solution (at the 6-g level) was administered. No sensations were experienced in response to an S solution. The sensations reported were warmth, weakness, tightness, and palpitation. In no case were the sensations accompanied by an appropriate objective sign. Blood pressure, monitored frequently throughout, remained stable.

These observations, taken in conjunction with earlier failures to demonstrate objective correlates of symptoms (3), are interpreted as indicative of sensations referred to the body surface from a viscus, rather than peripheral, area. When one compares the areas of the body where type B sensations occur (Fig. 1) to those where esophageal pain is referred (5), one finds a remarkable degree of coincidence. The area for referred esophageal pain "corresponds, as a rule, fairly well with the portion of the esophagus involved." Thus, heartburn, believed to arise by spasm or irritation of the gastroesophageal junction, is felt at the lower end of the sternum. A preferred site for the experience of upper esophageal pain is the midline at the upper sternal border. From here the sensation spreads to involve the face, the head and neck, the upper chest and back, the shoulders, and the upper arms. Those areas outside of the face correspond to the dermatomal distribution of the nerves of the 2nd, 3rd, 4th, and 5th cervical segments. Clinical experience indicates that pain arising in the esophagus is of two varieties: burning and pressure. Major causes of pain arising from the esophagus are spasm of the muscle coats and chemical stimulation of free nerve endings of the mucous surface by the backwashing of the strongly acid gastric juice from the stomach, giving rise to the sensation of heartburn.

If one accepts that tingling or warmth are lesser intensities of burning sensation and that tightness or stiffness may be varieties of pressure sensation, one is driven to the conclusion that the type B response observed on exposure to MSG solutions might be described accurately as a transient esophagalgia.

In order to test this hypothesis, some further experiments were undertaken. Four members of the reactor group who had experienced symptoms at the 6-g (3%) level but had no response at the 3-g (1.5%) level were selected. They were subjected to a double-blind administration of 100 ml solutions, one containing 3 g MSG. These subjects were thus exposed to a total dose of MSG that had previously failed to provoke symptoms, although the concentration of the material had been effective. Three of the four subjects experienced typical type B sensations.

The two females and one male who reacted at the level of 3 g MSG in the 200 ml of soft drink were given a 3-g dose of MSG in gelatin capsules with the objective of avoiding contact with nerve endings of the oropharynx and esophagus. A placebo control was provided by an equal number of capsules containing lactose. One of the three subjects experienced a typical reaction of burning in the upper chest and shoulders, whereas the two other subjects experienced nothing. The reacting subject was differentiated from the other two in that she reported regular heartburn "after eating her own cooking"; the others denied any experience of this sensation.

In an attempt to further localize the source of the type B sensations, a group of nine subjects, five nonreactors and four reactors used the soft-drink mixtures with or without MSG at the 3% concentration as a mouthwash and gargle, avoiding swallowing the solution. Each was given 100 ml of solution and instructed to spend 3 to 5 min using it. This procedure gave one report of a type B reaction: a reactor subject using a G solution. All other reports were of taste, aftertaste, salivation, or dryness. The possibility that the one reaction arose from the inadvertent swallowing of the solution cannot be excluded.

The experiments of Shaumberg et al. (6) leave no doubt that typical symptoms can be produced by MSG administered intravenously. Furthermore his studies demonstrate the ability of MSG to produce a rather general irritation of free nerve endings. These experiments, however, should not lead one to conclude that the customary production of symptoms relies on an elevated concentration of MSG in the blood since we have demonstrated no relationship between blood levels of glutamic acid and the occurrence of symptoms (3). Comparison may be made with the use of the bitter substance Decholin (dehydrochloric acid) to measure the arm-to-tongue circulation time where injection of Decholin into an arm vein is followed about 15 sec later by the perception of a bitter taste. The latency of the first appearance of sensation after MSG injection (17 to 20 sec) accords well with an arm-to-esophagus transit.

The nerve endings of the esophagus are accessible to an irritant substance in three ways: (a) by direct exposure following swallowing, (b) by reflux of the material back into the esophagus from the stomach, and (c) from the circulation. The latency (usually about 20 min) of the appearance of symptoms following ingestion of MSG has often been tacitly equated with the time needed for absorption. However, such a latency would be quite compatible with gastroesophageal reflux. Furthermore, tests of esophageal irritability undertaken to differentiate esophageal pain from angina pectoris (1) demonstrate that the development of sensation is dependent on two factors: (a) the concentration of the irritant material and (b) the length of time that the material is present at the esophageal mucosa. A lapse of 15 to 30 min from the start of superfusion of the esophagus with 0.1 N HCl and the appearance of pain is not uncommon (1).

From the perspective, therefore, of dietary exposure to MSG, the following conditions would predispose to the appearance of symptoms:

1. Ingestion of a solution with a high concentration of MSG. (Experience indicates that for the vast majority of individuals the threshold concentration is in excess of 1.5%.)
2. A lack of a good flow of saliva to dilute and wash away the agent.
3. A tendency to experience gastroesphageal reflux.

CONCLUSIONS

These three studies, the one cited (3) and the two reported here, are consistent in demonstrating that large doses, or high concentrations, of MSG will provoke a variety of sensations in approximately 33% of a test population. However, the new soft-drink vehicle has provided a better definition to the problem by sharply defining the MSG-attributable symptoms. It is now possible to say with confidence that with concentrations of MSG of the order of 0.75%, it is extremely unlikely that any of the symptoms will be experienced by even a demonstrably sensitive individual. Furthermore, at a level of 1.5%, only a few individuals will be affected.

Type A sensations, evoked by G and S samples with equal frequency, are of common experience (4) and indeed might be expected when a fasting subject drinks a solution of high carbohydrate and significant sodium content.

Type B sensations arise from the upper part of the esophagus where high concentrations of MSG appear to be a specific or nonspecific irritant.

It is interesting that although the anecdotal literature of the Chinese restaurant syndrome contains reports of individuals who believed themselves to be suffering cardiac pain (2) and sought attention for this condition, no mention is to be found of the differential diagnosis (1) of esophagalgia being considered.

ACKNOWLEDGMENTS

This work was supported by the International Glutamate Technical Committee. The author would also like to thank Dr. A. Genoni of NESTEC for the generous supply of sample materials, Ms. Barbara Schubert and Ms. Nancy Sheehan for technical assistance, Dr. Karl Folkers for the analysis of blood samples, and the volunteer subjects for their patient cooperation in these studies.

REFERENCES

1. Bernstein, L. M., Fruin, R. C., and Pacini, R. (1962): Differentiation of esophageal pain from angina pectoris: Role of the esophageal acid perfusion test. *Medicine*, 41:143–162.
2. Gordon, M. E. (1968): Chinese restaurant syndrome. *N. Eng. J. Med.*, 278:1123.
3. Kenney, R. A., and Tidball, C. S. (1972): Human susceptibility to oral monosodium L-glutamate. *Am. J. Clin. Nutr.*, 25:140–146.
4. Kerr, G. R., Wu-Lee, M., El-Lovy, M., McGandy, R., and Stare, F. J. (1977): Objectivity of food-symptomatology surveys. *J. Am. Diet. Assoc.*, 71:263–268.
5. Moersch, H. J. and Miller, J. R. (1943): Esophageal pain. *Gastroenterology*, 1:821–831.
6. Shaumberg, H. H., Byck, R., Gerstl, R., and Mashman, J. H. (1969): Monosodium L-glutamate: Its pharmacology and role in the Chinese restaurant syndrome. *Science*, 163:826–828.

Glutamic Acid: Advances in Biochemistry
and Physiology, edited by L. J. Filer, Jr., et al.
Raven Press, New York © 1979.

Food-Symptomatology Questionnaires: Risks of Demand-Bias Questions and Population-Biased Surveys

George R. Kerr,* Marion Wu-Lee, Mohamed El-Lozy, Robert McGandy, and Frederick J. Stare

School of Public Health, The University of Texas Health Science Center, Houston, Texas 77025; and School of Public Health, Harvard University, Boston, Massachusetts 02115

The "Chinese restaurant syndrome" (CRS) is reportedly characterized by a unique symptom complex consisting of sensations variously described as "burning," "tightness," and/or "numbness" in the upper chest, neck, and face, beginning shortly after the start of a meal in a Chinese restaurant and lasting less than 4 hr (1–4). Less characteristic symptoms include dizzyness, headache, chest pain, palpitation, weakness, nausea, and vomiting (2,7). The syndrome has been speculated to be caused by the flavor enhancer monosodium glutamate (MSG), and Reif-Lehrer (5,6) recently reported that some 25% of a population surveyed by questionnaire may have experienced this condition.

Being concerned that "demand-bias" in the Reif-Lehrer questionnaire resulting from use of the question "Do you think you get 'Chinese restaurant syndrome?'" might have led to an exaggerated estimate of its true prevalence, we recently attempted to clarify the issue through a two-part questionnaire (3). The first part attempted to identify unpleasant symptoms associated with particular foods and places of eating, but did *not* include the phrase "Chinese restaurant syndrome." When the first questionnaire was completed, the same respondents received a second asking if they had ever heard of a "Chinese restaurant syndrome," what symptoms were associated with it, and whether they had personally experienced it. The questionnaires were administered to students of the Harvard Summer School, faculty, students, and staff of the Harvard School of Public Health, and employees of the Children's Hospital Medical Center of Boston. The study revealed that 3 to 7% reported symptoms on the first questionnaire that could possibly represent the characteristic syndrome, yet once the syndrome was mentioned, 31% believed that they had experienced it. We concluded that nonspecific symptoms that occurred in association with eating in Chinese restaurants were erroneously believed to represent the "Chinese restaurant syndrome" by those familiar with its name.

Because of the additional concern that data derived from this "health-conscious" sample might not be representative of the general population, Market Research

Corporation of America (MRCA) was commissioned by Ajinomoto, U.S.A. to administer the same questionnaire sequence to its National Consumer Panel.

MATERIALS AND METHODS

Study Population

The National Consumer Panel is a panel of households that report to MRCA their purchases of grocery and textile products. The sample is maintained at a level of about 7,500 active reporting households, generally representative of the United States and stratified by demographic characteristics such as household size, region, age of housewife, and so on. For the present study, a subsample of 2,269 households was selected at random from within each strata. Each of the 4,729 adult members of the 2,269 households received both study questionnaires.

Questionnaire Design

Questionnaire I stated that we were collecting information on the symptoms and discomforts that some people associate with particular foods. Subjects were told that a second questionnaire dealing with their knowledge of a particular food-associated health problem would be sent to those who completed the first one. Questionnaire I listed 18 food-associated symptoms, 3 of which were characteristically associated with CRS (burning sensation in the face or chest, tight sensation around face, neck or chest, and numbness or loss of feeling). Six symptom options were nonspecific, but often associated with CRS (chest pain, dizziness or light headedness, headache, nausea or vomiting, palpitation, and weakness), and 9 options were even more nonspecific or not associated with CRS (abdominal cramps, chills, diarrhea, flushing sensation in face or chest, heartburn, unusual perspiration or sweating, unusual thirst, tingling, and others to be specified). Respondents were asked the time of onset of each symptom after the start of the meal (options: under 10 min, 10 min to 2 hr, over 2 hr) and its duration (options: less than 1 hr, 1 to 4 hr, over 4 hr).

Having identified the unpleasant symptoms associated with food, we then questioned whether each was notably associated with a particular food class from a list of 15 options that included beverages, cereal or grain products, chocolates or other "sweets," dairy products, desserts, eggs or egg products, fowl or poultry, fruits or fruit juices, meats, nuts, seafood and shellfish, soups, spices, vegetables, and others (specify). Respondents were then asked if each symptom was associated with a particular place of eating from a list of 11 options that included cafeterias, church suppers, delicatessens, fast-food restaurants, hotels, lunch counters, personal residence, picnics, residence of friends, restaurants, and vendors and vending machines.

We then asked if each symptom was notably associated with a particular ethnic style of food preparation from a list of 15 options that included American, Arabic, Chinese, French, German, Greek, Hungarian, Indian, Japanese, Jewish, Mexican-Spanish, Polynesian, Scandinavian, "soul food," and others (to be specified). And finally, we asked if each symptom was associated with a particular ethnic food course from a list of 15 options that included chow mein, curry, gefilte fish, goulash, hot dog, pizza, raw fish (sashimi), salad (type to be specified), shish kebab, soup (type to be specified), spaghetti, sweet-and-sour pork, tempura, taco, and others (to be specified).

Questionnaire II contained an initial series of questions that again related to unpleasant symptoms that respondents might associate with specific foods or eating environments. We then asked whether they personally ate foods prepared in Chinese restaurants (including "take-outs"); whether they purchased Chinese food in markets or prepared Chinese food at home; whether they had ever heard of a condition called "the Chinese restaurant syndrome"; which of 18 symptoms (the same symptoms options listed in Questionnaire I) were believed to be associated with the syndrome; what the temporal relationships were between the symptoms and consumption of food in Chinese restaurants (the same time options as in Questionnaire I); whether the respondent had personally experienced CRS (options: yes, no, don't know); and whether the syndrome was associated with a particular food additive from a list of nine options that included artificial food colors, artificial food flavors, artificial sweeteners, BHA, BHT, iodized salt, MSG—monosodium glutamate, sodium nitrate, sodium nitrite, spices (to be specified), and other (to be specified).

Data Analysis

Data were analyzed by computer, using a proprietary data processing system. In determining the prevalence of CRS, symptoms commencing sooner than 10 min, later than 2 hr after the start of the meal, or having a duration greater than 4 hr were not considered compatible with characteristic CRS. Each of the three characteristic CRS symptoms ("burning," "tightness," and "numbness") were given a score of from 1 to 3 depending on their presence, time of onset, and duration. If a correspondent reported a characteristic CRS symptom but failed to report times of onset and duration and indicated that it was an uncommon occurrence, we gave the benefit of the doubt and assumed that the times would have been those characteristic of CRS. If a respondent identified only one of the characteristic symptoms, with the appropriate times of onset and duration, a score of 3 was allocated; all three symptoms with correct times of onset and duration would receive a score of 9. The symptoms reported were then scored with a range from 0 to 9 according to the following system: 0 = not CRS; 1 to 2 = probably not CRS; 3 to 6 = possibly CRS; 7 to 8 = probably CRS; and 9 = definitely CRS. A response reporting one characteristic symptom with the correct temporal associations would be considered a "possible CRS."

RESULTS

Both questionnaires were completed by 3,222 respondents. Thirty-seven percent of the 1,411 male respondents were 18 to 34 years of age; 32% were 35 to 54; and 31% were over 55. Of the 1,811 female respondents 32, 30, and 38% were in the same age ranges.

TABLE 1. Unpleasant symptoms associated with food

Symptom	% Sample reporting symptom	% Male	% Female	% "Correct" onset time	% "Correct" duration
Abdominal cramps	10.7	8.6	12.3	56.1	30.8
Burning sensation in face or chest	2.3	2.4	2.2	36.5	12.2
Chest pain	3.2	3.6	2.9	44.7	23.3
Chills	1.2	0.8	1.6	17.5	5.0
Diarrhea	12.1	10.1	13.6	47.9	36.7
Dizziness	2.9	2.6	3.2	42.6	22.3
Flushing sensation in face or chest	1.9	1.2	2.5	24.2	9.7
Headache	4.8	3.8	5.6	31.6	38.7
Heartburn	24.5	24.9	24.2	64.6	31.9
Nausea or vomiting	6.4	5.1	7.3	43.4	31.7
Numbness or loss of feeling	1.2	1.0	1.3	21.1	15.8
Palpitation	2.0	1.0	2.8	36.9	18.5
Tight sensation around face, neck, or chest	1.3	0.9	1.5	34.1	19.5
Tingling	1.3	0.9	1.6	16.7	11.9
Unusual perspiration	3.6	4.2	3.1	30.4	12.2
Unusual thirst	10.4	9.6	11.0	53.3	38.9
Weakness	1.8	1.2	2.2	40.4	17.5
Other	4.0	3.0	4.8	39.5	24.8

Questionnaire I

A total of 1,369 respondents, 43% of the total sample, indicated that one or more unpleasant symptoms were associated with the consumption of food. Twenty percent reported experiencing only 1 of the 18 symptom options; 10% reported 2 symptoms; 5% reported 3 symptoms, with progressively smaller numbers of respondents reporting additional symptoms. Two respondents reported all 18 symptom options.

The frequency of individual symptoms is presented in Table 1. It is apparent that nonspecific symptoms of "heartburn," diarrhea, abdominal cramps, and unusual thirst are experienced by over 10% of the study population. The characteristic CRS symptoms of "burning," "numbness," and "tight" sensation were reported by 2, 1, and 1% of the population, respectively. Headache was reported by nearly 5%, unusual perspiration by 4%, and the other CRS-associated symptoms by smaller percentages of the study sample. A larger percent of females reported experiencing 14 of the 18 symptom options. Many respondents did not report times of onset and duration of symptoms; of those that reported these times, the percent falling within the "acceptable" limits of CRS is also indicated in Table 1. It is apparent that many of the symptoms—CRS-characteristic, CRS-associated, and nonspecific—are experienced within the same periods of time.

On the basis of our scoring system, none of the respondents reported all three characteristic CRS symptoms within the correct time limitations (score 9) nor did

TABLE 2. *Specific food classes associated with symptoms*

Food class	% Positive association	% Positive respondents reporting		
		Characteristic CRS symptoms (total of 3)	CRS-associated symptoms (total of 6)	Nonspecific symptoms (total of 9)
Beverage	10.5	4.4	32.9	81.5
Cereal products	2.0	7.4	14.8	90.7
Chocolate or "sweets"	6.9	5.4	32.5	79.8
Dairy products	3.7	3.7	21.1	89.0
Desserts	3.7	4.0	25.0	89.0
Egg products	3.4	7.4	18.9	88.4
Fowl/poultry	1.5	4.7	23.3	83.7
Fruit/fruit juice	6.4	5.3	8.0	95.7
Meats	8.8	2.7	20.4	90.2
Nuts	5.5	3.9	24.2	86.9
Seafood and shellfish	3.8	3.7	21.1	89.0
Soups	3.4	4.2	11.5	94.8
Spices	18.1	4.9	12.3	94.7
Vegetables	8.3	3.4	11.3	94.5
Others	6.6	4.6	17.5	90.2

Many respondents reported more than one symptom.

any report symptoms that met the criteria for "probable CRS" classification (scores 7 to 8). Fifty-seven respondents, 25 males and 32 females (1.8% of the total study population), reported symptoms that were classed as "possible CRS" (scores 3 to 6). Of these 57 respondents, 46 reported only one of the characteristic symptoms; they also reported one (17 of the 57) or more (40 of the 57) CRS-associated symptoms, and 50 of the 57 reported one or more nonspecific symptoms.

The food classes associated with specific symptoms are presented in Table 2. Eighteen percent of the respondents reported unpleasant experiences with "spices," but all other options were selected by 10% or less of respondents. Characteristic CRS symptoms were noted by from 2 to 8% of those respondents who reported unpleasant symptoms for each option. CRS-associated symptoms were indicated for each food class by from 8 to 32%, and 80% or more of the unpleasant symptoms associated with each food were of the nonspecific category. Symptoms associated with "spices" had the following frequency sequence: "heartburn," 65%; unusual thirst, 15%; abdominal cramps, 13%; diarrhea, 9%; and unusual perspiration, 7%. No particular food group was clearly associated with characteristic CRS symptoms.

The places of eating that were associated with unpleasant symptoms are presented in Table 3: 15% of respondents reported unpleasant symptoms after meals eaten in their own residence; 11% after meals in restaurants; 9% after eating in fast-food restaurants; and 7% after meals in the residence of friends. Other options were selected by progressively fewer respondents, with only 2% reporting unpleasant symptoms after church suppers. Over 90% of the symptoms associated with each option were of the nonspecific category; 18 to 25% of symptoms were related to CRS-associated symptoms; and 4 to 9% of symptoms were considered characteristic

TABLE 3. Eating places associated with unpleasant symptoms

Eating place	% Positive association	% Positive respondents reporting		
		Characteristic CRS symptoms	CRS-associated symptoms	Nonspecific symptoms
Cafeterias	4.5	5.3	22.9	92.4
Church suppers	1.9	8.8	21.1	91.2
Delicatessens	2.5	3.9	15.8	96.1
Fast-food restaurants	8.5	5.9	19.5	94.5
Hotels	2.0	6.5	22.6	93.5
Lunch counters	3.6	3.7	18.7	96.3
Personal residence	14.8	5.6	22.1	93.0
Picnics	4.0	5.1	17.8	94.1
Residence of friends	6.9	6.2	21.5	94.7
Restaurants	11.2	6.2	22.1	90.6
Vendors and vending machines	2.7	6.3	25.3	91.1

TABLE 4. Ethnic styles of food preparation associated with unpleasant symptoms

Food style	% Positive association	% Positive respondents reporting		
		Characteristic CRS symptoms	CRS-associated symptoms	Nonspecific symptoms
American	10.9	2.8	22.4	95.0
Arabic	0.3	0	10.0	100.0
Chinese	4.2	9.1	18.9	87.9
French	1.3	11.1	13.9	94.4
German	1.7	5.7	17.0	96.2
Greek	1.2	5.7	11.4	97.1
Hungarian	1.0	3.3	10.0	96.7
Indian	1.3	10.3	5.1	97.4
Italian[a]	5.1	6.6	13.8	96.1
Japanese	1.1	8.6	11.4	94.3
Jewish	0.9	4.2	8.3	100.0
Mexican-Spanish	14.9	5.5	11.5	96.7
Polynesian	1.0	17.2	6.9	89.7
Scandinavian	0.2	0	0	100.0
"Soul food"	1.3	5.9	17.6	91.2
Other	1.2	6.2	15.6	84.4

[a] Extracted from the category of "other."

of CRS. No particular place of eating was notably associated with characteristic CRS symptoms. Of the 57 respondents who were classified as "possible CRS," 40% reported CRS symptoms after eating in restaurants, and 40% reported CRS symptoms after eating in their personal residence.

The ethnic styles of food preparation associated with unpleasant symptoms are presented in Table 4. Some 15% of respondents associated unpleasant symptoms with Mexican-Spanish foods; 11% with American; 4% with Chinese food; and

TABLE 5. Symptoms most often reported after eating different ethnic styles of food

	1 (% Symptoms reported)	2 (% Symptoms reported)	3 (% Symptoms reported)
American	Heartburn (57.4)	Diarrhea (27.8)	Cramps (25.2)
Arabic	Heartburn (60.0)	Cramps (40.0)	Diarrhea, thirst, nausea, vomiting other (10.0 each)
Chinese	Heartburn (37.1)	Thirst (34.8)	Diarrhea (13.6)
French	Heartburn (66.7)	Cramps (25.0)	Diarrhea (22.2)
German	Heartburn (69.8)	Cramps (28.3)	Thirst (18.9)
Greek	Heartburn (62.9)	Cramps (31.4)	Diarrhea, thirst (17.1 each)
Hungarian	Heartburn (63.3)	Cramps (26.7)	Diarrhea (20.0)
Indian	Heartburn (51.3)	Thirst (28.2)	Cramps (17.9)
Italian	Heartburn (70.4)	Thirst (20.4)	Cramps (17.1)
Japanese	Heartburn (34.3)	Thirst (31.4)	Diarrhea (28.6)
Jewish	Heartburn (58.3)	Thirst (37.5)	Diarrhea (16.7)
Mexican-Spanish	Heartburn (68.1)	Thirst (19.2)	Cramps (18.1)
Polynesian	Heartburn (41.4)	Thirst, diarrhea (20.7 each)	Thirst, diarrhea (20.7 each)
Scandinavian	Thirst (66.7)	Cramps, heartburn, diarrhea (16.7 each)	Cramps, heartburn, diarrhea (16.7 each)
"Soul food"	Heartburn (55.9)	Cramps (32.4)	Diarrhea (26.5)
Other	Heartburn (40.6)	Thirst (21.9)	Diarrhea, other (12.5 each)

progressively fewer positive responses with other options, down to 0.2% with Scandinavian food. While Italian food had not been listed as an option, over 5% of respondents associated unpleasant symptoms with Italian food under the "other" option. Over 80% of unpleasant symptoms associated with each ethnic cuisine were of the nonspecific category, and from 0 to 22% of symptoms were CRS-associated. Characteristic CRS symptoms were reported by 17% of those reporting unpleasant symptoms after eating Polynesian food and by 11% of those experiencing unpleasant symptoms after either French or Indian foods. (The characteristic CRS symptom most often associated with these three ethnic styles of preparing food was a "burning sensation," but CRS symptoms were also reported by 5 to 10% of those reporting difficulty after Chinese, German, Greek, Italian, Japanese, Mexican-Spanish, "soul food," and other ethnic styles of food preparation, and in each case a "burning sensation" was the most common symptom.) Of the 57 "possible CRS" respondents, only 6 (0.019% of the study population) reported characteristic CRS symptoms after consuming Chinese food.

Table 5 presents the three unpleasant symptoms most commonly associated with each of the ethnic styles of preparing food. With the exception of Scandinavian food, where unusual thirst was the most common complaint, heartburn was the most frequent symptom in all cases. The next most common symptoms were those of abdominal cramps, diarrhea, and unusual thirst. No particular ethnic style of food preparation was associated with a uniquely different pattern of symptoms.

TABLE 6. Unpleasant symptoms associated with particular ethnic foods

Food	% Positive association	% Positive respondents reporting		
		Characteristic CRS symptoms	CRS-associated symptoms	Nonspecific symptoms
Chow mein	2.5	10.1	21.5	87.3
Curry	2.8	9.5	8.3	95.2
Gefilte fish	0.4	11.1	0	100.0
Goulash	3.0	5.4	8.6	96.8
Hot dogs	7.4	2.2	15.2	94.2
Pizza	12.0	3.9	10.5	96.1
Raw fish (sashimi)	0.2	0	71.4	28.6
Salad	3.4	1.0	13.6	94.2
Shish kebob	0.8	4.3	8.7	100.0
Soup	3.1	3.5	8.1	97.7
Spaghetti	8.3	4.9	11.3	96.0
Sweet-and-sour pork	2.8	4.8	22.9	90.4
Tempura	0.3	0	22.2	88.9
Taco	9.7	4.4	8.8	97.3
Other	4.0	5.6	13.7	92.7

Specific ethnic foods associated with unpleasant symptoms are presented in Table 6. Pizza was associated with unpleasant symptoms by 12% of respondents, tacos 10%, spaghetti 8%, hot dogs 7%, and the other ethnic food options were associated with unpleasant symptoms by progressively fewer respondents. With the exception of the "raw fish" option, over 80% of the symptoms reported were of the nonspecific variety. CRS-associated symptoms accounted for 71% of the "raw fish" symptoms (mainly nausea and vomiting) and for 0 to 23% of the symptoms associated with other ethnic foods. Characteristic CRS symptoms accounted for 10% of unpleasant symptoms reported after chow mein and gefilte fish, but for less than 6% of symptoms associated with other ethnic food options. Of those classified as "possible CRS" on the basis of their symptoms, eight associated characteristic CRS symptoms with pizza, six with tacos, five each with spaghetti and "other," and four each with chow mein, hot dogs, and goulash.

Table 7 reports the three unpleasant symptoms most commonly associated with specific ethnic foods. With the exception of sashimi (where 71% of positive responses reported nausea or vomiting) and chow mein (where thirst was the most prominent symptom), heartburn was the primary complaint in all instances. The next most common complaints were, in almost all instances, those of abdominal cramps, diarrhea, and thirst. As with the preceeding questions, and with the exception of the "raw fish" option, it did not appear that any of the ethnic food options was associated with a unique pattern of unpleasant symptoms.

Questionnaire II

After receiving the above information without use of the leading phrase "Chinese restaurant syndrome," the second questionnaire attempted to ascertain the number of respondents aware of it, its symptoms, and who had personally experienced it.

TABLE 7. Symptoms most often reported after eating different ethnic foods

Food	1 (% Symptoms reported)	2 (% Symptoms reported)	3 (% Symptoms reported)
Chow mein	Thirst (38.0)	Heartburn (35.4)	Cramps, diarrhea (15.2 each)
Curry	Heartburn (58.3)	Cramps, thirst (19.0 each)	Thirst, cramps (19.0 each)
Gefilte fish	Heartburn (44.4)	Cramps, diarrhea, thirst (22.2 each)	Cramps, diarrhea, thirst (22.2 each)
Goulash	Heartburn (74.2)	Cramps (16.1)	Diarrhea (12.9)
Hot dog	Heartburn (66.4)	Cramps (18.4)	Thirst (15.2)
Pizza	Heartburn (68.4)	Thirst (21.9)	Cramps (14.1)
Raw fish (sashimi)	Nausea, vomiting (71.4)	Cramps, diarrhea, unusual perspiration (14.3 each)	Cramps, diarrhea, unusual perspiration (14.3 each)
Salad	Heartburn (64.1)	Cramps (26.2)	Diarrhea (17.5)
Shish kebab	Heartburn (69.6)	Cramps (21.7)	Thirst (13.0)
Soup	Heartburn (60.5)	Cramps (20.9)	Thirst (17.4)
Spaghetti	Heartburn (73.7)	Cramps (17.4)	Thirst (15.4)
Sweet-and-sour pork	Heartburn (50.6)	Thirst (26.5)	Diarrhea (20.5)
Tempura	Heartburn, diarrhea (44.4 each)	Heartburn, diarrhea (44.4 each)	Cramps (33.3)
Taco	Heartburn (68.4)	Thirst (20.5)	Cramps (16.2)
Other	Heartburn (49.2)	Cramps (24.2)	Diarrhea, thirst (19.4 each)

Chinese food was consumed with a wide range of frequency, with 31% stating that they ate in Chinese restaurants more often than once or twice a year. About 20 to 22% of the respondents reported the purchase or home preparation of Chinese food more often than once or twice a year. Of "possible CRS" respondents, 3% stated they never or rarely ate in Chinese restaurants; 4% stated they ate there more often than once or twice a year. Of respondents reporting one or more characteristic CRS symptoms, 8% never or rarely ate at Chinese restaurants; 13% ate there more often than once or twice a year.

Only 8% of respondents reported that they had heard of a condition called the "Chinese restaurant syndrome," 86% were unfamiliar with the phrase, and 6% gave no response. The symptoms associated with CRS by those who were "aware" of it are reported in Table 8. Headache was the most common symptom (38%), followed by unusual thirst (31%), dizziness (29%), abdominal cramps (20%), diarrhea (21%), flushing (22%), and heartburn (18%). The characteristic CRS symptoms were identified by a smaller number of respondents: "burning" (11%); "numbness" (11%), and "tight" sensation (12%). Thirty-one percent of those aware of the syndrome stated that "burning," "tightness," and "numbness" were *not* associated with CRS, and approximately 60% of those "aware" reported they "didn't know" or gave no response to each symptom option. Of the total panel, 2% were "aware" of CRS and able to identify at least one of its characteristic symptoms; 5% were "aware," reporting CRS-associated symptoms; and 5% were "aware," reporting nonspecific symptoms.

When queried as to the food additive associated with the syndrome, 8% of the

TABLE 8. *Symptoms associated with CRS by respondents familiar with name of syndrome*

Symptom	Yes (%)	No (%)	Don't know (%)	No response (%)
Abdominal cramps	20.2	27.1	26.7	26.0
Burning	11.2	31.0	27.9	29.8
Chest pain	8.5	34.1	28.3	29.1
Chills	5.8	37.6	28.7	27.9
Diarrhea	20.5	26.7	26.4	26.4
Dizziness	29.1	23.6	22.5	24.8
Flushing	22.1	24.8	24.8	28.3
Headache	37.6	19.8	21.7	20.9
Heartburn	18.2	27.5	27.5	26.7
Nausea or vomiting	15.9	30.2	26.7	27.1
Numbness	10.9	31.0	27.5	30.6
Palpitations	15.1	30.2	25.6	29.1
Thirst	30.6	23.3	24.0	22.1
Tightness	12.4	31.0	27.5	29.1
Tingling	4.3	34.9	29.8	31.0
Unusual perspiration	14.0	29.8	27.1	29.1
Weakness	9.7	32.9	26.7	30.6
Other	1.9	15.5	18.6	64.0

panel associated it with MSG; other additives were identified by only 1 to 2% of respondents. The majority of respondents "didn't know" or were "unaware" of any association between specific food additives and CRS.

With regard to the prevalence of CRS, 65 respondents or 2.0% of the study population reported that they had personally experienced the syndrome. Forty percent of the respondents stated they had not experienced it; 43% "didn't know," and 15% failed to answer the question. Because of a concern about the legibility of the answer options to the "Have you personally experienced the Chinese restaurant syndrome?" question, a repeat questionnaire was sent to 98 respondents who had stated they were aware of the CRS and had *not* expressed it. Nine of the 98 subsequently reported personal experience with CRS, resulting in a *subjective* prevalence rate of 74/3222 or 2.3%. Of these 74, however, only 6 had been classified as "possible CRS" on the basis of their symptoms.

Finally, of respondents who reported personal experience with CRS, 69% stated that they ate in Chinese restaurants more often than once or twice a year; 29% stated that they never or rarely ate at Chinese restaurants. Of respondents who reported they did *not* experience CRS, the figures were almost reversed: 70% never or rarely ate in Chinese restaurants, and 28% ate there more often than once or twice a year.

DISCUSSION

Webster's New Collegiate Dictionary defines *syndrome* as "a group of signs or symptoms that occur together and characterize a particular abnormality." The original reports of CRS described a symptom complex that was characteristic and unique. Subsequent reports have included an increasing number and variety of symptoms, however, and the original definition of syndrome has been modified by Rief-Lehrer to include any symptoms named in response to the question "Do you get any of the symptoms below (20 options including 'other') after you eat Chinese restaurant food either in restaurants or 'take out?' " "Other" symptoms reported included "depression," "detachment," "feel emotionally variable; laughing, crying," "sense of fullness after a limited amount of food," "water retention," and 31 other new symptoms, all of which now become components of a newly defined CRS.

It would appear reasonable to expect that a syndrome postulated to result from the pharmacologic effects of an agent should have limited interindividual variance: inclusion of nonspecific or noncharacteristic symptoms as components of a syndrome will detect "atypical" cases, but will also result in a larger, and more likely erroneous, estimate of its prevalence in a population. If carried to the absurd, an individual who consistently suffers the headache, thirst, weakness, and nausea of hangovers from overconsumption of rice wine in Chinese restaurants might logically conclude that he was susceptible to CRS.

Although we do not dispute that some people have developed a dramatic and frightening syndrome after eating in Chinese restaurants, we believe that the prevalence rates of 25% reported by Reif-Lehrer (5,6) and the 3 to 7% reported by Kerr et al. (3) are probably both in excess of the true figure. The basis for the different estimates of prevalence involves important issues of questionnaire design and administration. The fact that Reif-Lehrer's reports were published as "Special Articles" in the prestigious *Federation Proceedings* suggests a need to review some of these issues.

It is clear, for example, that many people have strong feelings about particular foods and welcome the opportunity to report such feelings. It is also well established that individuals who return questionnaires are usually more interested in the subject matter than those who do not. Analyses of food-associated questionnaire data must consider whether those who returned the questionnaire "wanted" to participate and whether their responses are representative of the entire population. This is a difficult problem to resolve except through some form of remuneration or non-food-associated reward system.

A second concern is that of "demand-bias" from leading questions. Questionnaires should minimize the risk that respondents may become aware of the actual issue and want to "help" the study. Rief-Lehrer's questionnaire included "Do you think you get Chinese restaurant syndrome?" And we believe this question may have led some of her study population to report non-CRS symptoms as instances of the syndrome. Properly designed questionnaires must go to great lengths to prevent the researcher's own interest or biases from being communicated to respondents. By addressing these two concerns in questionnaires, we were able to reduce the apparent prevalence of the CRS from 25 to 3 to 7%.

A third concern involves population bias: both Reif-Lehrer's study and our previous one were conducted in a population strikingly different from the general population. In both cases the participants were more "health-aware" than the general population, more likely to be familiar with the expression "Chinese restaurant syndrome," and thus more likely to report an exaggerated estimate of its prevalence. For example, 92% of our Harvard population associated unpleasant symptoms with food, compared with 43% of the present study population. From 2 to 9 times as many Harvard respondents reported experiencing each of the symptom options, and 31% believed they had experienced CRS compared with 2% of the general adult population. And where our previous prevalence estimate (based on the symptoms reported by the Harvard population) was in the range of 3 to 7%, the symptoms reported by the present National Consumer Panel suggest that 1 to 2% of the adult population experience symptoms that might possibly represent CRS.

An additional methodologic concern involves "exposure." People who do not eat in Chinese restaurants may not believe that they had experienced CRS even though they may have developed the appropriate symptoms in another eating environment. Conversely, individuals who frequently eat in Chinese restaurants would be more liable to have experienced some unpleasant event, in some temporal relationship with the meal, that might suggest that it was CRS rather than gastroenteritis, overindulgence, or some discomfort associated with the event or the eating environment rather than the food consumed.

And finally, we believe that a syndrome, if it is a true entity, and particularly if it is caused by a pharmacologic agent, must show reasonable conformity to its characteristic symptom complex. It is apparent from both our previous and current questionnaires that many people associate unpleasant signs and symptoms with specific foods and eating environments. In the great majority of cases these symptoms are nonspecific, and any questionnaire attempting to document the prevalence of a food-associated health problem must minimize the risk of such nonspecific symptoms being given undue significance. While 74 respondents (3.2% of the study population) reported that they had personally experienced CRS, only 6 of them were classified as "possible CRS" on the basis of their symptoms. Accordingly, some attempt must be made, in food-symptomatology questionnaires, to modify prevalence estimates resulting from *subjective* feelings by some *objective* criteria.

There are still many unresolved issues in regard to the true prevalence of CRS, and we hope the present study will lead to further refinements in questionnaire design. For example, many of these symptoms are ambiguous and imprecise: it is possible that the characteristic CRS symptom of "numbness" may have been reported by some as nonspecific "tingling" and that the characteristic symptom of "burning" was interpreted as nonspecific "flushing" (or vice versa). For the present, however, we chose not to speculate on respondent's interpretation of these words, as both "burning" and "flushing" may possibly be confused with the *normal* feeling of "warmth" accompanying the ingestion and metabolism of food.

The data from this study suggest that the characteristic symptoms of CRS, which may (and more often do not) occur in association with food consumption in a Chinese restaurant, has a prevalence rate of somewhere closer to 1 to 2% of the general adult public than either of the previous estimates of 3 to 7 and 25%.

ACKNOWLEDGMENT

This work was supported in part by the Fund for Research and Training, Department of Nutrition, Harvard School of Public Health, Boston, Massachusetts.

REFERENCES

1. Ambos, M., Leavitt, N. R., Marmorek, L., and Wolschina, S. B. (1968): Sin-Cib-Syn: Accent on glutamate. *N. Engl. J. Med.*, 279:105 (correspondence).
2. Gordon, M. E. (1968): Chinese restaurant syndrome. *N. Engl. J. Med.*, 278:1122 (correspondence).
3. Kerr, G. R., Wu-Lee, M., El-Lozy, M., McGandy, R., and Stare, F. J. (1977): Objectivity of food-symptomatology questionnaires. *J. Am. Diet. Assoc.*, 71:263–268.
4. Kwok, R. H. M. (1968): Chinese restaurant syndrome. *N. Engl. J. Med.*, 278:796 (correspondence).
5. Reif, Lehrer, L. (1976): Possible significance of adverse reactions to glutamate in humans. *Fed. Proc.*, 35:2205–2211.
6. Rief-Lehrer, L. (1977): A questionnaire study of the prevalence of Chinese restaurant syndrome. *Fed. Proc.*, 36:1617–1623.
7. Schaumberg, H. H., and Byck, R. (1968): Sin-Cib-Syn: Accent on glutamate. *N. Engl. J. Med.*, 279:105 (correspondence).

Summary

R. J. Wurtman

Laboratory of Neuroendocrine Regulation, Department of Nutrition and Food Science, Massachusetts Institute of Technology, Cambridge, Massachusetts 02139

In the pages of this volume, several major aspects of the biochemistry and physiology of glutamic acid were discussed:

The metabolism of glutamate.
Its probable role as a neurotransmitter.
Its function as the precursor for another neurotransmitter (GABA).
The actions of synthetic glutamate analogs like kainic acid.
The sources and metabolic fate of circulating glutamate.
The barriers that exist within the body for keeping excess glutamate from crossing the placenta to the fetus or the newborn.
The responses of the fetus and the newborn to exogenous glutamate.

For the first time, evidence has been presented showing the direct efficacy of glutamate in stimulating taste receptor cells at the molecular level. The probability that some unfortunate individuals might develop characteristic symptoms after eating foods rich in glutamate and the possible mechanism of this response were put into perspective. We considered at some length the experimental uses of glutamate and related compounds to produce brain lesions when injected into the brain, given intraperitoneally, administered in massive doses by gavage, and given without foods or water. We discussed the relationships between such brain lesions and the body's response to glutamate as a food constituent, consumed in nonexperimental situations. In addition, we discussed some aspects of the food science of glutamate and its presence, free and protein bound, in natural and synthetic foods. The symposium on which this volume is based may well be the prototype of a series of symposia that will be held in the next few years to bring together scientists concerned with food, nutrition, and metabolism, on the one hand, and those concerned with neuropharmacology, neurochemistry, and toxicology, on the other. In Lausanne, in July 1978, a conference was held in which exogenous tryptophan and the brain were discussed;[1] In December 1978, another symposium was organized in which the effects of exogenous choline and lecithin on the brain were discussed.[2] We should, perhaps, perceive our discussions on dietary glutamate and

---

[1] July 6–7, 1978: *Transport Mechanisms of Tryptophan in Blood Cells, Nerve Cells and at the Blood-Brain Barrier.*
[2] December 4–6, 1978: *Uses of Choline and Lecithin in Neurologic and Psychiatric Diseases.*

brain function as part of a far broader phenomenon, namely, the growing interest of scientists concerning the interactions between food constituents and brain function. Even though we have focused on glutamate, many of the processes discussed have direct relevance to the actions of other nutrients on the brain.

From the above generalizations, I would like to expand on four specific problems. First, is glutamate's role as a neurotransmitter, especially the adequacy of the data supporting its function as such. Second, is the access of circulating glutamate to the brain, especially to those regions, such as the arcuate nucleus, that may lie outside the blood-brain barrier. Third, is the use of glutamate as an experimental neurotoxin. Fourth, is the safety of glutamate as a food additive.

GLUTAMATE AS A NEUROTRANSMITTER

If one compares what we know about glutamate, aspartate, GABA, or glycine, any of the nonessential amino acid neurotransmitters, to the available information on other neurotransmitters (e.g., catecholamines, serotonin, and acetylcholine), it must be concluded that the situation is far from comparable. Although we have considerable understanding of the electrophysiology of glutamate, especially about its effects when topically applied to neurons, we lack an understanding about the locations of specific synapses where glutamate is released. Advances in our understanding of glutamate as a neurotransmitter continue to be retarded by several problems that must be solved before real progress can be made. First, we need a method for distinguishing glutamate molecules in the "neurotransmitter pool," if it exists, from other glutamate molecules in the brain. Until we have some means of marking that particular pool of brain glutamate that functions as a neurotransmitter, it will be very difficult to examine the kinetics of its synthesis and release, i.e., glutaminergic transmission.

Second, we need to identify the enzymatic steps in the biosynthesis of glutamate neurotransmitter molecules. Circulating glucose is converted to brain glutamate in nerve terminals; however, we know very little about the specific pathways connecting glucose to the transmitter. The fact that brain glutamate levels are stable and unresponsive to various metabolic and pharmacologic manipulations suggests that these levels are regulated by some sort of a feedback mechanism. We are, however, uncertain if this stability applies to a specific neurotransmitter pool. It would be wonderful indeed if there was some compound that, on administration to experimental animals, would cause a specific increase in brain glutamate levels (especially *neurotransmitter* glutamate levels) so that we could thereafter examine the mechanisms that restore these levels to normal. All attempts to identify such compound(s) have been unsuccessful.

A third major problem in designing experiments on glutamate as a neurotransmitter is the lack of specific drugs affecting its synthesis or release, or its interactions with receptors. Such drugs have been of enormous utility in studying the monoamine neurotransmitters.

It is easy to identify problems that retard the development of knowledge about glutamate as a transmitter, but much more difficult to solve them. Meanwhile, I believe that glutamate *is* a CNS neurotransmitter, and I suggest that most other neurobiologists share this belief.

ACCESS OF CIRCULATING GLUTAMATE TO THE BRAIN

If glutamate is an excitatory neurotransmitter with specific receptor sites existing on most brain cells, we are presented with a very real theoretical problem: Glutamate is found in the circulation in very high concentrations; if, as we have discussed, it is a highly potent agent, and if there *are* any parts of the brain to which circulating glutamate normally gains access, then how does the brain protect its cells from continuous excitation by circulating glutamate? If circulating glutamate does have access to any region of the brain (e.g., the arcuate nucleus) lacking a recognizable blood-brain barrier (i.e., tight junctions between neighboring capillary endothelia) visible by electron microscopy, how does that region protect itself against the free entry and effects of glutamate? How do such brain cells maintain normal functional activity when, as normally occurs, postprandially, plasma glutamate levels rise? Plasma glutamate concentrations can vary over a twofold range, depending on what is eaten. If there are twofold variations in the amount of glutamate presented to those areas of the brain that lie outside the blood-brain barrier, and if glutamate is a potent excitatory neurotransmitter, then it should be possible to detect major functional changes in the neurons of the arcuate nucleus and other neurons thought to lie outside the blood-brain barrier when blood glutamate concentrations change postprandially. No one, to my knowledge, has ever demonstrated such functional, glutamate-dependent changes in the arcuate nucleus or in any other brain region. Until the time that such glutamate effects can be demonstrated, my guess is that there is *no* brain region that is not effectively shielded from circulating glutamate by either a visible or an invisible blood-brain barrier. I hope that someone will be able to determine whether physiologic variations in plasma glutamate levels do produce parallel changes in glutamate concentrations in the arcuate nucleus, and in the electrophysiologic or other parameters of arcuate nucleus function. In this volume, Garattini showed preliminary evidence that experimentally induced changes in plasma glutamate concentrations tend to be unassociated with alterations in arcuate nucleus glutamate levels. The burden of proof rests heavily on anyone who proposes that circulating glutamate normally has access to any portion of the brain, in the absence of marked plasma hyperosmolarity or of some other insult to the blood-brain barrier.

If circulating glutamate gains access to the arcuate nucleus or other brain regions, it becomes necessary to examine the consequences of such entry. Perhaps the brain uses circulating glutamate levels to sense alterations in peripheral metabolic phenomena, just as it has been shown to utilize similar changes in plasma neutral amino acid levels. Perhaps the brain decides whether or not we are hungry or sleepy, based on circulating glutamate levels. I do not really think this is the case; however, I cannot reject this possibility on the basis of the available data.

SUMMARY

GLUTAMATE AS AN EXPERIMENTAL NEUROTOXIN

Does glutamate produce brain lesions in the primate? Reports from one laboratory allege that it is possible to produce brain lesions in monkeys by giving very high, concentrated doses of glutamate. Reprints from several other laboratories failed to confirm lesion production. What are we to conclude from these conflicting reports? Although it is not unknown in the history of science that several laboratories do similar experiments and come up with opposite conclusions, a preponderance of data supporting one position or the other must hold sway, or the scientist must resolve this contradiction.

If, as appears clearly to be the case in rodents, exogenous glutamate is able to produce CNS lesions, should we be worried about it? One point that emerges from all of the data presented is that glutamate has never been shown to produce a CNS lesion in *any* experimental animal when the glutamate was ingested as a dietary constituent. Animals that are allowed to choose whether or not to eat food that contains free as well as bound glutamate (and that are also, one suspects, allowed to drink water *ad libitum*) simply do not develop brain lesions. This implies two things. First, I think we need not worry about damaging our brains by adding glutamate to our foods. Second, I think it tells us something about the mechanism whereby glutamate produces lesions in rodents: lesions occur only if glutamate is ingested in very large amounts mixed with very small quantitites of water (i.e., very high concentrations). To me, the most economical explanation of the mechanism by which glutamate produces lesions in rodents is a two-step process, of which both steps are essential. The first step is the production of a massive increase in plasma osmolarity, well beyond the level that produces coma in people (i.e., 340 to 350 mOsmoles/liter). This effect is sufficient to shrink the capillary endothelial cells that form the blood-brain barrier, thus opening gaps between these cells and allowing the circulatory contents free access to the brain. (Since cells do not shrink at a uniform rate, the loci of gaps will exhibit randomness, causing a random distribution in glutamate-induced brain lesions.) This massive hyperosmolarity is not specific to glutamate, but can be produced by administration of concentrated solutions of arabinose, sucrose, sodium chloride, or other solutes. The second step probably involves the more specific action of glutamate as an excitatory amino acid: if the blood-brain barrier is opened at a moment when blood glutamate concentration is high, significant quantities of glutamate will enter the brain at such loci to damage local neurons. *Both* processes must occur in order for CNS damage to be produced. Are there ever any circumstances in which a human might consume enough glutamate with food to produce hyperosmolarity, thereby opening the blood-brain barrier to produce brain lesions? I think that the chance of this happening is virtually nil. Our thirst mechanisms are so effectively activated when plasma osmolarity reaches 298 or 300 mOsmoles/liter that there is no way in which people would elect to consume enough of a glutamate solution to produce the degree of hyperosmolarity necessary to open the barrier (i.e., about 380 mOsmoles/liter). In general, people become comatose when plasma osmolarity rises above 340 mOsmoles/liter, and comatose people stop eating. I suspect that in none of the experimental studies on glutamate neuropathology have animals had free access to drinking water; if so, the induction of lesions by glutamate is simply an artifact of forced water restriction.

What about the neurotoxicity of glutamate analogs? Clearly the development of kainic acid as an experimental neurotoxin that destroys the cell bodies of neurons has considerable potential utility, for example, in characterizing glutamate-responsive receptors. In spite of this, I would like to reserve judgement concerning the mechanism of action and the specificity of such compounds. One reason for my doing so is the data that McGeer presented on the inability of kainic acid to produce caudate lesions after glutaminergic neuronal inputs (from the frontostriatal tract) have been interrupted. It seems to me that, based on analogies with the actions of indirect-acting sympathomimetic agents, kainic acid may work by *releasing* glutamate or some other substance from terminals of frontostriatal neurons (and not primarily or solely as a direct glutaminergic receptor agonist). The proposed interpretation of the experimental data as showing some sort of cooperative action on the postsynaptic receptor between kainic acid and glutamate released presynaptically is, of course, possible; however, to my knowledge it would be without precedence in neuropharmacology. Another area of kainic acid research where I would urge restraint is the assumption that the animal that has received a kainic acid injection in its basal ganglia constitutes a bona fide experimental model for a human disease, Huntington's chorea. Those who have treated patients with Huntington's chorea know that the manifestations of the disease are diverse, far transcending anything observed in the kainic acid-treated rat.

GLUTAMATE AS A FOOD ADDITIVE

Finally, there is the question of glutamate's safety as a food additive. Obviously, it is never possible to make a categorical statement that something is safe; one can only make the best evaluation of safety using the evidence currently available. Based on the information that we have heard, I personally have no reservations about the general use of glutamate as a food additive. I cannot conceive of any situation in which people could possibly consume glutamate in sufficient concentration to produce brain damage (if, in fact, primates, including man, are susceptible to such damage).

A limited number of persons appear to develop an unpleasant group of symptoms, sometimes called the "Chinese restaurant syndrome" (CRS), after eating at restaurants; these symptoms may be etiologically related to the free glutamate in their food. We do not yet know the mechanism responsible for these symptoms, assuming for the moment that a specific glutamate-related entity really exists; however, my best guess would be that they derive not from an effect on the brain, but from the activation of glutamate receptors in the upper gastrointestinal tract. Clearly more studies are needed on glutamate's relationship to CRS.

We are fortunate that, as documented in this volume, a wealth of information is available on the metabolic fate of dietary glutamate and on things that dietary glutamate does *not* do (for example, traverse the placenta to any significant extent). This data base does much to reassure us concerning the safety of glutamate.

Subject Index

A

Adrenal gland, glutamate effects on, 303
Afferent terminals, amino acid depolarization of, 166-167
D-Alanine, effects on taste receptors, 3
L-Alanine, binding of, to receptors, 3
Amino acids
　afferent terminal depolarization by, 166-167
　availability to brain of, 125-137
　binding of, to taste receptors, 2-3
　blood-brain barrier transport systems for, 130
　excitant antagonists of, 167-168
　excitotoxic, safety implications of, 287-319
　in gut, 31
α-Aminoadipate
　blockade of NMA activity by, 307-308
　taste similarity to MSG of, 48
Animals, glutamate toxicity studies of, 203-215
Aspartame
　metabolic loading effects of, 116
　structure of, 116
Aspartate
　as ablative neuroendocrine tool, 304
　free, in natural foods, 32
　in human milk, 117-119
　metabolic flux of, 63
　neurochemical pathway for, 169-170
D-Aspartic acid, binding studies on, 182-183
L-Aspartic acid
　binding studies on, 181-182
　multiplicity of sites of, 183
　excitatory receptors for, 177-179

B

Baby food
　glutamate addition to, 311-314
　MSG in, 31-32
Barbels, of catfish, taste receptors in, 2
Behavior, MSG effects on, 270-272
Blood-brain barrier
　for glutamate, 128-134
　transport systems for, 129
Body weight, MSG effects on, 236-240
Brain
　amino acid availability to, 125-137
　excitotoxin microinjection into, 291-297
　MSG toxicity to, 218
　　histopathology, 250-252, 255-260

C

Catfish, taste receptor studies on, 1-9
Central nervous system (CNS)
　glutamate metabolism by, 147-148
　glutamic receptors in, 177-185
Cerebral blood flow, calculations for, 125-126
Cheese, glutamic acid in, 30
"Chinese restaurant syndrome," glutamate and, 288, 373, 375-387, 393
Circumventricular organs (CVO), of brain, 132
　glutamate uptake by, 132-134, 288
CMP, synergism with MSG, 4, 6, 7
L-Cysteate, as ablative neuroendocrine tool, 304

395

SUBJECT INDEX

Cysteine-S-sulfonic acid, microinjection into brain, 291

D

Diabetes, glutamate-induced, 300
Dicarboxylic acid, in milk, factors affecting, 111-123
Diet, MSG-flavored, self-selection of, 18-20
Dog, glutamate toxicity, lack of, 204, 208, 210

E

Endocrine function, MSG effects on, 260-261, 278-284
Excitotoxic amino acids
  as food additives, 311-314
  as neuroendocrine probes, 297
  research applications and safety implications of, 287-319

F

Fertility, in mice, MSG effects on, 245-249
Fetus, glutamate transfer to, 103-110
  effects on hypothalamus, 217-229
Food
  flavor preference for MSG and, 51
  intake, MSG effects on, 240-242
  MSG-flavored, self-selection of, 11-23
Food-symptomatology questionnaires, MSG studies using, 375-387

G

GABA
  in brain, 153
  effect on luteinizing hormone release, 306-307
  glutamate as precursor of, 158-159

GABA-ergic neurons, identification of, 189-191
Glia, glutamate uptake by, 145-147
Glucose, as glutamate neurotransmitter precursor, 141-142
Glutamate (Glu), *see also* Monosodium glutamate
  analogs of, in neostriatum, 191-198
  in animal protein, 29
  annual production of, 27
  axon-sparing lesions from, 293-295
  biochemistry of, 69-84
  blood-brain barrier for, 128-134
  in brain, 153, 325-226
    plasma levels and, 391
  brain uptake of, 125-137
  effects on pituitary function, 277-285
  as food additives, 311-314, 393
    plasma levels of, 324-325
  free, in human milk, 32
  as GABA precursor, 158-159
  glutamine exchange with, 59-61
  intake and absorption of, 55-59
  intolerance syndromes of, 373, 375-387, 393
  kinetics of, in newborns and adults, 322-324
  levels in fetal tissue, 222
  metabolic pools of, 61-64
  metabolism of
    in adults, 341-349, 359
    in CNS, 147-148
    comparative, 85-102
    hormone role in, 64
    in infants, 353-358
    outline of, 69-70
    in pregnancy, 103-110
    regulation, 55-68
    responses to dietary changes, 64-66
  in natural products, 25-34
  neurochemical pathway for, 169-170
  neuronal and glial uptake of, 145-147
  neurotoxicity of, 288, 392-393
    ablation approach to, 297-299
    general features, 288-289
    to humans, relevance of, 329-330, 393

Glutamate
  neurotoxicity of (contd.)
    molecular specificity in, 289-291
    provocative approach, 304-308
    systemic, 287-288
  as neurotransmitter, 139-147, 151-161, 163-175, 390-391
    biochemical model, 139-141
  in organs, 25
  placental transfer of, 103-110
  in plant and animal protein, 29
  plasma levels of, 126-128, 321-322
    area under curve (AUC), 323-324
    biological variations, 334-335
    factors affecting, 336-348
    in human adults, 328-329
    in human premature newborns, 327-328
  safety levels for, 321-331
  in striatum, 187-201
  synergistic taste effect of, with 5'-ribonucleotides, 1-9
  taste receptors, biochemical studies of, 1-9
  toxicity in animals
    dog, 204
      lack of, 208, 210
    guinea pig, 207-208
    hamster, 208
    monkey, 210
      lack of, 211-213, 217-229
    mouse, 204, 205-206, 255-258
      lack of, 231-253, 258-261
    rabbit, 287
      lack of, 204
    rat, 203-204, 206-207
L-Glutamic acid
  binding sites on, 180-181
    multiplicity of, 183-184
  CNS receptors for, 177-185
  excitant agonists of, 177-178
  excitatory receptors for, 178-179
  excitotoxic structure analogs of, 294
  ligand-binding studies on, 179-180

  microinjection of analogs into brain, 291-293
    localization of, toxic mechanism of, 295-297
  as a transmitter precursor, 151-153
Glutaminase, in striatum, 198-200
Glutamine
  biochemistry of, 69-84
  in brain, 153
  as glutamate neurotransmitter precursor, 142-145
  glutamic acid exchange with, 59-61
  in vivo relationship with glutamate, 156-158
  metabolism of, 70-73
Glutaminergic neurons, identification of, 187-189
Glutaminergic system, neuroanatomy of, 153-155
Glutathione
  biochemistry of, 69-84
  metabolism of, 73-81
GMP, synergism with MSG, 4-7
Gonadotropin axis, glutamate effects on, 302-303
Gonadotropins, MSG effects on secretion of, 281-282
Growth, MSG effects on, 236-240, 262-263, 278
Growth hormone
  glutamate effects on, 305
  MSG effects on secretion of, 279-281
Guinea pig, glutamate toxicity to, 207-208

H

Hamster, glutamate toxicity to, 208
DL-Homocystate, taste similarity to MSG, 48
Homocysteic acid, microinjection into brain, 291
Hormones
  in glutamate metabolism, 64

Hormones (*contd.*)
  MSG effects on, 260-262, 279-284
Huntington's chorea, kainic acid
  proposed animal model, 192
Hypothalamus
  glutamate effects on, 303-304
    in fetal monkey, 217-229
  lesions of, MSG-induced, 256-258

I

L-Ibotenic acid, taste similarity to
  MSG, 48
IMP, synergism with MSG, 4, 6, 7,
  49-51
Infant(s)
  glutamate ingested by, 117-119
  glutamate metabolism in, 322-324, 353-362
  premature, glutamate intake and
    levels in, 354-355

K

Kainic acid
  binding studies on, 182
  microinjection into brain,
    291
  proposed animal model for
    Huntington's chorea, 192, 393
  role in cerebral neurotransmission,
    154, 393

L

Laboratory animals, glutamate toxicity to, 203-215
Lactation
  effects on glutamate content of
    milk, 117
  MSG intake during, 244-245
Litter size, of mice, MSG effects
  of, 249-250
Luteinizing hormone (LH), glutamate
  effects on, 305, 306-307

M

N-Methyl aspartic acid, microinjection
  into brain, 291
Milk
  dicarboxylic acids in
    factors affecting, 111-123
    lactation effects, 117
  free amino acids in, 33
  human, glutamate in, 33, 111-123,
    314, 328
Monkey
  glutamate effects on
    in utero, 104-109, 217-229
  glutamate metabolism in, 85-102
  toxicity, 288
    lack of, 204-205, 210-213, 217-229
Monosodium glutamate (MSG), *see
  also* Glutamate
  in baby food, 31-32
  consumption of
    effects on plasma levels, 338-348
    in various countries, 28
  daily intake, 120
    in meals, 341-348
    in water, 338-340
  effects on
    mouse reproduction, 255-264
    pituitary function, 277-285
    reproduction, 233-236
  flavor effects of, 39-48
  human reaction to
    food symptomatology questionnaire of, 375-389
    placebo-controlled studies on,
      363-373
  lack of
    effects on fetal hypothalamus,
      217-229
    long-range effects of, 261-272
    of mouse reproduction, 231-253
  properties of, 26
  ribonucleotide synergism with, 4-7
  role in flavor preference for food, 51
  safety of, 89-100, 103-109, 217-228,
    258-274, 321-331, 393
  self-selection of foods flavored with, 11-2

Monosodium glutamate (*contd.*)
  taste of
    compounds similar to, 48
    intensity of, 43-45
    interaction with other taste, 45-48
    nucleotide synergism with, 49-51
    psychometric studies on, 35-54
    taste threshold of, 42-43
Mouse
  glutamate metabolism in, 85-102
  glutamate toxicity to, 204, 205-206
    lack of, 231-253, 258-260, 273-274
    in utero, 231-253
MSG, *see* Monosodium glutamate
Muscles, amino acids in, 26

N

Natural products, free and bound glutamate in, 25-34
Neostriatum
  glutamate in, 187-191
  glutamate analogs, in neurotoxicity of, 191-198
Neurons
  amino acid depolarization of, 164-166
  glutamate uptake by, 145-147
Neurotransmitter, glutamate as, 139, 161
  biochemical model, 139-141
  problems of evaluation, 163-175
Nucleotides, synergistic taste effects with MSG, 1-9, 49-51
Nutrients, factors affecting choice of, 12-14

O

Obesity, glutamate-induced, 299-302
Organ weight, MSG effects on, 264-265
Organs, glutamate content of, 25

P

Parenteral foods, glutamate ingestion in, 358-360

Pituitary gland, glutamate effects on, 277-285, 299
Placenta, glutamate transfer by, 103-110
Plasma, glutamate levels in, *see entries under* Glutamate
Poisons, taste aversion of, 2
Pregnancy
  glutamate effects during, to monkeys, 212, 217-229
  glutamate metabolism in, 103-110
Prolactin, MSG effects on secretion of, 283-284
Proline, glutamate effects on levels of, 303, 305
Protein
  dietary, digestion of, 57
  glutamic acid in, 29
Psychometric studies, on taste of MSG, 35-54
  evaluation terms, 35-36
  flavor profile, 38-39

R

Rabbit, glutamate toxicity to, 287
  lack of, 204
Rat, glutamate toxicity to, 203-204, 255-275
Reproduction, glutamate effects on
  in mouse, 231-253, 278-279
  in rat, 265-269
5'-Ribonucleotides, synergistic effect of, with L-glutamic acid, 1-9, 49-51
Rodent, *see* Rat

S

Safety margin, studies of, 105-109, 321-331, 393
Somatotropin axis, glutamate effects on, 302
Striatum
  glutamate in, 187-201
  glutaminase in, 198-200

T

Taste, 18-21, 38-45
  glutamate taste receptors and, 1
  stimulus-receptor interaction in, 2
Taste receptors
  amino acid binding to, 2-3
  enhancement effects on, 3-4
  "hidden," 3-4
Taurine, effect on luteinizing hormone release, 306-307
DL-Threo-β-hydroxyglutamate, taste similarity to MSG, 48
Thyroid gland, glutamate effects on, 303
Thyroid hormones, MSG effects on secretion of, 282-283
L-Tricholomic acid, taste similarity to MSG, 48

U

Umami ("tastiness"), MSG and, 53
  nucleotide synergism in role of, 1
UMP, synergism with MSG, 4, 6, 7

W

Water, MSG-flavored, self-selection of, 11-23

THE LIBRARY
UNIVERISTY OF CALIFORNIA, SAN FRANCISCO
(415) 476-2335

THIS BOOK IS DUE ON THE LAST DATE STAMPED BELOW

Books not returned on time are subject to fines according to the Library Lending Code. A renewal may be made on certain materials. For details consult Lending Code.

14 DAY JUL 2 2 1992 RETURNED JUL 2 1 1992 **14 DAY** SEP 1 3 1993 **RETURNED** AUG 3 1 1993 RETURNED NOV 1 5 1993	**28 DAY** DEC 6 - 1995 RETURNED JAN 0 8 1995	

Series 4128